M000118214

B. A. Bolt W. L. Horn
G. A. Macdonald R. F. Scott

Geological Hazards

Earthquakes – Tsunamis – Volcanoes
Avalanches – Landslides – Floods

With 116 Figures

Springer-Verlag
Berlin Heidelberg GmbH 1975

B.A. Bolt, Professor of Seismology, Seismographic Stations, Department of Geology and Geophysics, University of California, Berkeley, USA

W.L. Horn, Consultant Engineer, formerly Chief, Flood Forecasting and Control Branch, California Department of Water Resources, Sacramento, USA

G.A. Macdonald, Professor of Geology, Hawaii Institute of Geophysics, University of Hawaii, Manoa, USA

R. F. Scott, Professor of Civil Engineering, Engineering Department, California Institute of Technology, Pasadena, USA

The dustcover depicts the eruption of the volcano Hekla/Iceland – a prize-winning "Agfacolor" photo by Rafn Hafufioro.

Additional material to this book can be downloaded from http://extras.springer.com

ISBN 978-3-642-86822-1 ISBN 978-3-642-86820-7 (eBook)
DOI 10.1007/978-3-642-86820-7

Originally published by Springer-Verlag Berlin Heidelberg New York in 1975
Softcover reprint of the hardcover 1st edition 1975

Library of Congress Cataloging in Publication Data. Main entry under title: Geological hazards. Bibliography: p. . Includes index. 1. Geodynamics. 2. Natural disasters. I. Bolt, Bruce A., 1930– .
QE501.3.G46 551 74–32049

Preface

Growth of population, communication and interdependence among countries has sharpened the impact of natural disasters. Not only have calamities and miseries been given wider publicity, but the realization has grown that through rational study and foresight much can be done to mitigate these hazards to life and social wellbeing.

In this book we present a summary account of hazards which nowadays are usually classified as geological: earthquakes, faulting, tsunamis, seiches, volcanoes, avalanches, rock and soil slides, differential settlement and liquefaction of soil, and inundation.

The book is aimed first at the general reader who is interested in studying the history of such hazards and examining ways that risk can be reduced even if all dangers cannot be eliminated. We also hope that the book will be useful to college students in introductory courses in geology, engineering, geography, country and urban planning, and in environmental studies. We have tried to bring out for the students the problems that remain to be solved.

The treatment is elementary and descriptive, rather than mathematical. Nevertheless, our approach has been analytical and critical; we have not tried to hide controversy and difference of opinion in dealing with problems of hazard control and planning. Those of us who would wish to mitigate environmental problems must face squarely the complexity of the natural environment and economic forces and the lack, in general, of thorough and tractable theoretical models (even with modern computers). Often, if a firm grip is taken at one place then we lose our hold at another, as with a flood control dam which permits housing developments on a downstream flood plain, but may heighten the danger if the dam fails through earthquake shaking or faulting.

Extensive study of geological hazards is currently going on in many parts of the world. In the last few years there has been a spate of special conferences, both national and international, on aspects of the subject. UNESCO, for one, has sponsored conferences on nearly every topic touched on here, from earthquake stimulation by water impounded behind large dams, to early-warning systems for tsunamis. We have tried to include the main results of these conferences in this book, with the hope that it will be valuable in many countries as a summary of the present knowledge.

In one important respect we have not gone as far as we would have liked. It has become clear that the traditional presentation to the public of the concept of risk and the process of decision making on risk reduction is becoming quite unsuited to the sharper demands of the present circumstances. The elaboration of unquantitative statements on "maximum possible", "credible", "allowable", and so on has become self defeating. The evasion of real statistical basis of risk, under cover of either a contempt for statistics or a belief that the public will not accept rational odds, is surely not justifiable. Only when students of the subject tackle questions of acceptable balance of risk will sounder and more practical methods of risk evaluation and decision be found. The serious student of geological hazards should be aware of statistical methods.

Something should be said about the division of knowledge and responsibility of the four authors. The first (B.A. BOLT), a Professor of Seismology at the University of California, Berkeley, who undertook the preparation of Chapters 1 and 3, has taught seismology courses including treatment of earthquake and tsunami hazards. The case histories on risk are ones on which he had made special studies as a consultant or field investigator. The second chapter, on volcanic hazards, was written by G.A. MACDONALD, who has for many years taken a close interest in such problems while at the Volcano Observatory in Hawaii and as a Professor of Geology at Hawaii University. The chapters on hazards from ground movements are the work of R.F.SCOTT, a Professor of Civil Engineering at California Institute of Technology, who has specialized in soils engineering, including both theoretical and field studies of landslides, settlement and liquefaction of soils in earthquakes. Chapter 7 on flood hazards was written by W.L. HORN, who as a Principal Engineer of the California Department of Water Resources, had the responsibility for flood forecasting and operations for over a decade. Chapter 8 is mainly the work of B.A. BOLT, who partook as a consultant in certain aspects of the California "Urban Geology" study.

Difficulties in the assessment of geological hazards arise when, so to speak, cobblers do not stick to their lasts, but progress depends on just that and we make no apologies for trying to integrate, at least to some extent, the separate disciplines dealt with here. We warn the student, however, to be prudent and open-minded in judging risk in disciplines away from his own special studies.

For ease of cross-referencing, the first integer of a Section, Figure or Plate refers to the Chapter number. Thus, Plate 3.2 may be found in Chapter 3. Metric values are used throughout the text. Some few Figures remain with English measurements. Appendix E gives conversion tables between English and metric systems. The world map at the front of the book shows the geographical locations of many places mentioned in the text.

We are indebted to a number of colleagues who gave invaluable criticism of various parts of the manuscript. Reviews or other contributions to particular chapters were made by Dr. R.D. ADAMS, Mr. W.K. CLOUD, Dr. A. EWART, Mr. L. JAMES, Mr. J. LEFTER, Dr. G. OAKESHOTT, Dr. S. OMOTE, Dr. J.P. SCHAER, Dr. J. SCHULZ, Mr. R.L. WIEGEL and Dr. H. WILLIAMS. Our thanks are also due to Dr. BEVERLEY BOLT who helped in many ways. R.F. SCOTT would like to acknowledge the hospitality of Churchill College, Cambridge, where part of his section was prepared.

March 1975 The Authors

Contents

Chapter 1
Hazards from Earthquakes

1.1. The Great Good Friday Alaska Earthquake and Tsunami, March 27, 1964

Eyewitnesses

During the late afternoon of Good Friday, 1964, at 17:36 h local time, a great earthquake struck the sparsely inhabited mountainous area of northern Prince William Sound in south-central Alaska (see Fig. 1.1). Waves from the earthquake source spread through the Earth and caused serious damage over more than 20,000 square km. In the area of significant damage, or *meizoseismal* area, the largest city affected was Anchorage, some 130 km from the earthquake's center (see Fig. 1.2).

In the well-to-do suburb of Turnagain Heights, on a high cliff overlooking Cook Inlet, Mr. Robert B. Atwood, editor of the Anchorage Daily Times, afterwards recorded his experience.

"I had just started to practice playing the trumpet when the earthquake occurred. In a few short moments it was obvious that this earthquake was no minor one: the chandelier, made from a ship's wheel, swayed too much. Things were falling that had never fallen before. I headed for the door. On the driveway I turned and watched my house squirm and groan. Tall trees were falling in our yard. I moved to a spot where I thought it would be safe, but as I moved I saw cracks appear in the earth. Pieces of ground in jigsaw puzzle shapes moved up and down, tilted at all angles. I tried to move away, but more appeared in every direction. I noticed that my house was moving away from me, fast. As I started to climb the fence to my neighbor's yard, the fence disappeared. Trees were falling in crazy patterns. Deep chasms opened up. Table-top pieces of earth moved upward, standing like toadstools, with great overhangs. Some would turn at crazy angles. A chasm opened beneath me. I tumbled down. I was quickly on the verge of being buried. I ducked pieces of trees, fence posts, mail boxes and other odds and ends. Then my neighbor's house collapsed and slid into the chasm. When the earth movement stopped I climbed to the top of the chasm. I found angular landscape in every direction."

The post-earthquake appearance of the landscape at Turnagain Heights so graphically described by Mr. Atwood is shown in Plate 1.1. The extreme damage was a direct result of the failure of the clay soil during the ground shaking. This led to a massive landslide down toward the sea (see Chapter 4).

In downtown Anchorage damage varied, depending on the strength of the foundation material and the type of building. In the Hillside Manor apartments, constructed of concrete block, Mr. John R. Williams, a geologist, was sitting

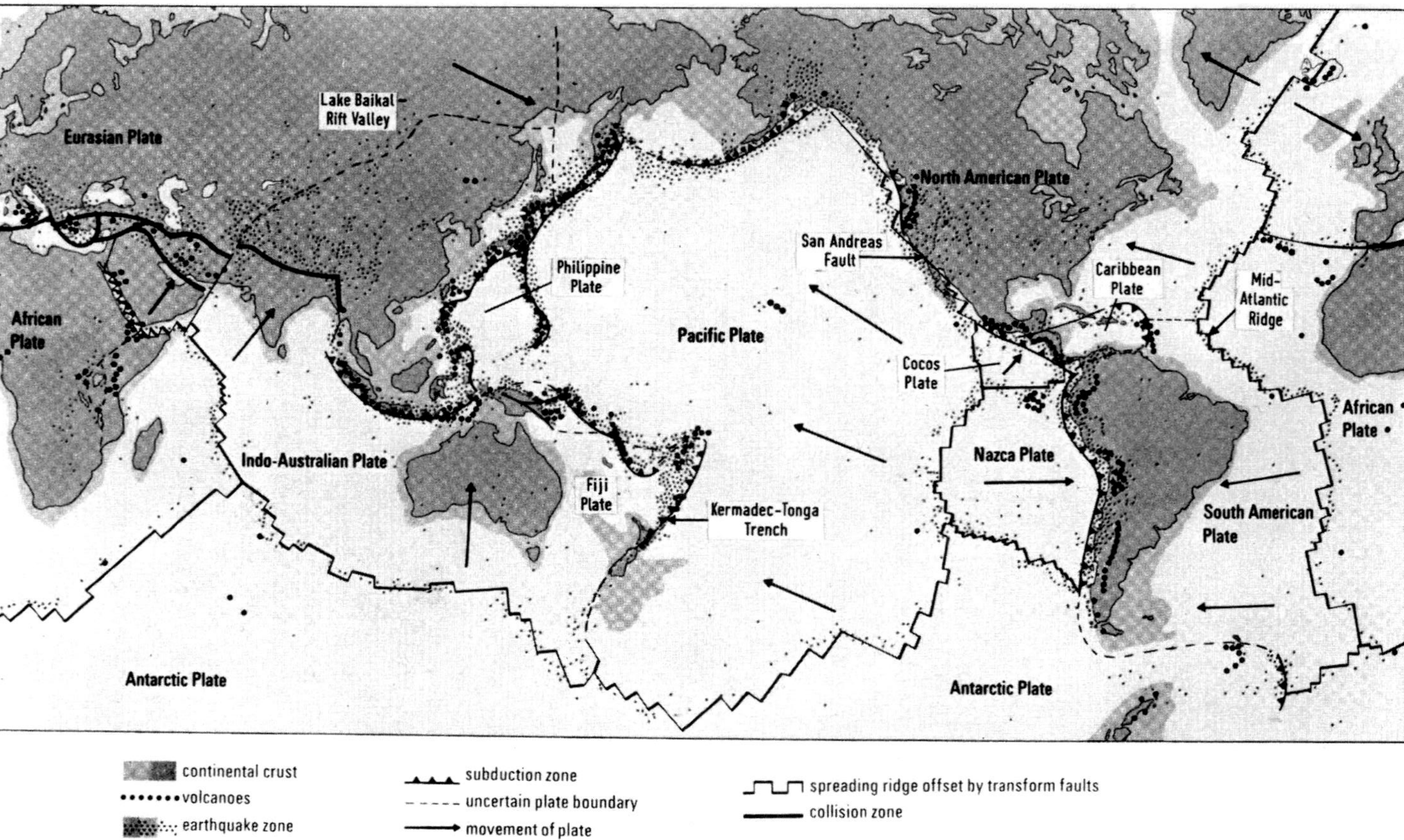

Fig. 1.1. Map of the world showing diagrammatically the distribution of volcanoes and earthquakes in relation to the major tectonic plates

Plate 1.1. View of damaged houses due to great slide at Turnagain in 1964 Alaskan earthquake. Soft clay bluffs about 22 m high failed in the shaking. The slide extended 2,800 m along the coast-line and regressed inland 300 m. (Courtesy of USGS)

on a couch in his living room. He recalled that, "At first we noticed a rattling of the building. The initial shaking lasted perhaps five to ten seconds. The first shaking was followed without any noticeable quiet period by a strong rolling motion which appeared to move from east to west.

"After a few seconds of the strong rolling motion, I took my son to the door leading to the hall, opened the door to prevent jamming, and stood in the doorway. I looked in the hallway and back in the apartment and noticed blocks working against one another in interior walls and saw some fall into the street and into the apartment and hall. I took my son and ran to a parked car. I looked at the building, which was swaying in an east-west direction. Blocks were toppling, ground heaving, trees and poles were swaying strongly. The Hillside apartment building was a total loss. Our apartment was one of the least damaged. The light day beds were in position; the portable TV on the wheel stand was upright; the plaster cast of a child's hand on the wall was in place. The stove did not move, but the ice-box slid from the wall."

In sharp contrast with the behavior of this type of construction was the frame house (poured concrete basement) at 1555 Eighth Street. Although the occupants were severely shaken, and lamps toppled and cupboards opened, the only apparent damage was a crack in the south wall!

The Great Sea Wave

The waters of Prince William Sound and the Gulf of Alaska were disturbed by the sudden vertical displacement of the ocean bottom so that a great water wave, or *tsunami*, crashed against the coastal regions of south-central Alaska and spread over the Pacific Ocean (see Chapter 3).

In the town of Valdez, 70 km from the earthquake center, sea waves in the harbor devastated the waterfront. Within seconds of the initial shaking, eye witnesses realized that something violent was occurring along the pier area. The most obvious action at first involved the steamer Chena, approximately 120 m long. Like a cork, it rose some 6 to 9 m, then dropped, struck bottom, shot forward, bottomed again, and was lifted clear. It was nothing short of a miracle that the ship survived. The men on the ship claim that it heeled over to the 50° mark and was righted by the waves. The bow rose until it could be seen well above the dock warehouses. Two men on the ship were killed by falling cargo and another died of a heart attack.

Almost immediately, the Valdez dock began to move violently and broke in two, sending warehouses flipping forward into the sea. Men, women and children staggered around the dock looking for something to hold on to. Very soon, a large water wave arose, smashing structures in its path; buildings were reduced to kindling wood, heavy trailers were thrown all over the waterfront and cars and trucks were smashed into twisted masses of metal, some people recalling a wave 9 m high crossing the waterfront. All these events took only a few minutes. Approximately ten minutes after the initial wave had receded, a second surge crossed the waterfront carrying large amounts of wreckage. There followed a lull of approximately five to six hours during which time search parties were able to look for possible survivors. There were none.

Deformation of the Earth's Crust

The eyewitness accounts from the great Good Friday earthquake are a fascinating record of both social and scientific value. They demonstrate how people react to sudden natural disaster and also contain clues necessary for scientific appraisal of the causes, nature, and consequences of the natural phenomenon itself. After the Alaskan earthquake, great effort was expended over a period of many years by geologists, geodesists, geophysicists, engineers and others to extract the maximum knowledge from the calamity. Because eyewitnesses, however, often give conflicting accounts, colored by the stress of the moment, after earthquakes the main attention is given to the systematic study of damage to buildings, to measurements of offsets along faults (if any), and to displacements of the land surface, particularly along beaches, where absolute measurements of motions relative to the sea level can often be obtained. Whenever they are available, records on seismographs of seismic shaking (both near to and far from the earthquake source), of atmospheric waves on microbarographs, and of tsunami waves in the ocean measured by sea level recorders and tide gages are analyzed.

Seismic waves from this great earthquake were recorded at seismographic stations all over the world, often with such amplitude that the most sensitive seismographs were driven off scale and little could be measured. The time of travel to various stations pinpointed the position of the first break or *focus* in the Earth at latitude 61°.1 north and longitude 147°.4 west, at a depth between 20 and 50 km under Prince William Sound. Measurements of wave amplitudes put the magnitude of the earthquake between 8 and 8.6, making it one of the greatest earthquakes yet recorded (see Appendix A). Unfortunately, no instruments built to record the complete strong ground motion in the meizoseismal zone were in place in Alaska at the time. For this reason, the detailed ground motion at Anchorage and other damaged towns that would provide the key for engineering assessments will never be known precisely.

Deformation of the Earth's crust was more extensive than any yet studied in a single earthquake (see Fig. 1.2). Vertical displacements along almost 1,000 km

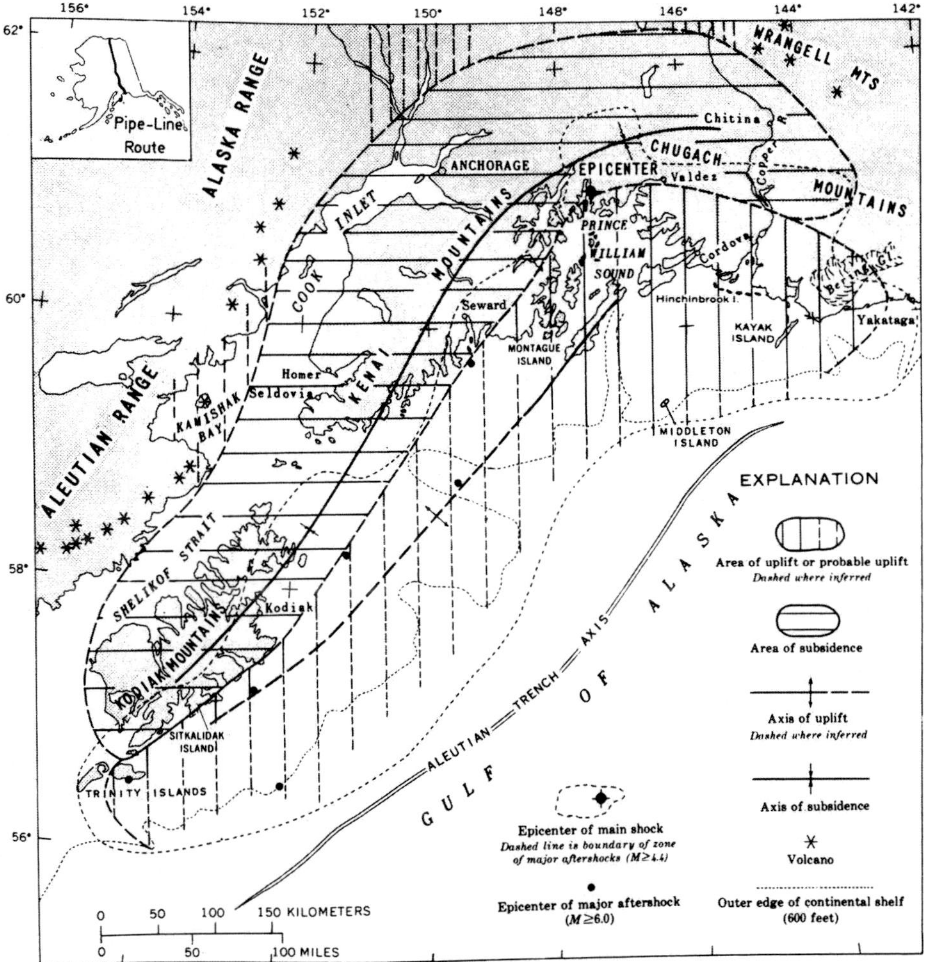

Fig. 1.2. Deformation of the crust in the great Good Friday 1964 Alaskan earthquake. Also shown is the area of aftershocks which followed the main shock (shown as full circle). (After National Ocean Survey)

of the continental margin, from the southwestern tip of the Kodiak Islands through Prince William Sound, were estimated from hundreds of measurements of upper growth limits of intertidal marine organisms relative to normal sea level. Many barnacles, for example, form a conspicuous band on rocky shores with a sharply-defined, easily-recognizable upper limit depending mainly on the ability of the barnacles to survive prolonged exposure to the air.

Land level changes in Alaska consisted essentially of (i) a broad zone of *subsidence* of as much as 2 m along the Kodiak-Kenai-Chugach Mountains; (ii) a major zone of *uplift* bordering it on the seaward side and extending from the coast to the sea floor; and (iii) a zone of minor *uplift* that borders it on the landward side and extends northward into the Alaskan and Aleutian ranges. The maximum vertical uplift in zone (ii) reached as much as 11 m in places. The reader will see from Fig. 1.2 that these crustal displacements are compatible with a major crustal break running along the coastal area from Prince William Sound southwest to the seaward side of Kodiak Island.

After the main earthquake, many thousands of aftershocks occurred in this rupture zone, 1,260 being recorded during the first four months following the main shock. This distribution of aftershocks may be taken to define the great extent of crustal fracturing involved in the 1964 earthquake.

Faults on land associated with the earthquake were found only on southwestern Montague Island in Prince William Sound and on the sub-sea continuation of one fault southwest of the island. On the longer of the two faults, a scarp was clearly visible with a maximum vertical component of displacement of 6 m. Other suspected instances of faulting, carefully checked in the field by geologists of the United States Geological Survey, turned out to be landslides or surficial cracks in unconsolidated soils.

The earthquake occurred in a seismically active zone that runs westward along the coast of Alaska, and follows the deep Aleutian trench to the south of the Aleutian Island arc (see Fig. 1.1), which is made up of a chain of late Cenozoic volcanoes. Careful radiocarbon dating of materials taken from uplifted wave-cut benches along the shore lines (see Plate 5.1) of the arc have shown that uplift of the kind that took place in 1964 is probably repeated once in about every thousand years. The 1964 earthquake was only the most recent pulse in a long period of deformation that began ten million years ago in late Pliocene time and has continued on and off up to the present day.

In 1912 and 1934 there had been two earthquakes of about magnitude 7.2 near the center of the 1964 earthquake, and quite large earthquakes occurred in central Alaska in 1937, 1943, 1947, and 1958. However, seismicity in the Kodiak Island region, according to available seismographic records, shows no significant increase or decrease just before the Good Friday earthquake in 1964.

Geological and seismological studies made after the earthquake have led to a model which explains most of the observed vertical and horizontal motions and extent of the shaking. This model requires the thrusting southward of Alaska at a shallow angle of perhaps 20° over a fracture lying along the submarine trench (see Fig. 1.4 for a generalized picture).

The large-scale faulting began near the center of the main shock in northern Prince William Sound, the rupture then spreading upward toward the Aleutian trench and horizontally both toward the southwest and the east within the area later defined by the aftershocks (see. Fig. 1.2). As the ruptures spread, seismic waves were generated in bursts of energy and these vibrations spread in all directions through the Earth.

The Cost

Three hundred people were killed in the earthquake, some from the effects of shaking and others from drowning in the tsunami. The violent shaking triggered numerous rockslides, snow avalanches and landslides throughout south-central Alaska. Fracturing, slumping, lurching and subsidence were common in unconsolidated deposits, and cracks and pressure ridges were observed in lake and river ice. In all, it was estimated in 1968 that $310,000,000 worth of property damage was sustained in Alaska. The dislocation to industry and to the life of the state was widespread with the destruction of harbors, docks, railroad tracks, bridges, highways, power facilities and structures of many kinds.

Alaska provides clear examples of geologic hazards from large earthquakes, tsunamis, landslides (see Chapter 3) and volcanic eruptions (see Chapter 2), although the risk is far from uniform across this great land area. Indeed, parts of Alaska have few geological hazards (see the discussion of the Alaska pipeline in Section 1.5) but the probable cost in life and damage from the distribution of hazards is an everchanging one.

The earliest permanent settlements were established by the Russians on Kodiak Island in 1783, and the oldest report of an earthquake is associated with the eruption of Pavlov Volcano in 1786. Early Russian settlers and indigenous Alaskans confined their activities to places where fish and furred animals were hunted readily, but after the Klondike gold-rush in the 1890s, the pattern of settlement changed to one of filling in the gaps in southern Alaska. As population growth turned inland the number of reported earthquakes increased sharply. Even today, if it were not for detection by seismographic stations, a very distorted picture of the seismicity of large uninhabited segments of northern Alaska and the Aleutian Islands would be obtained from reports of felt earthquakes. This distortion of seismicity patterns by unevenness in population has been found in many places, for example in the early earthquake reporting of New Zealand.

1.2. Seismicity of the World

Historical Record and Intensity

On the average, 10,000 people die each year from earthquakes. A UNESCO study gives 350,000 deaths from 1926 to 1950 from earthquakes, and damage losses amounting to $10,000,000,000. In Soviet Central Asia in this interval 2 towns and 200 villages were destroyed. Since then several towns including Ashkhabad (1948), Agadir (1959), Skopje (1963), Managua (1972) and hundreds of villages have been almost razed to the ground (see Plate 1.2). Historical writings testify to man's long concern about earthquake hazards.

Plate 1.2. The great variation in damage in the Managua 1972 earthquake. The torquezal dwellings (space between vertical wood studs is a "pocket", packed with mud and stone) collapsed (high intensity rating?) but the modern Banco de America building in the background stood with no substantial structural damage (low intensity rating?). (Courtesy of M. Sozen)

The longest catalog comes from the ancient Chinese civilization, as early as the Shang dynasty, more than 3,000 years ago. Records assembled by Chinese scholars list over 1,000 damaging earthquakes in the 2,750 years from 780 B.C. to the present time. An extensive catalog of Japanese earthquakes also exists since the era of Tokugawa Shogunate, about 1600 A.D. In the Mediterranean cradle of Western civilization, ancient Hebrew and Arab records make allusion to earthquakes since early times. The Biblical report of the ruin of Sodom and Gomorrah and of the throwing down of the walls of Jericho (about 1,100 B.C.) are two of the earliest probable allusions to earthquakes in the Bible. In the first case, a geologic theory is that a great earthquake along a boundary fault of the Dead Sea rift valley knocked down buildings and released natural gas and bitumen. Ignition from cooking fires then led to the Biblical conflagration that destroyed Sodom and Gomorrah.

The widely-used yardstick of the "strength" of an earthquake is earthquake *intensity*. Intensity is the measure of damage to the works of man, to the ground surface, and of human reaction to the shaking. Because earthquake intensity assessments do not depend on instruments, but on the actual observation of effects in the meizoseismal zone, intensities can be assigned even to historical earthquakes. In this way, the historical record becomes of utmost importance in modern estimates of seismological risk.

The first intensity scale was developed by de Rossi of Italy and Forel of Switzerland in the 1880s. This scale, with values from I to X, was used for reports of the intensity of the 1906 San Francisco earthquake, for example. A more refined scale was devised in 1902 by the Italian volcanologist and seismologist Mercalli with a twelve-degree range from I to XII. A version is given in Table 1.1, as modified by H.O. Wood to fit conditions in California. The descriptions in Table 1.1 allow the damage to places affected by an earthquake to be rated numerically. These spot intensity ratings can often be separated by lines which form an *isoseismal* map (see Fig. 1.3). Such intensity maps provide

Table 1.1. Modified Mercalli (MM) intensity scale of 1931

I	Not felt except by a very few under especially favorable circumstances.
II	Felt only by a few persons at rest, especially on upper floors of buildings. Delicately suspended objects may swing.
III	Felt quite noticeably indoors, especially on upper floors of buildings, but many people do not recognize it as an earthquake. Standing motor cars may rock slightly. Vibration like passing of truck. Duration estimated.
IV	During the day felt indoors by many, outdoors by few. At night some awakened. Dishes, windows, doors disturbed; walls make cracking sound. Sensation like heavy truck striking building. Standing motor cars rocked noticeably.
V	Felt by nearly everyone, many awakened. Some dishes, windows, etc., broken; a few instances of cracked plaster; unstable objects overturned. Disturbances of trees, poles, and other tall objects sometimes noticed. Pendulum clocks may stop.
VI	Felt by all, many frightened and run outdoors. Some heavy furniture moved; a few instances of fallen plaster or damaged chimneys. Damage slight.
VII	Everybody runs outdoors. Damage negligible in buildings of good design and construction; slight to moderate in well-built ordinary structures; considerable in poorly built or badly designed structures; some chimneys broken. Noticed by persons driving motor cars.
VIII	Damage slight in specially designed structures; considerable in ordinary substantial buildings, with partial collapse; great in poorly built structures. Panel walls thrown out of frame structures. Fall of chimneys, factory stacks, columns, monuments, walls. Heavy furniture overturned. Sand and mud ejected in small amounts. Changes in well water. Persons driving motor cars disturbed.
IX	Damage considerable in specially designed structures; well-designed frame structures thrown out of plumb; great in substantial buildings, with partial collapse. Buildings shifted off foundations. Ground cracked conspicuously. Underground pipes broken.
X	Some well-built wooden structures destroyed; most masonry and frame structures destroyed with foundations; ground badly cracked. Rails bent. Landslides considerable from river banks and steep slopes. Shifted sand and mud. Water splashed (slopped) over banks.
XI	Few, if any, (masonry) structures remain standing. Bridges destroyed. Broad fissures in ground. Underground pipelines completely out of service. Earth slumps and land slips in soft ground. Rails bent greatly.
XII	Damage total. Practically all works of construction are damaged greatly or destroyed. Waves seen on ground surface. Lines of sight and level are distorted. Objects are thrown upward into the air.

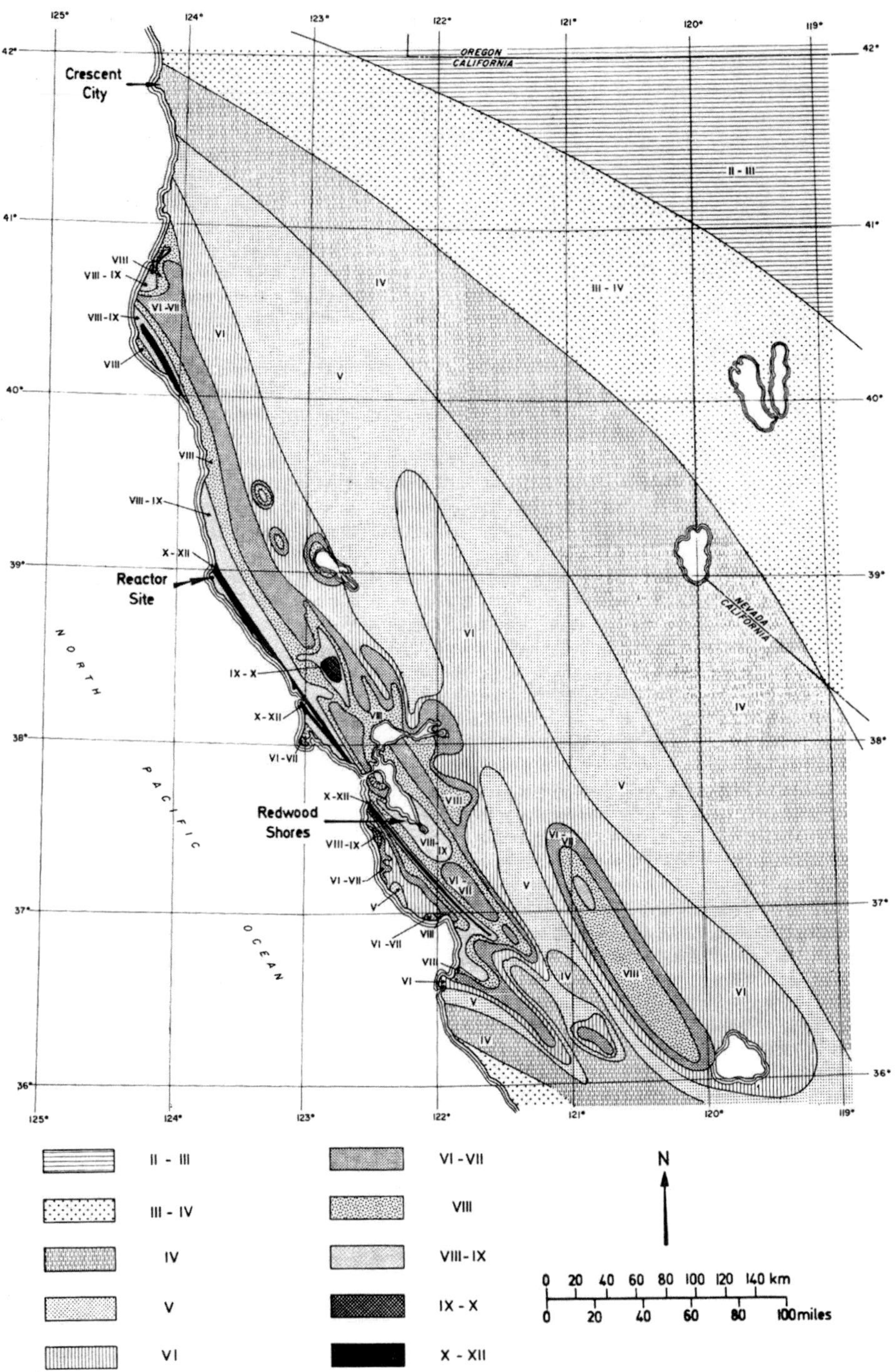

Fig. 1.3. Isoseismals for the great 1906 California earthquake. The contours separate areas of equal intensity of shaking as rated on the Modified Mercalli (MM) scale. (After the Report of the State Earthquake Commission, 1908)

crude, but valuable information on the distribution of strong ground shaking, on the effect of surficial soil and underlying geological strata, the extent of the source, and other matters pertinent to insurance and engineering problems (see Chapter 8).

Because intensity scales are subjective and depend upon the social and construction conditions of a country, they need revising from time to time. Regional effects must be accounted for and it is interesting to compare the Japanese scale (0 to VII) summarized in Table 1.2 with the Modified Mercalli descriptions.

Table 1.2. Japanese seismic intensity scale

0	Not felt; too weak to be felt by humans; registered only by seismographs.
I	Slight: felt only feebly by persons at rest or by those who are sensitive to an earthquake.
II	Weak: felt by most persons, causing light shaking of windows and Japanese latticed sliding doors (Shoji).
III	Rather strong: shaking of houses and buildings, heavy rattling of windows and Japanese latticed sliding doors, swinging of hanging objects, sometimes stopping pendulum clocks, and moving of liquids in vessels. Some persons are so frightened as to run out of doors.
IV	Strong: resulting in strong shaking of houses and buildings, overturning of unstable objects, spilling of liquid out of vessels.
V	Very strong: causing cracks in brick and plaster walls, overturning of stone lanterns and grave stones, etc. and damaging of chimneys and mud and plaster warehouses. Landslides in steep mountains are observed.
VI	Disastrous: causing demolition of more than 1% of Japanese wooden houses; landslides, fissures on flat ground accompanied sometimes by spouting of mud and water in low fields.
VII	Ruinous: causing demolition of almost all houses; large fissures and faults are observed.

Earthquake Location

Earthquakes from past centuries are located from intensity ratings, the center being placed near the middle of the isoseismal map. Since about 1900, knowledge of the distribution of earthquakes is fortunately not dependent only on felt reports, which depend greatly on the distribution of population, but is obtained by an objective detection of earthquakes by seismographs around the world. Today, there are some 1,000 seismographic stations operating continuously, with at least one observatory in even the smallest countries where there is a seismic risk. The usual seismographs consist of a suspended mass, like a pendulum, which is damped and attached to an amplifying device.

Modern seismographs can magnify a seismic wave in the ground with period of one second over one million times. From the times of arrival of these magnified waves the position of the source of the earthquake and its size can be calculated no matter where it originated around the globe.

The distribution of earthquakes is indicated broadly in Fig. 1.1. It can be seen that earthquakes, like volcanoes and high mountain ranges, are not randomly scattered, but are, for the most part, concentrated in narrow belts. Many earthquakes occur along mid-oceanic ridges and pose no hazard to humans. The greatest seismic activity is concentrated along the margins of *tectonic plates* (see Fig. 1.1), such as the Pacific plate, the interiors of which are almost aseismic. The Antarctic plate, mostly surrounded by spreading ridges, has the quietest margins and almost no interior earthquakes.

A current picture of the Earth's outer shell envisages it as made up of more than 15 relatively undistorted plates of lithosphere (crust and upper mantle) about 60 km thick, which move relative to one another. They spread from the mid-oceanic ridges (see Fig. 1.1) where, by means of the up-flow of magma, new lithospheric material is continually added. On the opposite margins of the plates there are usually deep submarine trenches, for example, along the island-arc systems of the Pacific plate and the Nazca plate. At these trenches, the plates converge from opposite directions (e.g. the Nazca and South American plates along the Andes), and one plate is consumed or *subducted* beneath the other into the deeper parts of the Earth. A generalized diagrammatic model for what is envisaged is given in Fig. 1.4. Earthquake foci are located at many depths in the downgoing plate. Elsewhere, as in the Caucasian-Himalayan belt, continents which are transported on the plates collide and here too, earthquake activity is high.

The foci of earthquakes, located from measurements of seismic waves, occur from quite close to the surface down to depths of about 700 km; no deeper earthquakes have ever been recorded. Over 75 per cent of the average annual seismic energy, however, is released by earthquakes with foci less than about 60 km deep. These are the earthquakes which constitute the significant hazard, and are known as *shallow-focus* earthquakes.

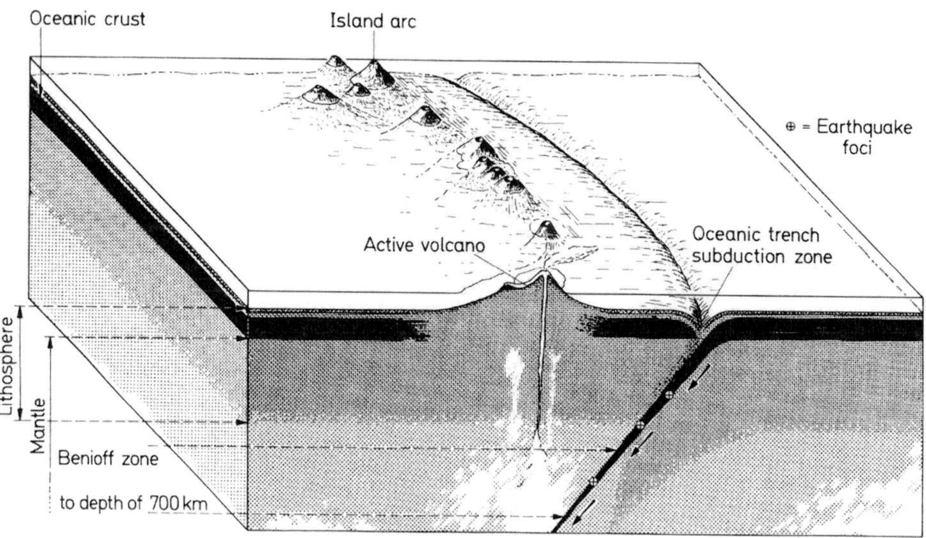

Fig. 1.4. Model of a lithospheric plate and *Benioff zone* moving down beneath an island arc. The less dense magma rises upwards and erupts as lava on the ocean floor where it builds an island arc

Earthquake Magnitude and Moment

Seismic waves generated by an earthquake source are of three main types. The fastest consist of *longitudinal* waves, called P waves, which travel through both solid and liquid parts of the Earth. Their particle motion is similar to sound waves and involves compression and expansion of the transmitting material. In the solid rock of the Earth secondary waves, called S waves, also propagate, whose particle motion is transverse to the direction in which they travel and involves shearing of the rock and whose speed is always less than that of P waves.

The P and S waves that move through the body of the Earth are followed by long trains of waves guided by the free surface of the Earth. These surface seismic waves are of two kinds, called Love and Rayleigh waves, both involving horizontal shaking but only Rayleigh waves having vertical displacements. As they travel they disperse into rather long wave trains and are responsible for much of the shaking felt from earthquakes beyond a few kilometers from the source itself. Because of their greater speed, the first waves to reach any point on the Earth are the P waves. The first P onset, of course, starts from the place where the earthquake originates, the point called the *focus* or *hypocenter*. Because the focus is always below the surface, the point on the Earth's surface straight above it is required for mapping earthquake location. This surface point is called the *epicenter*.

The relative severity of earthquakes, rated subjectively by the intensity, can be measured using seismographs by comparing particular features of the seismograms. These factors, or *earthquake parameters*, include the duration of the recorded motion, the amplitude of certain wave heights, and so on. The usual instrumental severity scale, originated by Charles F. Richter, depends upon a measurement of the amplitude of the largest wave recorded by a seismograph from a particular earthquake.

Because earthquake strength varies over a huge range, it is convenient to compress the measurement of wave amplitudes by use of logarithms. Richter's precise definition of *magnitude* is: "the logarithm (to base 10) of the maximum seismic wave amplitude (in thousandths of a millimeter) recorded on a special seismograph called the Wood-Anderson, at a distance of 100 km from the earthquake epicenter".

The definition has been extended so that any calibrated seismograph may be used at any distance. The scale is not limited, either at top or bottom, each magnitude step on the scale representing an increase of ten times in measured wave amplitude of the earthquake.

The *energy* of shaking in an earthquake is different but knowledge of seismic energy relates also to damage and hazard. An increase of one magnitude step has been found to correspond to an increase of 30 times the amount of energy released as seismic waves. Therefore, the energy of an 8.6 magnitude earthquake, such as the 1964 Alaskan earthquake, is not twice as large as that in a shock of magnitude 4.3, but rather, the magnitude 8.6 earthquake releases almost one million times as much energy as one of magnitude 4.3!

The largest earthquakes recorded (January 31, 1906 in Colombia-Ecuador and March 2, 1933 in Japan) had a Richter magnitude of about 8.9. Because rocks can only be strained so much before breaking, this magnitude is about the limit which can be produced by tectonic means in the Earth. At the other end of the magnitude scale, the most sensitive modern seismographs can measure earthquakes which have magnitudes below zero. A magnitude 2 is about the smallest earthquake which can be felt by human beings.

There is only a rough correlation between the magnitude of an earthquake and the intensity. It turns out that earthquakes of magnitude 5 roughly correspond to MM intensities near the source of the earthquake of VI to VII (see Table 1.1), while a great magnitude shallow-focus earthquake generates a large *area* of maximum intensity, as happened in the 1906 California earthquake and the 1964 Alaska earthquake (see Fig. 1.2).

The most recent global lists of earthquakes indicate that each year there are between 18,000 and 22,000 shallow-focus earthquakes of magnitude 2.5 or greater. Only a few of this prodigious number, fortunately, constitute any geological hazard to man (see Appendix A).

Another parameter used to rate the strength of an earthquake has come into theoretical vogue. If the earthquake is thought of as arising from the sudden relaxation of pairs of forces in the elastic rocks of the upper part of the Earth (say, by fault slip), then, just as in mechanics, the total moment of these forces is a measure of size. The *seismic moment* is defined as the rigidity of the rock times the area of fault face which moved times the amount of slip. The great 1906 San Francisco earthquake had a seismic moment of about 10^{28} dyne-cm, while the 1971 San Fernando shock had a moment of nearly 10^{26} dyne-cm.

Plate-Edge and Interplate Earthquakes

Moving plates of the Earth's surface (see Fig. 1.1) provide an explanation for a great deal of the seismic activity of the world. Collisions between adjacent lithospheric plates, destruction of the slab-like plate as it descends into the *subduction zone* beneath island arcs (see Fig. 1.4), and spreading along mid-oceanic ridges are all mechanisms likely to be associated with large-scale straining and fracturing of crustal rocks. Thus, the earthquakes in these tectonically active boundary regions are called *plate-edge earthquakes*. The very hazardous shallow earthquakes of Chile, Peru, the eastern Caribbean, Central America, southern Mexico, California, southern Alaska, the Aleutians, the Kuriles, Japan, Taiwan, the Philippines, Indonesia, New Zealand, the Alpine-Caucasian-Himalayan belt are of plate-edge type.

As the mechanics of the lithospheric plates become better understood, long-term predictions may be possible for plate-edge earthquakes. For example, many plates spread toward the subduction zones at rates of from 2 to 5 cm per year. Therefore in active arcs like the Aleutian and Japanese Islands, knowledge of the history of large earthquake occurrence might flag areas that currently lag in earthquake activity.

While the plate-tectonic theory is an important one for a general understanding of earthquakes and volcanoes, it does not tell the whole story, for within continental regions, away from boundaries, large devastating earthquakes sometimes occur. These *intraplate earthquakes* can be found on nearly every continent.

In the United States, the most famous are the major earthquake series of 1811–1812, that occurred in the New Madrid area of Missouri, along the Mississippi River. Another important group which seems to bear little relation to the present plate edges, occurs in northern China (see the section below).

Although the Australian plate (see Fig. 1.1) is often classified generally as almost aseismic, earthquakes also visit it; particularly the zone running from Spencer Gulf in South Australia north to 25°.5 S, 137°.0 E. Between the years 1938 and 1941, four moderately strong shocks occurred at the edge of the West Australian Pre-Cambrian Shield, the largest, of June 27, 1941, having a magnitude of 6.75. Fortunately, few people lived nearby. Another example, marginal to the Australian shield, is the damaging Meckering shock (magnitude 6.8) of October 14, 1968, which occurred in the far southwest corner, not far from Perth. This is the only Australian earthquake that was definitely associated with observed faulting. The fault was an arcuate compression rupture, concave to the east, with length of about 32 km and an uplift to the east of 1.5 m.

Such major internal seismic activity indicates that lithospheric plates are not rigid or free of internal rupture. The occurrence of intraplate earthquakes makes the prediction of earthquake occurrence more difficult and calls for different explanations from those that account for plate-edge earthquakes.

United States Earthquakes

The occurrence of earthquakes in the United States is very uneven, both in time and in space (see Fig. 1.5); the historical record goes back to 1638. A list of some of the more important ones is given in Appendix B.

One of the earliest earthquakes of importance was in the Three Rivers area (north of Quebec) of the lower St. Lawrence River region on February 5, 1663. From a summary of contemporary reports, a Modified Mercalli maximum intensity of X has been assessed by some seismologists. Large rock slides and changes in waterfalls were seen. Chimneys were broken as far away as Massachusetts Bay.

This 1663 earthquake, occurring 310 years ago, continues to have an important effect on the assessment of risk in the northeast United States and the adjacent Canadian provinces. Crustal readjustments following the removal of the ice load after the Ice Age that ended about 10,000 years ago have been suggested as the source of the strain energy. Largely on the basis of the 1663 shock, it has become almost traditional to draw a zone of high seismic risk (Zone 3 in Fig. 1.5 B; compare with Fig. 1.11) running along the St. Lawrence from the Great Lakes. Recent seismicity studies perhaps tend to change this conclusion. Southwest of Quebec, there is currently a seismicity gap bounded by instrumentally located epicenters concentrated in a zone running at *right angles* to the St. Lawrence between Ottawa and Montreal and trending from NW of Ottawa towards Boston. This northwest-southeast alignment is consistent with the Canadian map (Fig. 1.11) but not with Figs. 1.5 A and 1.5 B.

The western United States exhibits the greatest seismic activity in the whole country except Alaska. Along the Pacific margin, the San Andreas fault and

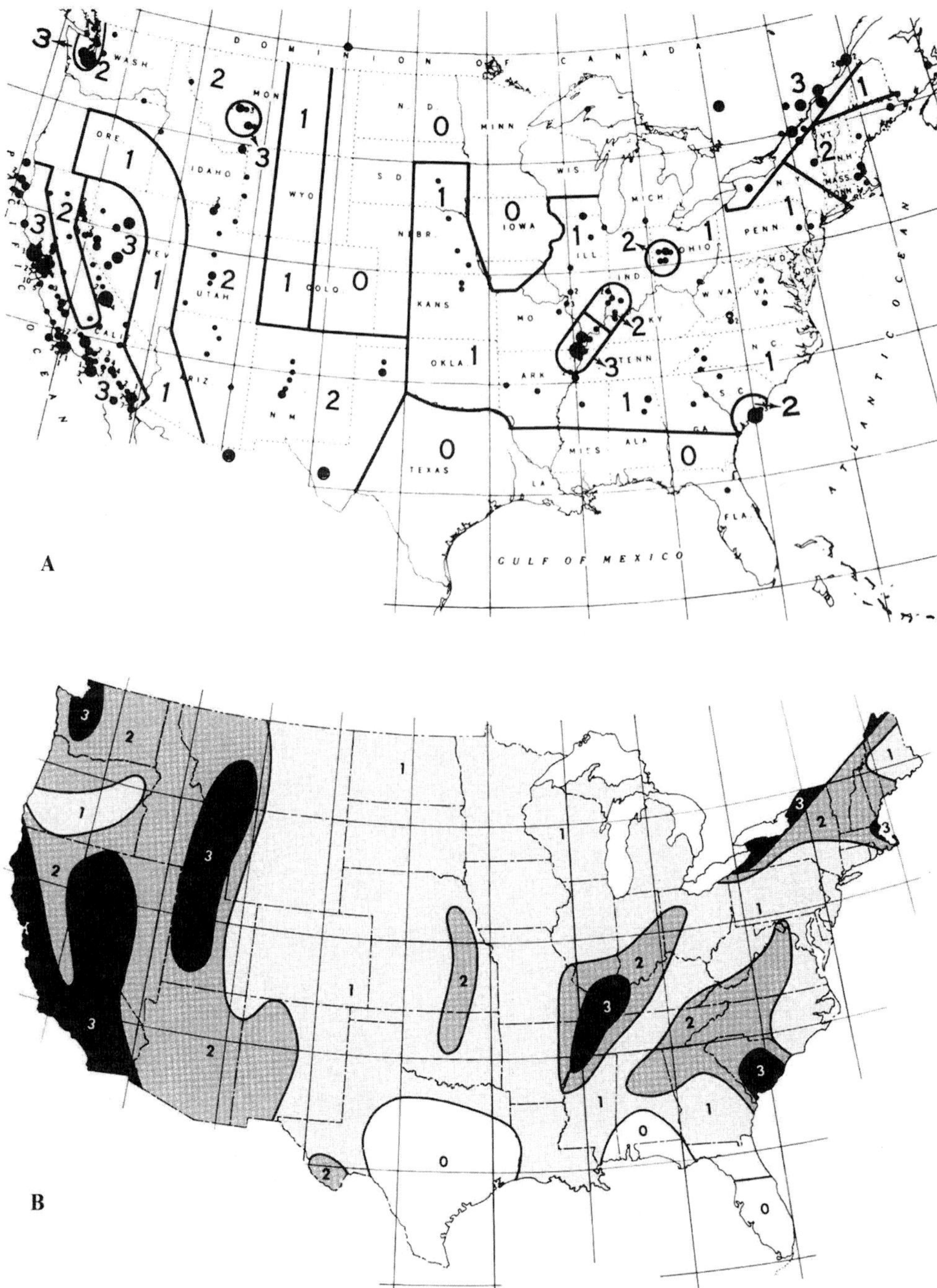

Fig. 1.5. Two forms of seismic risk maps for the United States. The top map is one compiled by the U.S. Coast and Geodetic Survey (as revised 1949) and shows, as dots, epicenters of some of the major historical earthquakes. The bottom map is a rather different and modified version originating from S.T. Algermissen (National Ocean Survey) in 1969

its subsidiaries appear to be the northeast boundary of the Pacific plate and to connect the spreading of the lithosphere in the Gulf of California with spreading along the Gorda ridge under the ocean off northern California (see Fig. 1.1).

The historical record goes back to about 1800 and we are indebted to Franciscan missionaries for the early reports. In 1812, the "year of earthquakes", the churches at both San Juan Capistrano and Purisima were destroyed in separate shocks. In 1836 and 1838, large earthquakes originated near San Francisco and on January 9, 1857 a major earthquake in central California near Fort Tejon was accompanied by rupture of the San Andreas fault. On October 21, 1868, a large earthquake occurred with rupture on the Hayward fault, which branches northward from the San Andreas fault in the vicinity of Hollister.

The California earthquake of April 18, 1906 is referred to a number of times in this book. A State Earthquake Commission was set up under the chairmanship of Andrew Lawson and its Report provided one of the classical studies of a single earthquake, studies by engineers unfortunately not being included but appearing separately. The epicenter of the earthquake was on the San Andreas fault, just to the west of the Golden Gate. The magnitude of the shock was 8.25 and the area over which it was felt was some 780,000 square km—not large for an earthquake of this magnitude. Fortunately, only some 500 people were killed. The greatest damage was done by the great conflagration following the earthquake in San Francisco. Of key importance was the formulation by H.F. Reid of the *elastic rebound theory* of earthquake genesis.

In the Report of the State Earthquake Commission the cause of the earthquake was said to be "the sudden rupture of the Earth's crust along the line, or lines, extending from the vicinity of Point Delgada to a point in San Benito County near San Juan; a distance of nearly straight course of nearly 430 km. For a distance of 300 km, from Point Arena to San Juan, the fissure caused by this rupture is known to be practically continuous. Beyond Point Arena, it passes out to sea, so its continuity with a similar crack near Point Delgada is open to doubt."

On this last comment, recent oceanographic evidence indicates that the San Andreas fault continues as a submarine feature from Point Arena to Point Delgada, suggesting that the rupture was continuous from San Juan Bautista to near Cape Mendocino (see Fig. 1.3).

Although the Central Valley and Sierra Nevada have been historically free from significantly hazardous earthquake sources, all structures in this region are subject to palpable shaking from large earthquakes of the San Andreas system and also from large earthquakes to the east in the Owens Valley and along the Honey Lake escarpment and Mohawk Valley of the Basin-and-Range province. The centennial of perhaps the largest of California's historic earthquakes, the Owens Valley shock of March 26, 1872, has now passed. Extensive fault ruptures appeared for a distance of 150 km along the valley. Across the valley, the width of ground disturbances ranged from less than one to 15 km. In geological terms, a century is but an instant and even today the curious can see clearly the largest rupture from Haiwee to Big Pine. The shaking was felt as far east as Salt Lake City, and at least 25 persons died, mainly from the destruction of adobe houses in Lone Pine.

Oregon is largely aseismic, while moderately deep earthquakes have caused damage and loss of life in the Puget Sound region of Washington. Moderate earthquakes with historic fault rupture have occurred in Nevada, Utah and Montana.

Active fault zones in the central United States are not well delineated, in contrast to those in the western region. There is no definite evidence of surficial fault breakage for any central or eastern United States earthquake.

The upper part of the Mississippi Embayment, southwest Kentucky, southern Illinois, and southeastern Missouri, is one of the most seismically active regions in the eastern United States. It has been a region of downwarp (perhaps due to compressional forces) since Cretaceous times. Complex fault structures have been mapped near the head of the Embayment near New Madrid, the site of the 1811–1812 earthquakes (MM intensity XII). It is of interest to note that if the settlement of North America by Europeans had taken place after 1812, even the New Madrid region would be classified as a minor seismic one. It would be hard not to conclude, contrary to what the compilers of Fig. 1.5 have done, that the infrequent occurrence of earthquakes in central and eastern United States indicates that the seismic risk is generally much lower than that in much of California.

Apart from the New Madrid and the northeast regions, concern is usually expressed about the seismic risk in one area on the eastern seaboard (see Fig. 1.5). On August 31, 1886, an earthquake (Intensity MM IX–X) occurred near Charleston, South Carolina, which did considerable damage in Charleston (much of which was built on filled land) and killed about 60 persons. It was felt as far away as New York, Boston, Cuba and Bermuda. No associated surface faulting has ever been detected.

An explanation of the great distances to which earthquakes in central and eastern United States are perceptible has been recently worked out. Seismograph recordings of contemporary earthquakes in Illinois show that seismic waves are not as severely attenuated compared with California as they travel through the Earth's crust in the eastern United States. Thus, an earthquake of the same magnitude would be felt at a greater distance than in California. This finding suggests, incidentally, that the magnitude of some of the historical eastern earthquakes may have been overestimated, so that the largest of the New Madrid sequence may have had a magnitude of only about 7.6.

Chinese Earthquakes

The Academy of Sciences of the People's Republic of China has been studying the dynastic and ancient literary works, temple records, and other sources for earthquakes in ancient China. The oldest earthquake that has been traced occurred in 1,100 B.C.; however, the record can be considered fairly complete only from about 780 B.C., the period of the Chou dynasty in northern China.

The intensity distribution of many can be assessed from preserved reports. For example, the San-ho earthquake (the greatest known near Peking) of September 2, 1679 is mentioned in the records of 121 cities with data on building damage, ground cracks and other geological features near the source, together with felt reports from distant places. From such isoseismal information, by comparison with earthquakes of recent times, the magnitude can be roughly assessed at 8.

The earthquake which cost the greatest known loss of life anywhere in the world occurred on January 23, 1556 in Shensi near the city of Hsiăn. Dynastic records are available which give an estimate of 830,000 people who died from all causes in this earthquake, a death toll so great that one might question the validity of the figures. The explanation is that the earthquake struck a densely

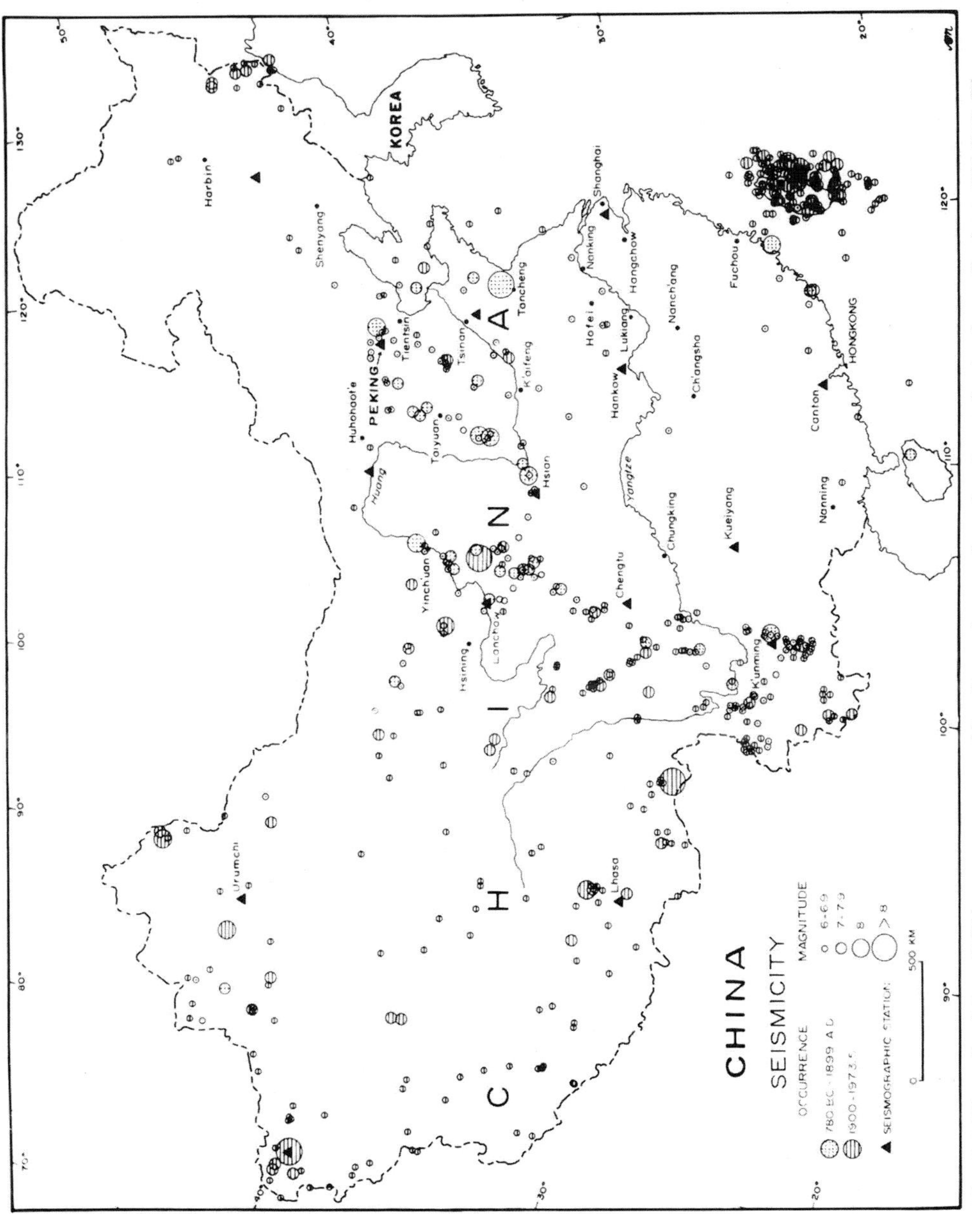

Fig. 1.6. The seismicity of China as represented by damaging earthquakes (magnitude above 6) since 780 B.C. The map is copied from one generously supplied by the Geophysical Institute, Peking

populated region, where the peasants lived mostly in caves in the loess hillsides. (*Loess* is windblown dust that compacts into thick layers (see Section 6.6).)

The 1556 earthquake occurred at 5 o'clock in the morning, when the families were asleep indoors and dwellings collapsed on them. In addition, demoralization, famine, and disease which can follow such a great disaster no doubt accounted for a significant number of deaths.

The seismicity of China since 780 B.C. is shown in Fig. 1.6, which has been redrawn from a colored map kindly made available by the Geophysical Institute, Academia Sinica, Peking. Three features of the map are outstanding: (i) the highly seismic province of Taiwan is associated with a subduction zone (see Fig. 1.1); (ii) the southeastern region of the country, including Kwangtung, Fukien, Hunan and Kiangsi provinces, is relatively aseismic and presents no great earthquake hazard; (iii) large, damaging intraplate earthquakes occur along the borders of the Western Mountains, running from the Himalayan region along the edge of the Tibetan Plateau up to the vicinity of Peking. The fracture zone runs almost north-south at longitude 103° from Kunming to the east of Lanchow, then at 35° latitude, branches eastward along the northern edge of the Tsingling Shan mountains past Hsiăn. Earthquake foci then follow a bifurcation into two northerly trends, one along the Shansi *geosyncline* (a trough hundreds of kms long in which several thousand meters of sediment have accumulated) east of the Hwo Shan mountains through Taiyuan and then northeastward to the west of Peking. The second follows the east flank of the Taihang Shan along the west edge of the North China Plain to at least 40° north. An unusual sequence of moderate earthquakes occurred in 1966 in this zone near Shintai (about 200 km northwest of Tsinan in Fig. 1.6), the largest of which, magnitude 7.2, occurred on March 22 and was noted for widespread liquefaction of soil with failures of embankments and canals.

On the basis of the historical seismicity, seismic risk maps have been drawn for China and seismic resistant building codes prepared. In 1973, there were some 60 strong-motion accelerometers in various seismic areas of the country (compared with about 1,000 in the USA, 700 in Japan, and 90 in New Zealand), so that in the next few decades important records of strong ground shaking should be available from these instruments.

1.3. Causes of Earthquakes

Tectonic Earthquakes

In the time of the Greeks it was natural to link the Aegean volcanoes with the earthquakes of the Mediterranean. As time went on it became clear that most damaging earthquakes were in fact not caused by volcanoes, but that the majority of earthquakes called *tectonic earthquakes* are associated with deformation of the Earth's outermost shell, particularly the crust, which in continental areas is about 35 km thick. These are generated by the rapid release of strain energy stored within the elastic rocks.

Tectonic earthquakes often occur in regions of great change in elevation, such as the earthquakes along the Bocono fault in the Venezuelan Andes, the earthquakes of the Hindu Kush and Himalayas and, for that matter, earthquakes along the high submarine ridges under the oceans (see Fig. 1.1). So, too, as we have seen in Section 1.2, earthquakes occur along the deep oceanic trenches that border tectonic plates.

To complete the picture, many tectonic earthquakes occur where there are no great mountains or changes of elevation. Examples are the 1811–1812 New Madrid intraplate earthquakes in central United States and the Dasht-e Bayāz earthquake of August 31, 1968 in northeastern Iran.

Elastic Rebound Theory

Field evidence in the 1906 California earthquake showed clearly that the strained rocks immediately west of the San Andreas fault had moved northwest relative to the rocks to the east. Displacements of adjacent points along the fault reached a maximum of 6 m near Olema in the Point Reyes region.

H.F. Reid studied the triangulation surveys made by the U.S. Coast and Geodetic Survey across the region traversed by the 1906 fault break. These surveys made in 1851–1865, 1874–1892 and just after the earthquake, showed that (i) there were small inconsistent changes in *elevation* along the San Andreas fault; (ii) significant horizontal displacements parallel to the fault trace had occurred; and (iii) distant points on opposite sides of the fault had moved 3.2 m over the 50-year period, the west side moving north.

Based on geological evidence, geodetic surveys, and his own laboratory experiments, Reid put forth the *elastic rebound theory* for an earthquake mechanism. This supposes that the crust of the Earth in many places is being slowly displaced by underlying forces. Differential displacements set up elastic strains that reach levels greater than can be endured by the rock. Ruptures (faults) then occur, and the strained rock rebounds along the fault under the elastic stresses until the strain is partly or wholly relieved. This theory of earthquake mechanism has been verified under many circumstances and has required only minor modification.

The strain slowly accumulating in the crust builds a reservoir of elastic energy, just as, for example, a coiled spring, so that at some place, the *focal point*, within the strained zone, rupture suddenly commences, and spreads in all directions along the fault surface in a series of erratic movements due to the uneven strength of the rocks along the tear. This uneven propagation of the dislocation leads to bursts of high-frequency waves which travel into the Earth to produce the seismic shaking that causes the damage to buildings. The fault rupture moves with a typical velocity of two to three km per second and the irregular steps of rupture occur in fractions of a second. Ground shaking away from the fault consists of all types of wave vibrations (P, S, and surface waves) with different frequencies and amplitudes.

Types of Fault

Fault displacement in an earthquake may not be almost entirely horizontal, as it was in the 1906 San Francisco earthquake along the San Andreas fault,

Plate 1.3. Normal fault scarp (4.5 m) associated with the Dixie Valley-Fairview Peak earthquake (Nevada) of December 1954. (Courtesy of P. Byerly)

but often large vertical motions occur, such as were evident in the 1954 Dixie Valley, Nevada earthquake (Plate 1.3) and in California in the 1971 San Fernando earthquake, where an elevation change of three meters occurred across the fault in some places.

The classification of faults is straightforward and depends only on the geometry and direction of relative slip. Various types are sketched in Fig. 1.7. The *dip* of a fault is the angle that the fault surface makes with a horizontal plane and the *strike* is the direction of the fault line exposed at the ground surface relative to the north.

A *strike-slip* fault, sometimes called a *transcurrent* fault, involves displacements of rock laterally, parallel to the strike. If when we stand on one side of a fault and see that motion on the other side is from left to right, the fault is *right-lateral strike-slip*. Similarly, we can identify *left-lateral strike-slip*.

A *dip-slip* fault is one in which the motion is largely parallel to the dip of the fault and thus has vertical components of displacement. A *normal* fault is one in which the rock above the inclined fault surface moves downward relative to the underlying crust. Faults with almost vertical slip are also included in this category.

A *reverse* fault is one in which the crust above the inclined fault surface moves upward relative to the block below the fault. Thrust faults are included in this category but are generally restricted to cases when the dip angle is small.

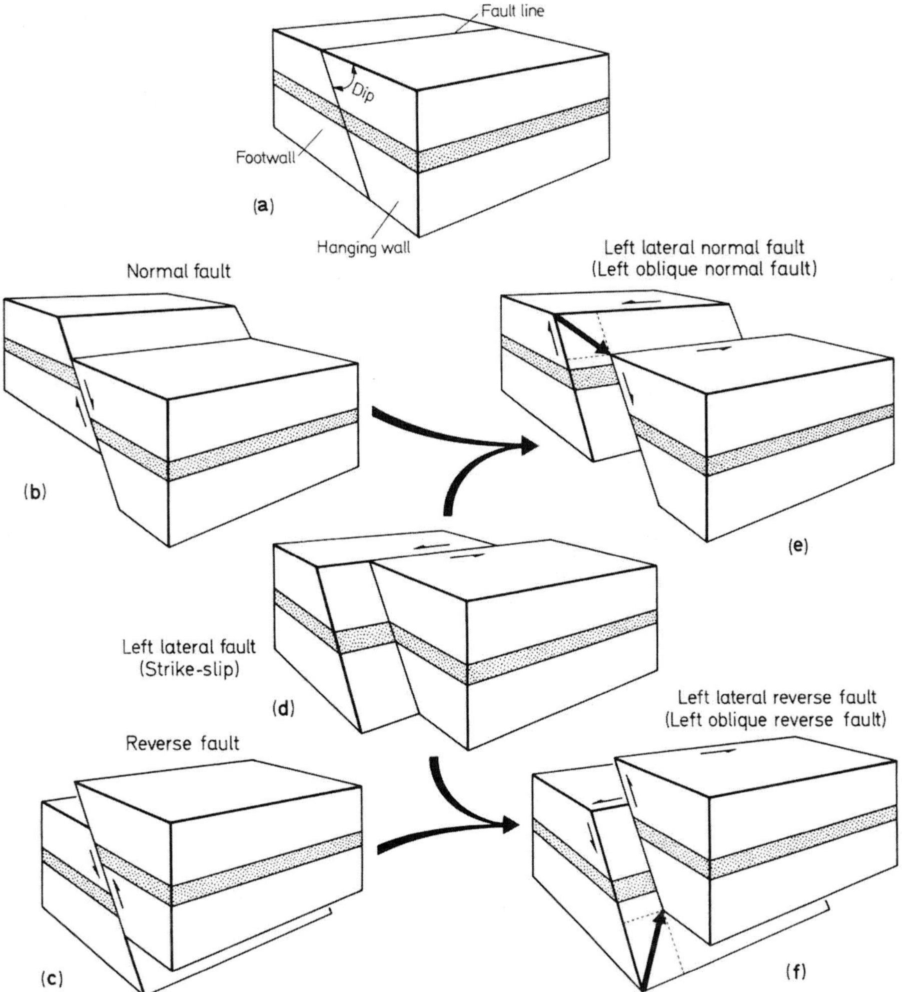

Fig. 1.7a–f. Diagrammatic sketches of fault types (a) names of components, (b) normal fault, (c) reverse fault, (d) left-lateral strike-slip fault, (e) left-lateral normal fault, (f) left-lateral reverse fault. (After California Geology, November 1971)

Geological hazards vary with the fault type, in general. The main effects are illustrated in Fig. 1.7. Along a strike-slip fault the zone of disturbance is usually the narrowest, whereas in normal or thrust faulting grabens and slides along the fault scarp may extend the motion over a considerable width.

Field studies have brought to light cases where fault fractures do not penetrate from the bedrock through overburden. Weathered rock and soil near the surface sometimes absorb the differential slip, a spectacular example of which comes from the 1964 Alaskan earthquake. Reverse faulting on the Patton Bay fault produced a scarp 2.5 m high in the gravel-covered bedrock at the level of the beach, but at the top of a sea cliff traversed by the fault no comparable scarp could be observed. Thus, the weathered rocks, through a distance of about 20 m, absorbed over 2 m of total displacement.

Despite such examples, fault rupture often does penetrate through hundreds of meters of unconsolidated deposits, indeed some amplification of the displacement has been found in soft deposits. Such contradictions have led to the development of trenching techniques across old fault traces to determine whether surface measurements of slip are typical.

For over a decade it has been known that displacement in fault zones occurs not only by sudden rupture in an earthquake but also by slow differential slippage of the sides of the fault. The fault is said to be undergoing *tectonic creep*. Slippage rates range from a few millimeters to a centimeter or so per year.

The best examples come from the San Andreas zone near Hollister, California, where a winery built straddling the fault trace is being slowly deformed and, in the town, sidewalks, curbs, fences and homes are being offset. On the Hayward fault, on the east side of San Francisco Bay, many structures are being deformed and even seriously damaged by slow slip, including a large water supply tunnel, a drainage culvert and railroad tracks that intersect the zone.

Creep apparently occurs when a zone of gouge extends to depth along the fault. Gouge is finely crushed rock, commonly altered to clay; it is of low strength and is often water saturated, thus ensuring that the effective friction coefficient is small. Below the zone of gouge, more competent rocks are presumably welded across the fault plane and elastic strain energy is accumulated. But near the surface the gouge material is carried along by the adjacent stronger surface rocks, with more-or-less continuous slippage.

The most recent geodetic studies of strain accumulation along the San Andreas fault system in central and northern California (see Fig. 1.3) show that relative right-lateral motion is now occurring at the rate of about 3.2 cm each year. Specially precise measurements of horizontal displacement are made from mountain top to mountain top across the fault zone using laser-beam geodimeters. If air temperature variations along the line of sight are measured by aircraft the precision of measurement is better than 1 cm in 10 km, or 10^{-6}. In the central part of the Coast Ranges, most of the motion is concentrated within a few tens of kilometers of the fault trace. Northwards to San Francisco Bay, it is concentrated over the trifault system of the San Andreas, Hayward and Calaveras, about 20 km apart, while north of San Francisco Bay the elastic deformation appears to be spread over more than 40 km. Some of this motion is undoubtedly reflected in slow fault slip but, according to the model mentioned above, we cannot be sure that where slow slip is occurring strain is not accumulating in the stronger rocks 5 to 20 km down in the crust. Thus, faults where fault creep is observed cannot at present be taken as zones of low risk.

A glance at any detailed geologic map shows that most areas of the crust contain a great many faults. Some are small and can be traced only for short distances, merging into mere rock fractures. Others have long been inactive. Geological field work has established that many long, ancient faults (such as the Melones fault in California and the Newark fault in the Appalachians) have had no movements across them for millions of years. Various processes may have healed the ruptures and the strength across the mapped fault may be almost as great as outside the zone. Thus, most known faults do not contribute to geological hazards.

It is difficult, as in the case of volcanoes, to say whether a fault is *active, inactive* or merely *dormant*. For example, no fresh rupture has occurred on the San Andreas fault near Fort Tejon since the great earthquake of 1857. The Dead Sea fault in Palestine, once an active locus of earthquakes according to Biblical and historical records, has had no rupture on it or been associated with large earthquakes for many hundreds of years. In a rough way, an active fault can be defined as one that has moved in the recent past and the geological and seismological evidence is that it may move in the near future. The recent past includes the historical period and sometimes extends to include at least the *Holocene epoch*, stretching back about 10,000 years.

Small earthquakes, down to even *microearthquake* size (magnitude <2), may provide such indication, but contrary cases cannot be ruled out with present knowledge. It is hardly ever clear whether the occurrence of microearthquakes along a fault really indicates if sufficient strain is present for a large earthquake to occur on it in the future. The microearthquakes may be merely remnants of ancient strains.

In recent years it has become easier to determine if movement has in fact taken place on faults and to date this movement. Methods involve careful field mapping, trenching and the use of radiocarbon dating to find the age of carbonaceous fossils along the fault, geodetic measurements and side-angle aerial photography.

Dilatancy in the Crustal Rocks

At a depth in the crust of 5 km or so, the lithostatic pressure (due to the weight of the overlying rocks) is already about equal to the strength of uncracked typical rock samples at the temperature (500° C) and pressure appropriate for that depth. If no other factors entered, the shearing forces required to bring about sudden brittle failure and frictional slip along a crack would never be attained; rather, the rock would deform plastically. A way around this problem was the discovery that the presence of water provides a mechanism for sudden rupture by reduction of the effective friction along crack boundaries.

Soviet seismologists in the early 1960's, working with earthquakes in the Garm region, reported that P wave velocities in the crust appeared to decrease, then increase, in a region before a sizeable earthquake occurred in it. Similar observations have recently been obtained in a number of places in the United States.

Studies of the time of travel of P and S waves before the 1971 San Fernando earthquake (see Section 1.4) indicated that four years before it occurred the ratio of the velocity of the P to the velocity of the S waves decreased rather suddenly by 10 percent from its average value of 1.75. There was, thereafter, a steady increase in this ratio back to a more normal value. One explanation is the *dilatancy* model. We may suppose plausibly that as the crustal rocks become strained, cracking occurs locally and the volume of rock increases or "dilates".

Cracking may occur too quickly for ground water to flow into the dilated volume to fill the spaces so the cracks become vapor-filled. The consequent fall in pore pressure would lead to a reduction mainly in P velocities. Subsequent diffusion of groundwater into the dry cracks would increase the pore pressure, and provide water for lubrication along the walls of the cracks, while at the same time, the P velocity would increase again.

The full picture of the dilatancy theory of earthquake genesis is not yet clear, but the hypothesis is attractive in that it is consistent with precursory changes in ground levels, electrical conductivity and other physical properties which have been noted in the past before earthquakes. The theory has a potential for forecasting earthquakes under certain circumstances. For example, measurement of the P velocity in the vicinity of large reservoirs before and after impounding of water (see below in this Section) might provide a more direct method of indicating an approaching seismic crisis near dams than is now available.

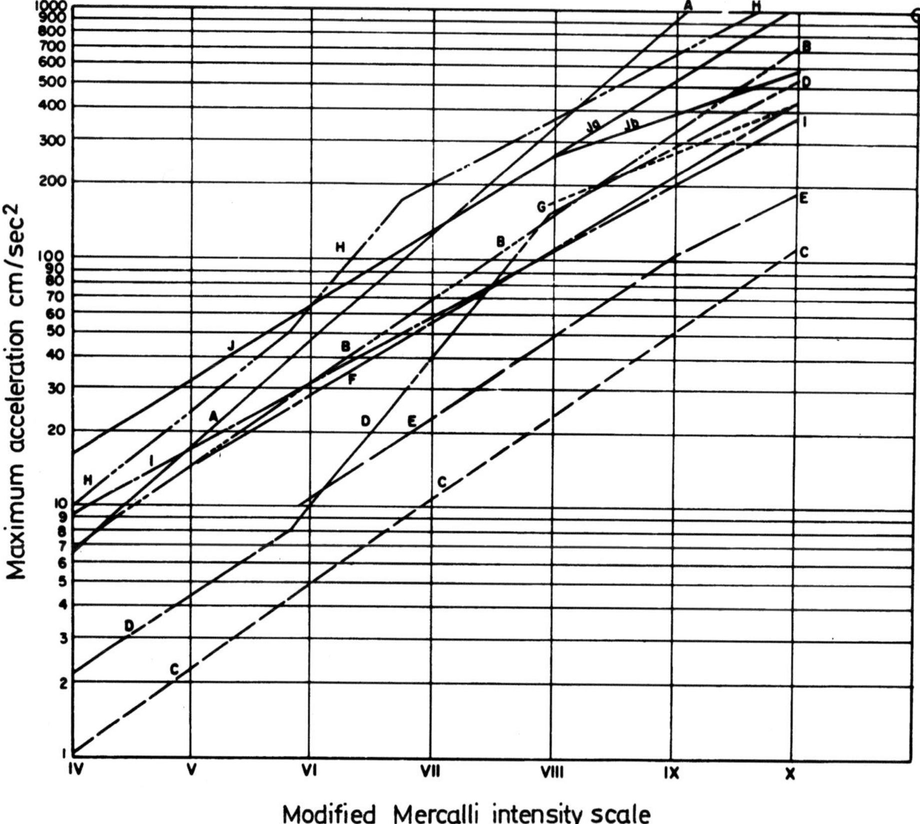

Fig. 1.8. Some empirical correlations of MM intensity with peak ground acceleration measured in earthquakes. *A* Hershberger; *B* Gutenberg-Richter (1942); *C* Cancani (1904); *D* Ishimoto (1932); *E* Savarensky-Kirnos (1955); *F* Medvedev et al. (1963); *G* N.Z. Draft By-Law; *H* Kawasumi (1951); *I* Cornwell; *J* Neumann (1954). O San Fernando earthquake, Feb. 1971

Prediction and Control of Earthquakes

The forecasting of earthquakes, like that of phases of the moon, weather and other natural phenomena, is, clearly enough, an integral part of science. Aspects that tend to receive most publicity are: prediction of the *place*, prediction of the *size*, and prediction of the *time* of the earthquake. For most people, prediction of earthquakes means prediction of the time of occurrence. A more important aspect for mitigation of hazard is the prediction of the *strong ground motion* likely at a particular site (see Section 1.5). This type of prediction is often based upon correlations between the observed Modified Mercalli intensity and recorded accelerations of the ground (see Fig. 1.8).

As noted in Section 1.2, prediction of the region where earthquakes are likely to occur has now been largely achieved by seismicity studies using earthquake observatories. Because empirical relations between the magnitude of an earthquake and the length of observed fault rupture have also been constructed (see Table 1.3), some limits can be placed on the size of earthquakes for a particular region.

Table 1.3. Earthquake magnitude versus fault rupture length

Magnitude (Richter)	Rupture length (km)
5.5	5–10
6.0	10–15
6.5	15–30
7.0	30–60
7.5	60–100
8.0	100–200
8.5	200–400

Many attempts have been made to find clues for forewarning. Recently, Chinese seismologists, using purely statistical methods, have claimed limited success in predicting moderate-sized earthquakes in an area where there is a long run of historical data. Elsewhere, emphasis has been placed on geodetic data, such as geodimeter measurements of deformation of the Californian crust along the San Andreas fault.

An *ex post facto* premonitory change in ground level was found after the Niigata earthquake (see Section 1.4), which if it had been discovered beforehand, might have served as one indication of the coming earthquake.

Another scheme is based on detecting spatial and temporal gaps in the seismicity of a tectonic region. In 1973, a prediction was made by seismologists of the U.S. Geological Survey that an earthquake with a magnitude of 4.5 would occur along the San Andreas fault south of Hollister within the next six months. The prediction was based on four principal shocks which had occurred within 3 years on both ends of a 25-km-long stretch of the San Andreas fault, bracketing a 6-km-long section free from earthquakes in that time interval. The assumption was that the mid-section was still stressed but locked, ready to release the elastic energy in an earthquake. However, no earthquake occurred in the six months predicted. One difficulty with such methods is the assessment of a zero epoch with which to compare the average background occurrence rate.

A more promising method is based on the theoretical and experimental hypothesis involved in the dilatational theory mentioned above. Unfortunately, as indicated by the dilatation cycle time given in Table 1.4, the forecasting interval is rather long for the more hazardous earthquakes.

Table 1.4. Duration of precursory changes in velocities of P and S waves

Earthquake magnitude	Duration
2	1 day
3	1 week
4	1 month
5	3 months
6	1 year
7	6 years

It has often been pointed out that even if the ability to predict the time and size of an earthquake was achieved by seismologists, many problems remain on the hazard side. Suppose that an announcement were made that there was a chance of one in two of a destructive earthquake occurring within a month. What would be the public response? Would the major industrial and commercial work in the area cease for a time, thus dislocating large segments of the local economy? Even with shorter-term prediction, there are difficulties if work is postponed until the earthquake period is over. Suppose that the predictive time came to an end and no earthquake occurred; who would take the responsibility of reopening schools and resuming other activities?

An allied approach to reduction of seismic danger may some day turn out to be rewarding. Is there a way in which earthquakes can be controlled, like floods and tsunami run-up? A chance occurrence near Denver, Colorado suggests one possibility. There, when waste liquid was being pumped down a deep borehole for disposal, earthquakes which correlated quite closely in number with the amount of water injected began to occur in the area. Perhaps the water reduced friction along pre-existing faults around the disposal well to a point where residual strain in the crust could be reduced by sudden slip, or perhaps an aquifer was inflated with consequent adjustments. As a follow-up, small earthquakes were produced by water injection by seismologists of the U.S. Geological Survey at the Rangely oil field.

The experiment has led to the speculative proposal that the magnitude of earthquakes along an active fault might be restricted to, say, magnitude 5 or less by deliberately inducing rupture by pumping water down holes bored into fault zones. The length of fault rupture (see Table 1.3) might be restricted by bounding the region with "seismovalves" at which points water would be pumped from the fault zone in order to "lock" it. The scheme supposes a better knowledge of rate and direction of groundwater movement than is now available for most regions.

Nuclear Earthquakes

Engineering projects have been undertaken, in both the USA and the USSR, using large underground nuclear explosions. The program in the United States to harness energy released in atomic bombs for construction purposes is part of the Plowshare Project. Some Plowshare experiments have been to stimulate production of natural gas by fracturing gas-bearing strata. In one experiment on September 10, 1969, a 40-kiloton nuclear bomb was fired 2,570 m underground in Garfield County, 70 km northeast of Grand Junction, Colorado, fracturing interbedded sandstone and shale. In the months following the explosion, well-head pressure rose by a factor of 6 and the value of the additional gas extracted is expected to reach $1,570,000.

In the Soviet Union, a large-scale program of water management has been carried out for years around the Caspian Sea. In the last 15 years the groundwater level had dropped by 2.5 m; by the turn of the century a further fall of 1.7 m is predicted if no corrective steps are taken. The Russians have planned to intercept the northward flowing rivers by dams and divert the water southward. One canal of 112 km length was proposed for excavation by nuclear explosions. Since 1966, overseas seismographic stations have detected seismic waves from large sources in the region near the Caspian Sea. Indications are that these are unannounced underground nuclear explosions which have been used for this water-management program.

In all such applications to mining and construction, of nuclear explosions with yields comparable with moderate earthquakes, guarantees have to be met concerning seismic safety. Ground vibrations generated by underground detonation may be strong enough to cause structural damage. Tests on this danger have been carried out at the Nevada Test Site by the U.S. Atomic Energy Commission and Soviet investigators have published similar studies of underground nuclear explosions.

Large Reservoir-Induced Earthquakes

Before Lake Mead, on the Colorado River, was filled in 1935, there was no historical record of earthquake activity in the area; afterwards small earthquakes became frequent. Moreover, local seismographic stations, established by the U.S. Coast and Geodetic Survey, showed that the number of shocks correlated rather closely with changes in reservoir loading. Since the Lake Mead experience, about ten to fifteen additional documented cases (see Table 1.5) are known where large reservoirs have probably triggered earthquakes. However, as Table 1.5 shows, there is as yet no clear-cut tectonic pattern in these earthquake occurrences. Sometimes the triggering occurs in known seismic areas (such as Kremasta, Greece), sometimes in regions of very low seismicity (such as Lake Mead and Koyna). The foundation geology for the listed reservoirs varies greatly from granitic and metamorphic rock to shales and sandstones. The fault mechanisms

Table 1.5. Man-made reservoirs with induced seismicity

Location (dam-country)	Dam height (m)	Capacity ($m^3 \times 10^9$)	Basement geology	Date impounded	Date of first earthquake	Seismic effect
L'Oued Fodda, Algeria	101	0.0002	Dolomitic marl	1932	1/33	Felt
Hoover, USA	221	38.3	Granites and Pre-cambrian shales	1935	9/36	Noticeable (M=5)
Talbingo, Australia	176	0.92				Seismic (M<3.5)
Hsinfengkiang, China	105	11.5	Granites	1959		High activity (M=6.1)
Grandval, France	78	0.29		1959/60	1961	MM intensity V in 1963
Monteynard, France	130	0.27	Limestone	1962	4/63	M=4.9
Kariba, Rhodesia	128	160	Archean gneiss and Karoo sediments	1958	7/61	Seismic (M<6)
Vogorno, Switzerland	230	0.08		8/64	5/65	
Koyna, India	103	2.78	Basalt flows of Deccan Trap	1962	1963	Strong (M=6.5) 177 people killed
Benmore, N. Zealand	110	2.04	Greywackes and argillites	12/64	2/65	Significant (M=5.0)
Kremasta, Greece	160	4.75	Flysch	1965	12/65	Strong (M=6.2) 1 death, 60 injuries
Nuzek Tadzik, USSR	300	10.5		1972 (to 100 m)		Increased activity (M=4.5)
Kurobe, Japan	186			1960/69		Seismic (M=4.9)

that have been determined for triggered earthquakes vary from strike-slip to dip-slip (see Fig. 1.7) in character. The largest load ever placed on the Earth's crust by man is Lake Kariba on the Zambezi River, East Africa. Impounding began in December 1958 and storage capacity is over 150 cubic km of water. There are, unfortunately, no seismic records before loading, but sporadic activity has been recorded from the beginning of operation of seismographs near the lake in 1961.

A great many more large reservoirs have not induced nearby earthquakes, although some are built in tectonically active areas (such as San Luis, California). In the United States, by 1973 there were 52 completed dams over 100 m high. Eighty-two percent of these are in seismic zones 2 and 3 (see Fig. 1.5), but only eighteen have seismic activity in the vicinity. Perhaps ten cases of seismicity may be related to water impounding. Lake Mead remains the one reservoir in the United States with sufficient seismicity to provide a high statistical significance between the positive correlation of water level and numbers of earthquakes.

Microseismicity at Lake Mead continues, and up to 1973 over 10,000 earthquakes have been recorded near the lake and Hoover dam, about 1,000 of which have been felt by residents in the area. The largest recorded shock occurred on May 4, 1939, with an intensity of Modified Mercalli VI.

Under the California State Water Plan a series of large dams has been constructed, linked by aqueducts and canals (see Chapter 7). By 1962, before completion of the dams, special networks of seismographic stations were established around them to determine local seismicity; the dams were at the same time instrumented with strong-motion accelerometers. The sensitive vertical-component seismographs have more recently been connected by telephone telemetry to recorders in Sacramento where foci are quickly located.

The Oroville earthfill dam on the west flank of the Sierra Nevada is 236 m high and its capacity is 4,365,000,000 cubic m (maximum operating level). The reservoir commenced to fill in November 1967 and was completed in September 1968. The seismic networks showed that from 1963 to 1967 no local earthquakes above magnitude 2 (above the detection threshold) occurred. Since the reservoir was loaded, no local seismicity has been detected within 8 km of the dam.

There are two likely causes for the triggering effect of a large reservoir, both of which may be partly responsible. The triggered earthquakes generally occur at shallow depths under the reservoir where the strain in the rock would be increased by the extra load of water placed on the Earth's crust. This enhanced strain may exceed the value needed for local faulting to occur at some sites. The additional load, however, turns out to be relatively small compared with the energy in the earthquakes. A more effective mechanism may be reservoir water percolating through fractures below the newly filled reservoir. The head of groundwater would increase the pore pressure in the rocks by a few tens of bars which would be sufficient to reduce the effective frictional resistance along fractures and lines of weakness, thus allowing sudden slip to take place.

The only effective method of decreasing any hazard to a dam structure from induced earthquakes is to lower the reservoir level.

Volcanic Earthquakes

As Fig. 1.1 shows, volcanoes and earthquakes often occur together along the margins of plates around the world. Like earthquakes, there are also intraplate volcanic regions, such as the Hawaiian volcanoes.

Despite these tectonic connections between volcanoes and earthquakes, there is no evidence that moderate to major shallow earthquakes are not essentially all of tectonic, elastic-rebound type. Those earthquakes that can be reasonably associated with volcanoes fall into three categories: (i) volcanic explosions, (ii) shallow earthquakes arising from magma movements, and (iii) sympathetic tectonic earthquakes.

As explained in Section 2.2, violent explosions are common as gas dissolved in the magma under pressure escapes, and these give rise to elastic waves which may be detected by seismographs in the vicinity. It is also likely that as the magma moves from one storage place to another along connecting tubes the heated crustal rock is subject to temperature gradients which lead to straining and cracking and inflation and deflation of the volcano volume. The presence of superheated steam and gases is favorable for local rupture along the cracks. This mechanism probably explains many shallow volcanic earthquakes reported during eruptions (see Chapter 2). Perhaps the clearest verification of this process comes from a recent eruption of Mauna Loa. A network of seismographs around the volcano detected small earthquakes starting at depths of about 15 km. Over a period of some months, the focal depths of the earthquake activity were found to decrease, as magma moved closer to the surface. The episode ended with a lava flow from the volcano. In such circumstances, the seismographic network might be used to predict eruption.

Category (iii), tectonically associated with volcanoes, is more difficult to tie down, as cases which may fit this category are rare (see Chapter 2). There is no report of significantly increased volcanic activity in the great 1964 Alaska earthquake (see Section 1.1), but Puyehue Volcano in the Andes erupted 48 hours after the great 1960 Chilean earthquake. We can at least suggest a mechanism for the stimulation of volcanoes by strong ground seismic shaking. One might suppose that in a large earthquake the ground shaking would set up waves in reservoirs of magma; the general compression and dilatation of the gaseous liquid melt may trigger volcanic activity.

One case of a coincident tectonic earthquake has been put forward in connection with a large eruption of Sakurajima Volcano in Japan. In January 1914, after the start of the eruption, a destructive earthquake occurred near the city of Kagoshima, not far from the volcano. The earthquake, of magnitude 7, had an estimated depth of 50 km. In 1868, a moderate tectonic earthquake on the south end of the Island of Hawaii was accompanied by fault movement of about a meter, a big local tsunami, and eruptions of both Mauna Loa and Kilauea Volcanoes.

1.4. Earthquake Case Histories

Hsinfengkiang Dam, China, March 19, 1962

Construction of a large concrete dam was begun in 1958 about 160 km northeast of Canton, with a maximum height of 105 m, and a crest length of 440 m and reservoir storage capacity of 11.5 cubic km (see Table 1.5). This dam was constructed in an area for which there was no historical record of destructive earthquakes and which would be rated low on any risk scale (see Fig. 1.6). Filling of the reservoir in 1959 was responsible for the outbreak of a long series of earthquakes that continued for the next six years at a rather high rate, so that by 1972, over 250,000 earthquakes of magnitude greater than 0.2 and very shallow focal depth (less than 10 km) had been recorded.

The reservoir lies on faulted granitic rocks of Late Mesozoic age. The faulting is complex with zones of crushed rock and fault gouge, and there is some geological and geodetic evidence for recent movement on the fault system.

After the reservoir level reached a maximum (115 m) for the first time on September 23, 1961, an increase occurred both in number and size of earthquakes. From February 1962 seismic activity decreased and no shock of magnitude greater than 3 occurred for 20 days, when on March 19, 1962, a principal shock of magnitude 6.1 and strike-slip mechanism occurred which caused sufficient shaking to produce a crack 82 m long in the upper dam structure. This remarkable case of the induction of earthquakes by large reservoirs has been studied in detail by Chinese seismologists and engineers.

Subsequently, the dam was strengthened and strong-motion instruments were placed at different elevations on the dam structure. Aftershocks provided excellent strong-motion records. For one earthquake of magnitude 4.5 a peak horizontal acceleration of 5 per cent that of gravity (0.05 g) recorded at the base of the dam had amplified tenfold at the crest. The engineers were able to check the dynamic behavior of the strengthened dam, using computer analyses based on recorded accelerograms.

Skopje, Yugoslavia, July 26, 1963

Earthquakes do not have to be of large magnitude to produce serious damage in regions where buildings are weak and devoid of earthquake resistive capability. When a great natural disaster strikes, a key question is whether the population should move to a less hazardous site, for example the experience of Skopje, a Macedonian city of 200,000 people in Yugoslavia.

The forerunner of modern Skopje, the Roman colony of Scupi, was situated at the confluence of the Lepenac and Vardar rivers. In 518 A.D., Scupi was completely destroyed by an earthquake, which led to the abandonment of the site, and a new city was built to the east of the present site by the emperor Justinian. Exaggerated accounts of the 518 A.D. earthquake relate that Scupi sank into the Earth with all its inhabitants, trees disappeared up to their tops and huge cracks formed in the ground, one measuring 4 m in width!

Since 518 A.D., records indicate that about 25 earthquakes of intensity VI or greater have occurred. An earthquake on August 10, 1921, is said to have formed a new fault rupture north of the city of Skopje. In the 1963 earthquake, on the other hand, no extensive fresh faulting was reported, but minor dislocations, crevices and sand blows were observed north and south of the city.

The 1963 earthquake struck in the early morning hours (05:17 h local time) with no warning foreshocks reported. It had a moderate magnitude of 6.0 and was centered only a few kilometers northeast of Skopje. Ten per cent of the buildings in the city collapsed and an additional 65 per cent were damaged beyond repair. The number of victims was in the vicinity of 1,200, but would have been much greater if the origin-time had been later, when school and administrative buildings (many of which were destroyed) were occupied. An unusual feature of the earthquake was the lack of fire damage, as on the summer morning no artificial heat was being used and electric power was switched off immediately after the earthquake struck. Only minor breaks in the city's water lines occurred and the sewer system remained operative. Of particular interest for the subject of this book, many earthquake casualties resulted from serious floods of the Vardar river in 1962 and 1963. These floods cost over $ 3,000,000, weakened foundations, and led to homeless people being crowded into flats which collapsed in the earthquake.

Severe damage was restricted almost entirely to within the city. The same circumstance was noted in the Agadir, Morocco earthquake of February 29, 1960 of magnitude 5.5. Observations suggest that ground motion was concentrated into a few pulses. In Nicaragua, a similar case perhaps occurred in the Managua earthquake of December 23, 1972, for which a peak acceleration of 0.39 g and bracketed duration (see Section 1.5) of 10 seconds were recorded (see Fig. 1.12).

Lack of knowledge of the geomorphology and stratigraphy of the Quaternary deposits of Macedonia made analyses of the seismotectonics in the post-earthquake studies less than complete. Faults occur in the mountains surrounding the Skopje basin, which itself contains zones of warping and faulting detected by geological and geophysical methods. The majority of fracture zones are classified as steep vertical or reverse faulting.

Some correlations were found between the type and depth of alluvium and the damage wrought, but lack of information on design and strength of the older buildings complicates the matter, as no earthquake building code was in use up to 1963. Engineering, seismological, and geophysical studies of the disaster indicated that there was no strong reason for re-siting the city, the seismic history and tectonics giving no clear-cut indication of places in the vicinity where the seismic hazard would be significantly less. The main requirement was to reconstruct using an enforced anti-seismic building code.

Inangahua, New Zealand, May 24, 1968

New Zealand is a seismic area (see Fig. 1.1) where moderately severe earthquakes occur somewhere every decade. The damaging Inangahua earthquake of 1968 occurred at 05:24 h on May 24 (local time) and had a Richter magnitude of 7.1 and a focal depth of 10 to 20 km. It had its epicenter about 15 km north of the town of Inangahua on the west side of the South Island. It was the sixteenth New Zealand earthquake to reach magnitude 7 or greater since reporting began about 1848. In 1929, the magnitude 7.6 Buller shock occurred 25 km to the east.

The 1968 earthquake caused widespread damage to buildings, railroads, roads and bridges, and produced rock falls, landslips, slumps and sand boils. The MM intensity reached X in Inangahua, while 45 km south, at Reefton, an accelerograph indicated a maximum ground acceleration of 0.35 g.

Although the meizoseismal region was sparsely populated (a few hundred people) and there were only two fatalities, estimates of the earthquake damage were large. For example, the Railways Department required $474,000 to restore the railroads, buildings and bridges; the National Roads Board, $515,000 for road repairs; and school repairs cost the Education Department $121,000. After the 1968 earthquake the government earthquake insurance scheme (see Section 8.7) paid out $2,500,000 on 10,500 claims.

The earthquake and its aftershocks occurred in a region of late Quaternary deformation in a structural depression of Cenozoic strata north-west of the north-easterly striking Alpine fault, which, while active in Holocene time, has not been associated with any significant earthquake since European settlement about 150 years ago.

Geological field work showed that surface rupture in the 1968 earthquake occurred on the Inangahua and Rotokohu faults. On the former both dip-slip and left-lateral strike-slip displacements were identified. The rupture extended for about 1 km with maximum offsets of 30 cm vertically and 33 cm horizontally. Geodetic measurements and mapping showed one meter change in height between bench marks across the Rotokohu fault at some places. The staggered fault traces extended for about 1.5 km with right-lateral offsets also reaching about 1 m; the overall deformation on the two faults suggests compressional forces and crustal shortening.

The greatest weakness in timber buildings was in the foundations, where free-standing piers, with little lateral bracing, allowed structures to shift and topple during the shaking. Masonry chimneys were also vulnerable, damaging tile roofs and leading to rain-water damage to walls and ceilings. In spite of repeated damage to masonry chimneys, in New Zealand, as in many other seismically active countries, little effort has been made to produce earthquake-resistant chimneys, although the New Zealand National Society for Earthquake Engineering is now studying the problem. Recently-built structures which complied with current New Zealand building codes generally behaved satisfactorily.

Niigata, Japan Earthquake, June 16, 1964

Niigata city is situated on the northwest coast of Honshu Island, Japan. According to its seismic history, well-known since at least 830 A.D., the Niigata area is in one of the relatively lower seismic risk zones of Japan. On June 16, 1964, just after 13:00 h local time, an earthquake of magnitude 7.5 occurred 60 km north of Niigata and caused significant damage extending about 200 km along the coast. One sizeable earthquake, in 1833, is known from the old literature at nearly the location of the 1964 earthquake and with similar damage distribution. Earlier in this century, only an earthquake in 1928 caused appreciable damage in Niigata city itself. These shallow earthquakes are located along the inner Honshu arc of Japan. Other destructive earthquakes that have taken place in the last half century along this general arc are the Tango earthquake of March 7, 1927 (magnitude 8.0), the Tottori earthquake of September 10, 1943 (magnitude 7.2), and the Fukui earthquake of June 28, 1948 (magnitude 7.3). All caused many fatalities because of their location near the densely populated coastline.

The Niigata earthquake included several remarkable episodes. A small island, Awashima, 5 km northeast of the epicenter, was uplifted from 80 to 160 cm with a tilt along a NNE striking axis and relative sinking on the northwesterly side. In the channel between the mainland and island, echo sounding picked out fresh dip-slip faulting striking NNE. The submarine *Yomiuri* with viewing windows was used by seismologists to observe the fault scarp.

Since 1965, the Japanese government has supported a long-term earthquake prediction program, in which emphasis has been given to geodetic surveys of crustal strain, particularly precise leveling. In the Niigata region, level lines had been measured in 1898, 1930, 1955, 1958, 1961 and just prior to the shock in 1964. These showed the coastline rising at a rate of about 1 mm per year over the 60-year interval. There was a suggestion of an accelerated rate of uplift for five or six years up to 1961 and then little change through 1964 until a sudden subsidence associated with the earthquake itself. Bench mark levels, in favorable circumstances, may thus give precursory signals for large earthquakes (see Section 1.3).

A further special feature of the Niigata earthquake was the liquefaction effects which occurred within the city, where two strong-motion acceleration seismographs recorded the waves. On the basis of these records, it is estimated that the maximum ground acceleration was approximately 0.16 g which, considering the amount of damage, is not high. Expansion of the modern city of Niigata had involved recent reclamation of land along the Shinano River, so that the present city is located over both the new and old deposits, including sand dunes and reclaimed areas. In the newly deposited and reclaimed land areas many big buildings tilted or subsided as a result of the earthquake (see Plate 1.4).

Plate 1.4. Liquefaction of saturated sands led to tilting of apartment buildings in Niigata in 1964 earthquake. (Courtesy of T. Mikumo)

Reports give 3,018 houses destroyed and 9,750 moderately or severely damaged in Niigata prefecture alone, most of which damage was caused by cracking and unequal settlement of the ground. About 15,000 houses in Niigata city were inundated by the collapse of a protective embankment along the Shinano River. The number of deaths was only 26.

The disaster was compounded by a tsunami. After the earthquake was registered, the Japan Meteorological Agency issued a tsunami warning, but only 15 minutes after the earthquake the sea wave deluged the coastal villages, locally causing flooding to depths of more than 4 m. Heights of from 1 to 2 m were observed along the coast as far south as 150 km away.

Peru, May 31, 1970

A well-defined belt of tectonic activity, characterized by high seismicity, young mountains, volcanoes and an ocean trench follows the Peru-Chile coast for 7,000 km, from Venezuela to southern Chile (see Fig. 1.1).

On May 31, 1970, at 15:23 h local time, about 25 km west of the coast city of Chimbote, a fault rupture began at a depth of about 50 km below the trench, producing an earthquake of magnitude 7.75, which led to the most catastrophic seismological disaster yet experienced in the Western Hemisphere. Its extent did not become fully known for weeks; rescue and relief were seriously hampered by landslides and rock avalanches which disrupted communications and blocked roads in the Andes. In an area of 75,000 square km in west-central Peru there were more than 50,000 deaths, 50,000 injuries, roughly 200,000 homes and buildings destroyed, and 800,000 people homeless. Within the meizoseismal region, roughly 100,000 square km, numerous villages were almost totally demolished. At least 18,000 people were buried beneath the great rock avalanche from Mt. Huascaran that covered the towns of Ranrahirca and most of Yungay.

Eyewitnesses said that the earthquake began with a gentle swaying, then vibrations became more intense, lasting variously from 30 to 50 or more seconds. No strong-motion accelerographs were in the meizoseismal area, but in Lima, perhaps as far as 200 km from the ruptured fault, a strong-motion instrument measured high-frequency vibrations with the surprisingly large value of 0.42 g (see Table 1.6) for one horizontal component.

The grimmest result was an enormous debris avalanche from the north peak of Huascaran Mountain (see Plate 1.5). This and the earlier avalanche of 1962, are discussed in Chapter 6. The avalanche amounted to 50,000,000 or more cubic m of rock and snow, ice and soil, traveling 15 km from the mountain to the town of Yungay with an estimated speed of 320 km per hour. Ridges as high as 140 m were overridden and boulders weighing several tons were projected 1,000 m beyond the avalanche margins.

A most graphic account of the Huascaran avalanche was later given by Senor Mateo Casaverde, a geophysicist with the Instituto Geofisico del Peru, who, by chance, prior to the earthquake was taking a French couple on a tour of Yungay:

"As we drove past the cemetery the car began to shake. It was not until I had stopped the car that I realized that we were experiencing an earthquake.

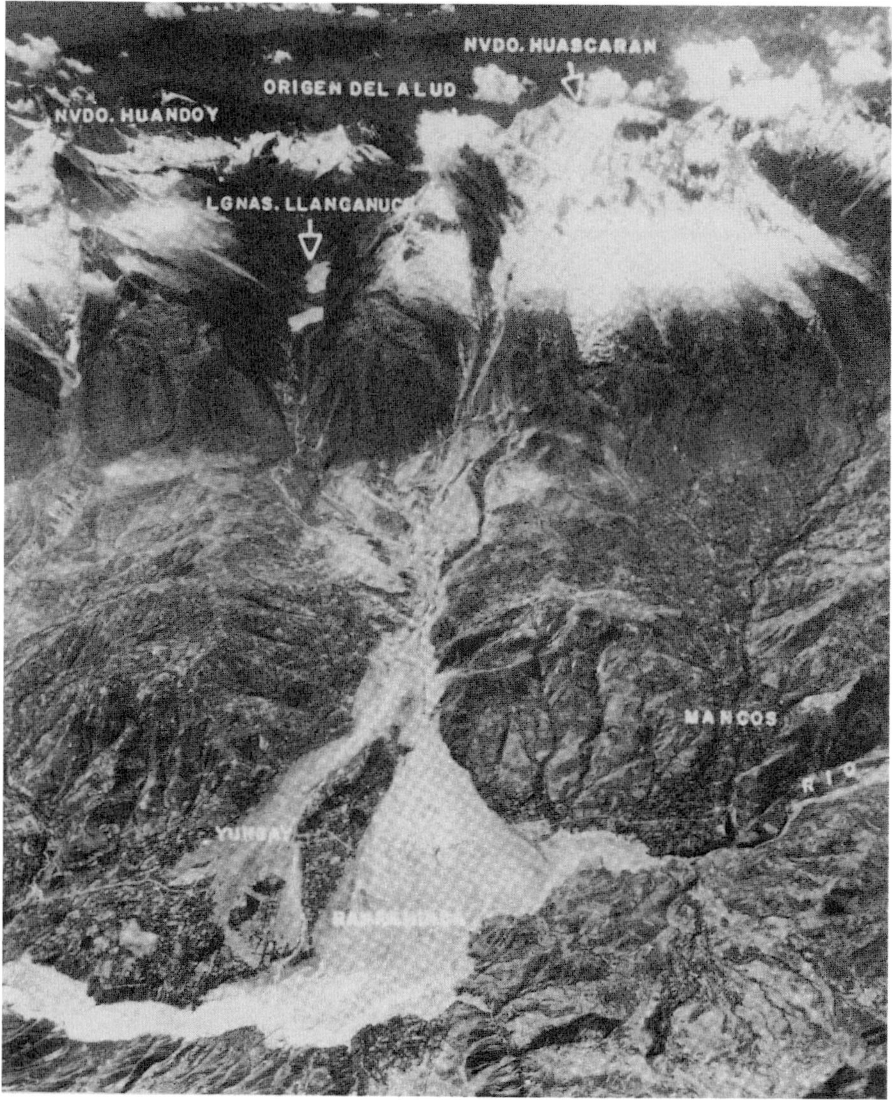

Plate 1.5. Aerial view of Mt. Huascaran and the debris avalanche that destroyed Yungay and Ranrahirca in the May 1970 Peru earthquake. (Courtesy of Servicio Aerofotografico Nacional de Peru and L. Cluff)

We immediately got out of the car and observed the effects of the earthquake around us. I saw several homes as well as a small bridge crossing a creek near Cemetery Hill collapse. It was, I suppose, after about one-half to three quarters of a minute when the earthquake shaking began to subside. At that time I heard a great roar coming from Huascaran. Looking up, I saw what appeared to be a cloud of dust and it looked as though a large mass of rock and ice was breaking loose from the north peak. My immediate reaction was to run for the high ground of Cemetery Hill, situated about 150 to 200 m away. I began running and noticed that there were many others in Yungay who were also running toward Cemetery Hill. About half to three-quarters of the way up the hill, the wife of my friend stumbled and fell and I turned to help her back to her feet."

"The crest of the wave had a curl, like a huge breaker coming in from the ocean. I estimated the wave to be at least 80 m high. I observed hundreds of people in Yungay running in all directions and many of them towards Cemetery Hill. All the while, there was a continuous load roar and rumble. I reached the upper level of the cemetery near the top just as the debris flow struck the base of the hill and I was probably only 10 sec ahead of it.

At about the same time, I saw a man just a few meters down hill who was carrying two small children toward the hilltop. The debris flow caught him and he threw the two children towards the hilltop, out of the path of the flow, to safety, although the debris flow swept him down the valley, never to be seen again. I also remember two women who were no more than a few meters behind me and I never did see them again. Looking around, I counted 92 persons who had also saved themselves by running to the top of the hill. It was the most horrible thing I have ever experienced and I will never forget it."

Considerable damage also was caused from shaking, in particular along the coast at Chimbote and Casma and in cities and villages up to 150 km inland, where most houses and buildings of adobe construction were destroyed or greatly damaged. Along the coastline, fortunately, no tsunami was generated.

San Fernando Valley, California, USA, February 9, 1971

The principal shock struck at 06:42 h Pacific Standard Time (14:00:41.6 Greenwich Time). Its focus was at latitude 34°24′0 N and longitude 118°23′7 W, at a depth of between 6 and 15 km, and its Richter magnitude was 6.5.

This earthquake is of first importance for any study of seismic hazards in a modern urban environment, being the first of significant size to occur in the Los Angeles basin (population 5,000,000) since the Long Beach earthquake of March 10, 1933 with a magnitude of 6.3. The kind and extent of damage which follows an earthquake is not a simple function of the magnitude (see Section 1.5). A magnitude 6.5 earthquake is often classed as "moderate" compared with a magnitude 8.25 shock, such as the 1906 San Francisco earthquake. Nevertheless, under appropriate conditions, including ground faulting or hazardous geological and soil conditions in densely populated areas, a magnitude 6.5 earthquake can cause severe damage locally. The Long Beach earthquake (at 17:54 h), for example, killed just over 100 people and did an estimated $40,000,000 to $50,000,000 damage (at 1933 values). Many school buildings were wrecked, leading to the enlightened Field Act regulating school construction in California.

The San Fernando earthquake is another example of a very damaging "moderate" shock, with 65 deaths and damage estimated to exceed $500,000,000. New categories of hazard information from this earthquake come from the major faulting in suburban housing areas, the effects of severe ground motion on modern

Plate 1.6. Damage to reinforced concrete columns at the new Olive View hospital in the 9 February 1971 California earthquake. (Photo by B.A. Bolt)

reinforced concrete structures, such as the Olive View Hospital (Plate 1.6) and highway overpasses, and the shaking of high-rise (up to 52 stories) buildings in downtown Los Angeles.

Surface faulting of length 12 to 15 km occurred in the San Fernando Valley and along the foothills of the San Gabriel Mountains. This faulting indicates both thrusting and left-lateral motion, the strike varying from place to place with a mean value of N72° W and dip of 45° N. The thrusting was such that the San Gabriel Mountains moved southward over the San Fernando Valley, the floor of which consists of more than 3,500 m of alluvium.

Thrust-faulting in this area was no geological surprise, although there has been a tendency to think mainly of strike-slip motion in California earthquakes. The actual faulting in the February 9 shock did not coincide with mapped faults, which indicates the difficulties of precise earthquake prediction. Mapped faults along the San Gabriel Mountains, which might in any pre-earthquake seismic hazard study have been suspected as active, did not rupture. The lesson is that, while clearly active faults (that is, associated with breakage in historical earthquakes) must be assigned appropriate risk, zones of risk should not be delineated too sharply on the basis of faults drawn on existing maps. (After the earthquake it was learned that some evidence of a fault zone through San Fernando had been obtained from ground water studies and borings at an earlier time.)

Aftershocks followed the earthquake for at least four years. By a fortunate accident, a strong-motion accelerometer at Pacoima continued to record for about seven minutes after being triggered by the main shock. It was found that in these first *seven minutes* some ten aftershocks were recorded.

Of great importance was the number of strong-motion accelerometers (over 200) which triggered. Several of the peak accelerations measured in this earthquake are given in Table 1.6. Acceleration of the ground measured near the abutment of the Pacoima dam was the highest ever recorded in any earthquake, reaching 0.7 g on the vertical component and on the horizontal components approximately 1.25 g (see Table 1.6). Reports had occurred in the seismological literature since at least the time of R.D. Oldham's account of the great Assam earthquake of June 12, 1897 that accelerations of the ground exceeded that of gravity. (In other words, if the acceleration were in a vertical direction objects would leave the ground surface.) The Pacoima recording put the matter beyond doubt so far as local motion is concerned. It must be noted, however, that the high-frequency peaks occurring in this record have brief duration and there is some theoretical evidence that the steep topography and dam structure provided some resonant amplification. The concrete dam was not damaged nor was the chimney of the caretaker's residence (about 2 km from the dam) cracked!

The social response of residents in the meizoseismal area was most commendable. There was shock, distress and sadness rising from the shaking and homes and possessions destroyed; but there was generally calmness, helpfulness and resourcefulness. Residents quickly organized road signs and barricades and parties to turn off gas heaters. The particularly warm, clear, sunny weather helped; people ate and slept on lawns and in cars. Bad weather would have required mass evacuation. Earthquake insurance policies were most uncommon in the area (see Section 8.7).

The behavior of wood-frame, one- or two-story houses in San Fernando and Sylmar was heartening. Many houses in the zone of faulting partially collapsed, were thrown off foundations or were damaged to the point of collapse and later required demolition. Nevertheless, injury to residents was not great, and severe damage from shaking was often the result of poor design or deterioration with age. A few blocks from the band of severe ground displacement the intensity of damage to homes sharply diminished, with spotty heavy damage interspersed with little or no structural damage (chimneys standing, etc.).

In new housing tracts, partial collapse of split-level houses with a large garage under the two-story section was common (Plate 1.7). The garage door eliminated one shear wall of the structure. Many new homes did not have sufficient timber cross-bracing in garages and basement walls. Strong lining such as plywood was not common. There were few masonry buildings in the meizoseismal area but those that were not reinforced were severely damaged.

The seismic hazard from broken utilities in the area did not turn out to be great. Perhaps because a major gas line ruptured in San Fernando, releasing the gas pressure, gas leakage did not become a serious problem and because electric power failed at the main distribution stations, broken power lines led to no calamities. Water supplies stopped, partly through rupture of the main water pipes where they crossed the fault traces. If there had been a high wind

Plate 1.7. Two-story wing of a new split-level house collapsed in the 1971 San Fernando earthquake crushing the automobile parked in the ground-level garage. The garage construction used reduces the lateral bracing. (Photo by B.A. Bolt)

the fire risk in these areas might have been severe. Portable battery radios provided an important link outside the damage area for most families.

The public schools had another important test of the efficacy of the 1933 Field Act. Over 5,000 school buildings in the Los Angeles school district met the seismic resistant requirements of the Act. Almost all behaved well structurally, with little chance that children would have been seriously injured if schools had been in session. Some 578 older buildings did not satisfy the Act requirements and, of these, at least 50 were so badly damaged they had to be demolished. These experiences helped, to some extent, in the urgent task of replacing substandard public school buildings, required by a recent California law to be complete by 1975.

Earth-filled dams of both the upper and lower Van Norman reservoirs were badly damaged (Plate 1.8). Fortunately, the water was not at the allowable maximum. A major earth slide in the interior portion of the lower dam left only a meter or so of earth dam to prevent catastrophe (see Section 4.6).

Major damage to freeway overpasses and road surfaces occurred. Under the severe vibrations and ground displacement, a number of key reinforced concrete columns failed. Fundamental design weaknesses were shown up, such as concrete spans falling from too narrow supporting ledges. Although changes have

Plate 1.8. Slumping of Van Norman dam in the 9 February 1971 California earthquake. (Photo by B.A. Bolt)

been made in design requirements since the earthquake, the question might well be asked: should an unusually complicated system of overpasses, many with elevations of 50 m or more, have been built in the first place in a seismic region of extensive faulting at the boundary between mountains and valley?

Managua, Nicaragua, December 23, 1972

This earthquake is of special social interest because it struck the administrative center of a nation. Its effects, therefore, give an indication of the breakdown in normal operations which can accompany an earthquake.

Two light earthquakes were felt in the city of Managua at 22:10 and 22:20 h, local time, on December 22, 1972. These were followed by the principal shock, magnitude 6.2, at 0:30 h on December 23. Many aftershocks followed (see Fig. 1.9). The two largest (magnitude about 5.0) occurred within an hour and caused substantial additional damage.

The focus of the earthquake lay beneath the center of Managua at a depth of less than 8 km. Of particular interest, location of the focus was estimated using the characteristics of the waves recorded by a strong-motion accelerometer at the ESSO refinery (see Fig. 1.12) on deep alluvium 17 km from the center of the city. The shaking led to considerable destruction in the central part of the city, encompassing 295 blocks, and significant damage extended as far as 15 km outside the central city limits.

Nicaragua is the largest country in Central America, with a population of 1,500,000. The population of Managua was 420,000. During the earthquake, approximately 5,000 persons died, 20,000 were injured and 250,000 were left homeless. The economic loss was severe for such a small and relatively poor country: 50,000 out of 70,000 homes collapsed; 95 per cent of the commercial and small shops were destroyed, as were 11 large factories; 4 hospitals with a total of 750 beds were destroyed, and 740 schoolrooms. The total damage amounted to over $1 thousand million.

The earthquake source was apparently fault rupture under the city, which is built on unconsolidated alluvial and pyroclastic (volcanic) materials south of Lake Managua. Four faults were identified by field geologic studies in Managua, along which displacement occurred during the earthquake sequences (see Fig. 1.9). Mapping the displacement was complicated by the urban build-up but the slip was mostly horizontal with left-lateral motion.

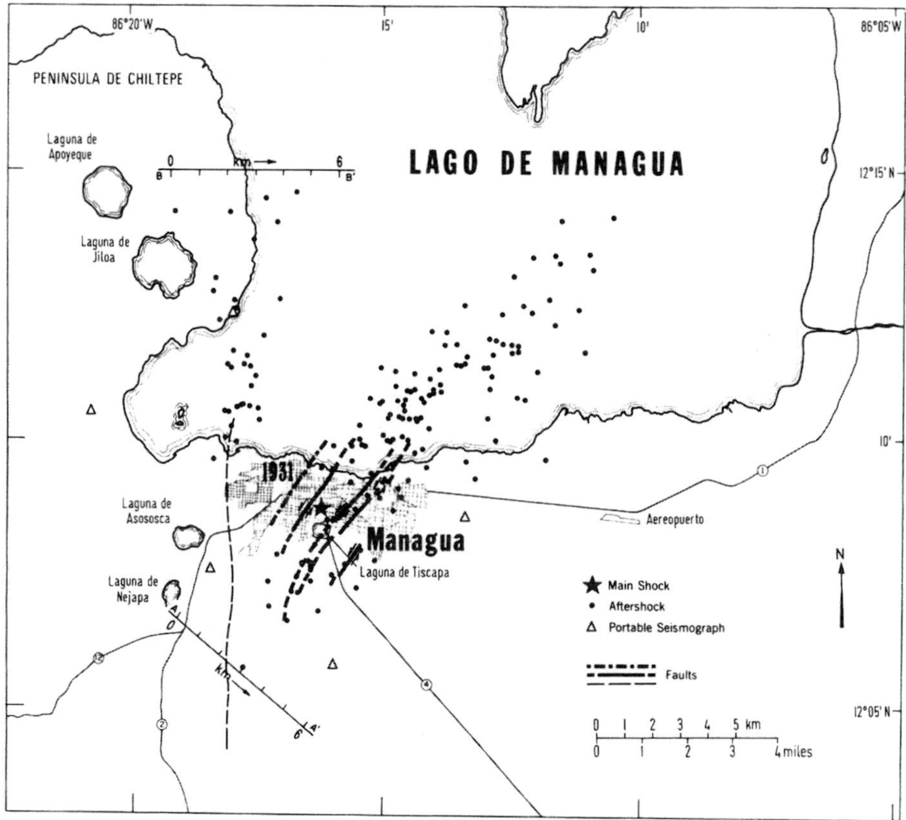

Fig. 1.9. Map of faulting observed after the 1972 Managua earthquake. The epicenters of the aftershocks (shown as dots) follow the general strike of the faults into Lago de Managua. Note, however, the separate group to the west. (Courtesy of J. Dewey, NOAA)

Maximum aggregate displacement on two of the faults reached about 30 cm. Fault rupture was responsible for severe localized damage to buildings straddling the fault, streets and underground utilities, and virtually all water lines across the faults were disrupted. Where the fault traces ran through unreinforced concrete block or older tarquezal (wood frame and adobe) construction they were marked by a swath of heavy damage.

Nicaragua is located (see Fig. 1.1) at the edge of the Pacific plate and over 450 local shocks have been cataloged since 1920. The country has also been subject to many volcanic eruptions, perhaps the most spectacular occurring in January 1835, when part of the volcanic cone of Coseguina collapsed following tremendous eruptions of ash (see Fig. 2.15). On March 31, 1931, an earthquake similar to but smaller (magnitude 5.5) than the 1972 event occurred in Managua. There was surface faulting and 400 persons were killed out of 60,000. Quite evidently, this warning had not been adequately followed up.

After the 1972 shock most government buildings in Managua were damaged. The ranch, El Retiro, of General A. Somoza, President of the Emergency Committee, was established as the central government headquarters outside the city and all military, police and other governmental administration was centered there. Five hours after the earthquake, 70 per cent of the police force had reported for duty; 12 members had been killed. The federal prison was destroyed with 39 lives lost; the remainder of the prisoners fled. Within hours after the earthquake, 300,000 residents had become refugees, streaming out of the city, causing traffic problems and clogging roads and crossings. Looting in the downtown area began, particularly in supermarkets, and martial law was declared about 38 hours after the earthquake.

The Managua earthquake provides the most recent example of *fire hazards* associated with earthquakes. Almost fifty years previously, perhaps the most famous case of fire occurred in the Kwanto earthquake of September 1, 1923 that devastated Tokyo, Yokohama and environs, when almost 140,000 Japanese were killed in the shaking, fire and tsunamis. Because of the combustible nature of the city and open cooking fires used, after the earthquake fire-storms broke out in parts of Tokyo that burned to death some 38,000 persons. Irreplaceable historical material in libraries, museums, art collections and archives was destroyed.

The Managua central fire station, built in 1964 to withstand earthquake damage, was occupied by 20 firemen, 8 fire trucks and 4 rescue ambulances. The main shock collapsed the second floor, crushing fire apparatus, killing 2 firemen and injuring others. The communications radio was destroyed and no emergency electric power was available. Fires soon began to break out in the city, where hose lines were laid from the lake and pumps put into place because the local water system failed. Some fires resulted from the earthquake, some from arson in order to collect insurance, and some from looting. All fire-fighting equipment and personnel continued operations on a 24-hour basis for seven days.

On Saturday, December 29, fires were still burning in the downtown section, but fortunately outside help had arrived, fourteen fire trucks and 135 personnel arriving within seven hours of the earthquake from other towns in Nicaragua. 132 men and 9 fire trucks were sent from other countries, and a United States Army Engineer Corps unit, arriving on December 26, played an important role in constructing a firebreak around the major fire area with its heavy equipment and explosives.

1.5. Reduction of Earthquake Hazards

Types of Hazards

We may classify seismic hazards into four classes:

i) Ground shaking
ii) Fault rupture
iii) Tsunamis and seiches
iv) Secondary hazards including:
   a) Avalanches, land and mud slides, differential ground settlement, ground lurching, soil liquefaction and ground failures.
   b) Floods from dam and levee failures.
   c) Fires.

All these hazards are considered, with examples, at various places in this book. As discussed in Section 1.3, different types of fault slip may lead to rather different hazards (see Fig. 1.7). Disturbances of the ground may be localized in a narrow zone along the surface rupture, as at many places along the strike-slip San Andreas faulting in 1906, or be spread over a zone many meters wide as is more usual in thrust motions such as the 1971 San Fernando earthquake. Special dangers, such as an active fault passing through a dam or across key life-lines such as water and gas mains, can be minimized by geologic zoning.

So little fault zone planning has been the rule, however, that schools, hospitals and other high population density and vital structures are to be found straddling active faults in California, Japan, Turkey and elsewhere. By contrast, in some places, natural fault features such as sag ponds might be utilized beneficially as *small* reservoirs. Offsets through them might disrupt water storage very little. Tsunami hazards under class (iii) receive special attention in Chapter 3.

Damage from vibrational effects is the most widespread and pervasive of all and the matter is treated in detail in the Sections which follow. Instances of ground failures listed in (iv) (a) and their avoidance are given in Sections 1.1, 1.4 and in the special studies discussed later in this Section. The behavior of these geologic hazards is discussed in detail in Chapters 4 and 5. In many earthquakes, field work has indicated that predominant damage has come from weak foundation conditions (structurally poor ground such as San Francisco Bay muds). Not only do such conditions lead, under certain conditions, to amplification of the ground shaking but filled material on slopes may move, increasing seismic risk to life and property.

Associated secondary but acute hazards are floods and fires. Floods arise not only from inundation of tsunamis along coastlines, bays and estuaries, but from large-scale seiches in lakes, canals, and also from the failure of dams due to ground shaking, when ensuing floods may be most devastating. A major calamity was prevented by only about a meter of earth-fill dam at the Van Norman reservoir in the 1971 San Fernando earthquake (see Plate 1.8).

Fire hazards continue to haunt the treatment of earthquakes. Vivid memories remain of the great conflagrations that followed the San Francisco 1906 earthquake and Tokyo's 1923 earthquake. In the reconstruction period after 1906 it was the vogue in California to speak of the catastrophic fire and to downgrade the effects of the earthquake shaking so as not to frighten off immigrants. With the cooling of emotions, restudy indicates that as much as 20 per cent of the total loss in San Francisco was due to ground motions. However, the fire, which in three days burned 12 square km and 521 blocks of the downtown district of San Francisco, was the major property hazard.

The great fire was no surprise. The National Board of Fire Underwriters had already reported in 1905: "In view of the exceptionally large areas, great heights, numerous unprotected openings, general absence of firebreaks and stops and the highly combustible nature of the buildings, the potential hazard is very severe. The above features, combined with the almost total lack of sprinklers and modern protective devices generally, make the probability alarmingly severe. In fact, San Francisco has violated all underwriting traditions by not burning up."

Although the main storage reservoirs on and near the ruptured San Andreas fault south of the city survived, the main water conduits were damaged where they crossed the fault or marshy ground, though contrary to general belief, the distribution reservoirs near the city were never empty. The problem was the damage to the water *distribution* system. Today, the overall fire risk in San Francisco is less, due to improved fire-fighting equipment, bay water pumps, under-street water storage and fire resistant construction.

Seismic Site Evaluation—Design Earthquakes

Regional seismicity or risk maps such as Figures 1.5 and 1.11 usually do not attempt to reflect geological conditions nor to take into account variations due to soil properties. It is necessary, therefore, for major vital construction in populated regions to make special geological-engineering studies for each site, the detail and level of concern which is used depending on the density of occupancy as well as the proposed structural type. In inhabited areas, more casualties are likely to result from a damaged nuclear reactor, for example, than from a damaged oil pipe line.

The factors which must be considered in assessment of seismic risk of a site have been well-defined in recent times.

i) Geological Input. Any of the following investigations may be required.

a) Provision of a structural geologic map (in color) of the region, together with an account of recent tectonic movements.

b) Compilation of active faults in the region and the type of displacement (e.g., left-lateral, strike-slip, etc.). Field work is sometimes necessary here. Of particular importance are geological criteria for fault movements in Holocene time (the past 10,000 years) such as displacements in recent gravels, dating by radio-carbon methods of organic material in trenches across the fault, and other methods.

c) Mapping of the structural geology around the site, with attention to scarps in bed rock, effects of differential erosion and offsets in overlying deposits. Such maps must show rock types, surface structures and local faults, and include assessments of the probable length, continuity and type of movement on such faults.

d) In the case of through-going faults near the site, geophysical exploration to define the location of recent fault ruptures and other lineaments. Geophysical work sometimes found useful includes measurement of electrical resistivity and gravity along a profile normal to the fault. Such data are often affected by the water table, which may change in elevation across the fault due to impermeable gouge barriers.

e) Reports of landslides, major settlements, ground warping or inundation from floods or tsunamis at the site.

f) Checks of ground water levels in the vicinity to determine if ground water barriers are present which may be associated with faults or affect the soil response to the earthquake shaking.

ii) Seismological Input. Procedures for the estimation of ground shaking parameters for optimum engineering design are still in the early stages and many are untested. It is important, therefore, to state the uncertainties and assumptions employed in the following methods.

a) Documentation of the earthquake history of the region around the site in detail. Seismicity catalogs of historical events are particularly needed in preparing lists of felt earthquakes. The lists should show the locations, magnitudes and maximum Modified Mercalli intensities for each earthquake. This information should be illustrated by means of regional maps.

b) Construction, where the record permits, of recurrence curves of the frequency of regional earthquakes down to even small magnitudes. Estimates of the frequency of occurrence of damaging earthquakes can then be based on these statistics.

c) A review of available historic records of ground shaking, damage, and other intensity information near the site.

d) Estimation of the maximum Modified Mercalli intensities on firm ground near the site from felt reports from each earthquake of significance.

e) Definition of the *design earthquakes*. The geological and seismological evidence assembled in the above sections should then be used to predict the earthquakes which would give the most severe ground shaking at the site. (Several such design earthquakes might be necessary and prudent.) Where possible, specific faults on which rupture occurred should be stated, together with the likely mechanism (strike-slip, thrust, and so on). Likely focal depth and length of rupture and estimated amount of fault displacement should be determined, with their uncertainties. These values are useful in estimating the possible Richter magnitude of damaging earthquakes from standard curves that relate fault rupture to magnitude (see Table 1.3).

iii) Soils Engineering Input. When there is geological indication of the presence of structurally poor foundation material (such as in flood plains and filled tidelands), a field report on the surficial strata underlying the site is advisable. In addition, areas of subsidence and settlement (either natural or from groundwater withdrawal) and the stability of nearby slopes must be studied. Because these surficial hazards are discussed in Chapters 4 and 5, we mention here only three factors that may require special scrutiny.

a) Study of engineering properties of foundation soils to the extent warranted for the type of building. Borings, trenchings and excavations are important for such analyses, as well as a search for the presence of sand layers which may lead to liquefaction.

b) Measurements (density, water content, shear strength, behavior under cyclic loading, attenuation values) of the physical properties of the soil *in situ* or by laboratory tests of borehole core samples.

c) Determination of attenuation values and P and S wave speeds in the overburden layers by geophysical prospecting methods.

Peak Accelerations of Strong Ground Shaking

For many design purposes, engineers require a quantitative estimate of the ground shaking that the structure is required to sustain. This estimate is the *design earthquake*. Two defining parameters which are frequently employed are the *maximum* (or peak) *acceleration* and the *duration* of strong shaking.

The maximum acceleration has now been measured from some hundreds of strong-motion records (see Table 1.6 for some of the most significant values), and the information has been summarized in curves showing acceleration plotted against Richter magnitude, distance from the ruptured fault and other variables. One rule-of-thumb that has emerged is that the peak vertical acceleration of ground is about half of the mean horizontal ground acceleration. (Close to dip-slip faulting, the fraction may be higher.)

The magnitude of the earthquake is not as strong an influence on peak acceleration as was once thought. As Table 1.6 shows, the 1972 earthquake on the San Andreas fault of magnitude 4.7 yielded a peak ground acceleration of 0.69 g about 10 km from the source; earthquakes with magnitudes as low as 4.5 at Ancona, Italy have given recorded peak accelerations of 0.6 g. Such large accelerations occur usually as only two or three high frequency peaks (probably corresponding to the arrival of an S wave) carry little energy, and have insignificant effect on substantial structures.

An obviously important variation is the rate of decrease of acceleration with distance away from the center of the earthquake disturbance (with allowance for fluctuations due to local amplifications). The correlations given in Fig. 1.10 give the general trend for recorded peak accelerations (frequencies less than 8 Hz) on firm ground. Rather than correlate peak accelerations with source magnitude, as is sometimes done, it seems preferable, for the reason given above, to separate the observed values into just two classes, called I and II in Fig. 1.10.

The first class consists of fault rupture that, because of the mode of dislocation, rock type, depth of rupture and so on, produces a high intensity of surface

Table 1.6. Earthquakes with strong recorded ground acceleration

Recording station	Horizontal distance Epicenter (E) Fault (F) (km)	Component	Maximum acceleration % Gravity	Remarks
May 16, 1968, Japan. Magnitude = 7.9				
Hachinohe	ca. 200 (E)	N–S E–W	24 19	Port area. Small shed. Soft soil
July 21, 1952, Kern County, California. Magnitude = 7.7				
Taft	40 (E)	N 21° E S 69° E	15 18	In service tunnel between buildings. Alluvium
October 17, 1966, Peru. Magnitude = 7.5				
Lima	200 (E)	N 08° E N 82° W	42 27	Small building. Coarse dense gravel and boulders
April 13, 1949, Puget Sound, Washington. Magnitude = 7.1				
Olympia	16 (E)	S 04° E S 86° W	16 27	Small building. Filled land at edge of Sound. Focal depth h = 50 km
December 11, 1967, India. Magnitude = 6.5				
Koyna Dam	8 (E)	Along dam axis; Normal dam axis	63 49	Dam gallery
January 21, 1970, Japan. Magnitude = 6.8				
Hiroo	18 (E)	E–W N–S	44 ca. 30	Focal depth h = 60 km
December 21, 1964, Eureka, California. Magnitude = 6.6				
Eureka	24 (E)	N 79° E N 11° W	27 17	2-story building. Alluvium
August 6, 1968, Japan. Magnitude = 6.6				
Uwajima	11 (E)	Transverse Longitudinal	44 36	Itashima Bridge site. Soft alluvium. Focal depth h = ca. 40 km
May 18, 1940, El Centro, California. Magnitude = 6.5				
El Centro	6 (F)	N–S E–W	32 21	2-story heavy reinforced concrete building with massive concrete engine pier. Alluvium
February 9, 1971, San Fernando, California. Magnitude = 6.5[a]				
Pacoima Dam Abutment	3 (F)	S 14° W N 76° W	115 105	Small building on rocky spine adjacent to dam abutment. Highly jointed diorite gneiss

(continued over)

[a] Maximum acceleration ≧ 0.15 g on 31 records within 42 km of the faulted zone during the February 9, 1971 San Fernando, California earthquake.

Table 1.6. (continued)

Recording station	Horizontal distance Epicenter (E) Fault (F) (km)	Component	Maximum acceleration % Gravity	Remarks
Lake Hughes Station No. 12	25 (E)	N 21° E N 69° W	37 28	Small building. 3-meter layer of alluvium over sandstone
Castaic Dam Abutment	29 (E)	N 21° E N 69° W	39 32	Small building. Sandstone
March 10, 1933, Long Beach, California. Magnitude = 6.3				
Vernon	16 (E)	N 08° E S 82° E	13 15	Basement of 6-story building. Alluvium
December 23, 1973, Nicaragua. Magnitude = 6.2				
Managua	5 (F)	E–W N–S	39 34	Esso Refinery. Alluvium
June 30, 1941, Santa Barbara, California. Magnitude = 5.9				
Santa Barbara	16 km (E)	N 45° E S 45° E	24 23	2-story building. Alluvium
June 27, 1966, Parkfield, California. Magnitude = 5.6				
C–H No. 2	0.08 (F)	N 65° E N 25° W	48 Failed	Small building. Alluvium
September 4, 1972, Bear Valley, California. Magnitude = 4.7				
Melendy Ranch	8.5 km (E)	N 29° W N 61° E	69 47	Ca. 19 meters from San Andreas fault. Small building. Alluvium. (No damage)
June 21, 1972, Italy. Magnitude = 4.5				
Ancona	Ca. 5 (E)	N–S E–W	61 45	Rock

waves, for example the 1971 San Fernando shock (see Castaic, 1971, on Fig. 1.10). The second class summarizes the vibrational intensity radiated from fault dislocation of a less efficient kind, as in the Imperial Valley earthquake of 1940 (El Centro). The peak acceleration near to a high intensity fault, on the average, would seem to be close to 0.6 g.

Currently, the sample of actual field records is small, so that estimates of uncertainty are crude. The indication at present, as indicated by the 90 per cent expectation ranges in Fig. 1.10, is that 9 times out of 10 the peak acceleration reached near to the class II source will be less than 0.4 g. Higher accelerations may be present for high frequencies, but generally, frequencies of engineering interest are no greater than 8 Hz (cycles per second). For distances over 100 km from the earthquake source, peak accelerations on country rock are unlikely to exceed 0.1 g.

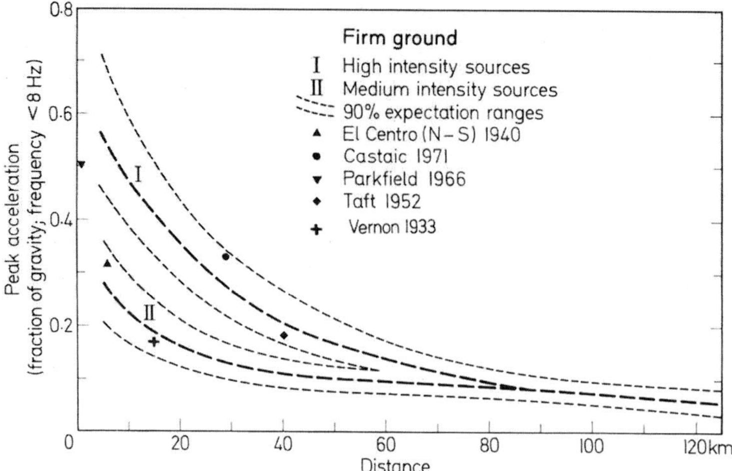

Fig. 1.10. Curve illustrating the general trend of the attenuation of maximum acceleration of ground shaking (frequency < 8 Hz) with distance from the source (fault rupture). The ground conditions are assumed to be rock or firm overburden. The symbols refer to strong-motion records described in Table 1.6. (After Bolt, 1973)

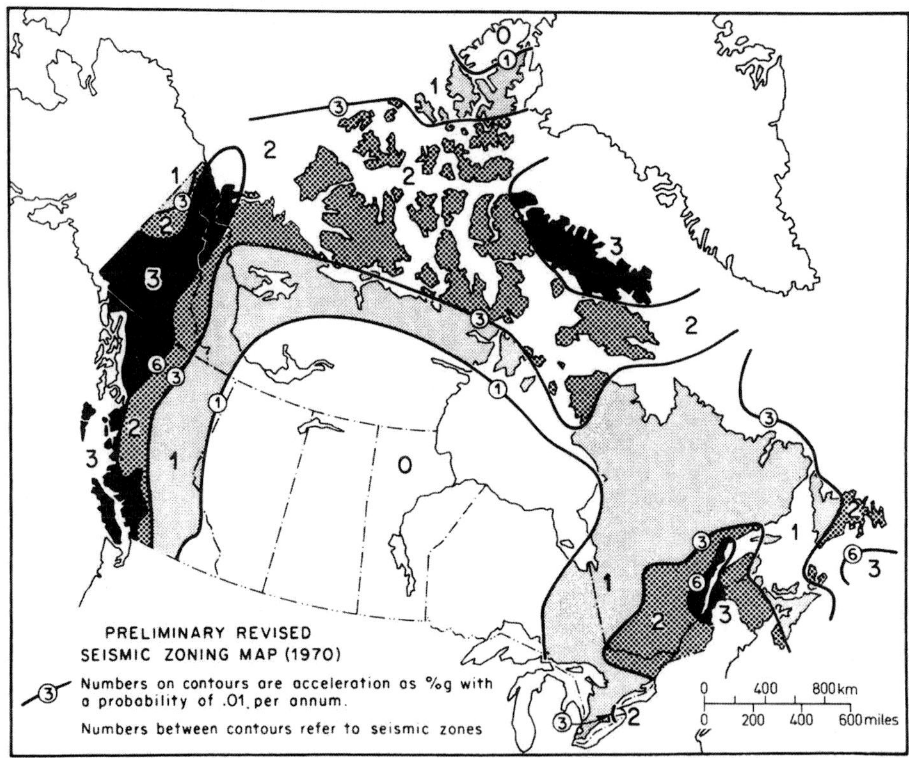

Fig. 1.11. One form of seismic zoning map (1970) developed for Canada which specifies the zones in terms of probabilistic risk. Numbers on the contours are accelerations as per cent gravity with a probability of 0.01 per annum. (After Whitham, Milne and Smith, 1970)

The probability of occurrence of ground accelerations exceeding certain values has already been used in national seismic zoning maps (see Fig. 1.11). It is of interest to compare the Canadian map in Fig. 1.11 with the risk maps used in the United States (Fig. 1.5), particularly along the border region.

Durations of Strong Ground Shaking

Many engineers and seismologists believe that duration is the single most important factor in producing damage and failure in some kinds of structures, soils and slopes. Because humans can feel shaking with amplitudes down to accelerations of one-thousandth of gravity (0.001 g), it is not surprising that people report feeling earthquake shaking up to one hundred seconds or more. However, what is important for hazard assessment is the prediction of the duration of seismic shaking above a critical ground acceleration threshold. Study of available strong-motion records, recorded on firm ground at different distances from the earthquake center, provides an average curve for duration as a function of the wave frequency, the amplitude threshold, and the Richter magnitude. As Fig. 1.12 shows, the significant duration of the shaking at Managua was only 10 sec. The magnitude of the earthquake affects the duration much more than it affects the peak acceleration because the larger the magnitude the greater the length of ruptured fault (see Table 1.3) and hence, the more extended the area from which the seismic waves are progressively emitted. The prolongation of shaking is mitigated to some extent by wave attenuation; the further a site is away from the rupture front on the extended fault the more will the wave amplitudes decay, the higher frequencies being attenuated more severely than the low ones.

An apt specification for a design earthquake is the *bracketed duration*. This is the elapsed time (for a particular frequency range) between the first and last acceleration excursions on the record greater than a given amplitude level (for example, 0.05 g). For instance, an earthquake time history may be called for in engineering design with 20 sec of 0.05 g or larger and 5 sec of 0.20 g or larger, with a maximum excursion of 0.30 g. Estimates (Table 1.7) have been made by B.A. Bolt from available strong motion records and assumed attenuation. Table 1.7 shows that 25 km from a magnitude 8 earthquake the bracketed duration

Table 1.7. Bracketed duration (seconds) (acceleration >0.05 g; frequency >2 Hz)

Distance (km)	Magnitude						
	5.5	6.0	6.5	7.0	7.5	8.0	8.5
10	8	12	19	26	31	34	35
25	4	9	15	24	28	30	32
50	2	3	10	22	26	28	29
75	1	1	5	10	14	16	17
100	0	0	1	4	5	6	7
125	0	0	1	2	2	3	3
150	0	0	0	1	2	2	3
175	0	0	0	0	1	2	2
200	0	0	0	0	0	1	2

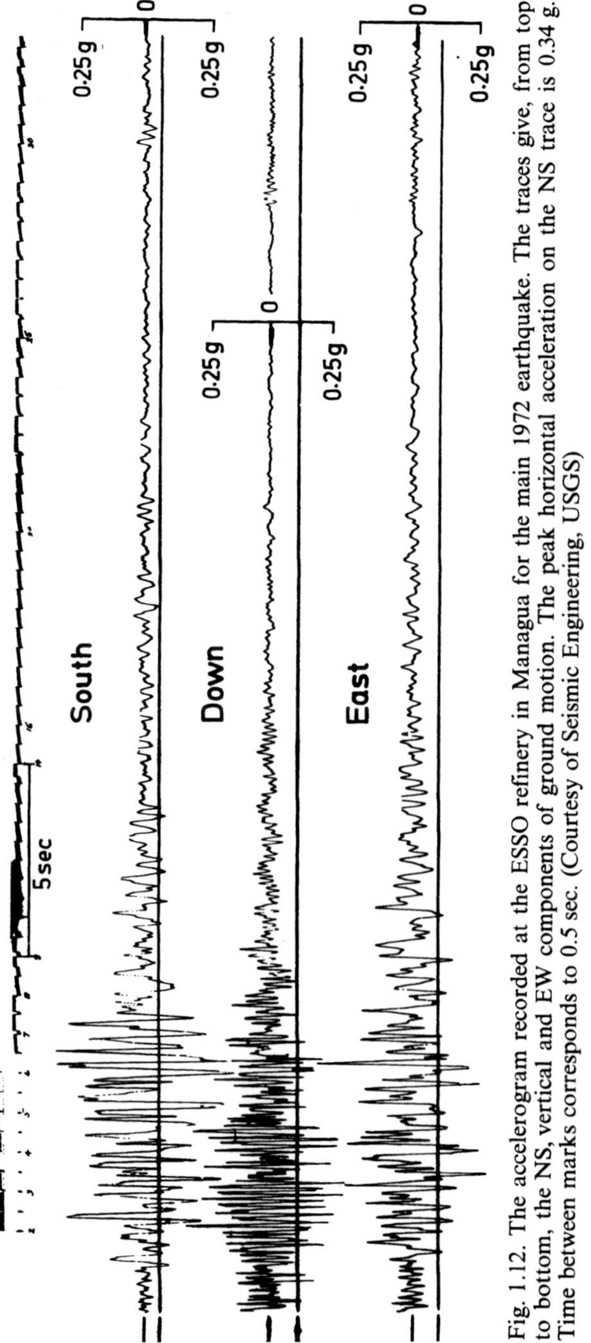

Fig. 1.12. The accelerogram recorded at the ESSO refinery in Managua for the main 1972 earthquake. The traces give, from top to bottom, the NS, vertical and EW components of ground motion. The peak horizontal acceleration on the NS trace is 0.34 g. Time between marks corresponds to 0.5 sec. (Courtesy of Seismic Engineering, USGS)

for the higher frequencies (above 2 Hz) and accelerations above 0.05 g would be about 30 sec. The reader can confirm that the duration of shaking at Managua (Fig. 1.12) is consistent with Table 1.7.

Not only are values of peak acceleration and duration required in best practice, but also some estimates of the order in which the peaks arrive, the number of pulses, and the general content of the frequency components (*spectra*) in the shaking. Unfortunately, not enough strong-motion records are available for average spectra to be computed with high confidence for all important geological situations. Synthetic seismograms have been computed based on statistical addition of available records and may be used so long as caution is employed.

The Redwood Shores Urban Development Study, California

In the 1960s Redwood City, California participated in the modification of 1,800 hectares of land on the shores of San Francisco Bay (see Fig. Fig. 1.3). The land was mainly an undeveloped peninsula, subject to flooding at high tide. First, a system of levees was raised to provide a freeboard above the maximum astronomical tide of 1 m. The surface of the peninsula, which was largely composed of bay mud, was then reworked and the level of the area was raised by rock fill from surrounding hills. During the development of this site and others nearby considerable public attention was focused on the special geological hazards. The city, therefore, set up an Advisory Board to examine the site conditions and advise how development should proceed in a manner consistent with requirements for public safety.

The flat terrain is underlain with shale, sandstone and other rocks of the Franciscan Formation, and these basement rocks are covered by alluvium and bay deposits of Quaternary (Pleistocene and Holocene) age to thicknesses of a few meters to more than 200 m. The buried basement surface is irregular, probably due to stream erosion before deposition of the sediments. Over 200 exploratory borings were made on the site, providing data on soil properties. The borings indicated discontinuous sand and gravel layers, with thicknesses generally less than 2 m but ranging up to 8 m. These sandy deposits were generally more than 12 m below the bay muds. Tests of the samples of sand and gravel suggested they would remain stable under seismic shaking but minor liquefaction could occur in the most intense earthquakes. However, the combination of size, shape and deep burial indicated that nowhere would the critical stability of the surrounding levees be affected and differential settlement might amount to no more than a few inches. Tide gage records in San Francisco Bay show that tsunami waves (see Section 1.6) entering the Golden Gate are reduced in amplitude by at least 75 per cent as they move towards Redwood City.

There was no evidence of late Quaternary fault movements through the site and the seismicity record, both historical and after locations using seismographs were available, showed no foci in the area. However, like most developed areas around San Francisco Bay, Redwood Shores lies between the San Andreas and Hayward faults, so that during a lifetime of 100 years for structures on the site at least one earthquake of magnitude 7 to 8.25 could be expected on one of the faults.

Water-saturated alluvium overlying the basement rocks is likely to have a significant influence on the vibration characteristics of the earthquakes. So far as is now feasible, therefore, it is advisable to ascertain the effect of the alluvial deposits on seismic motions from, say, a repetition of the 1906 earthquake. A mathematical analysis of plausible models of the mud and rock strata indicated that the short-period components of ground motion, which are most likely to cause damage to one- or two-story buildings, may not be significantly different from those to be expected in older cities built around the Bay. On the other hand, the mud cover is likely to increase the relative amplitude of long period components of strong motion somewhat, so that taller buildings would require more stringent design criteria than at sites away from the Bay shores.

After a consideration of the technical matters mentioned above, an appraisal of the geological risk at Redwood Shores was made by the Advisory Board, the main thrust of which was as follows: first, if reasonable construction costs are to be allowed, no *absolute* guarantees against injury and damage are possible in any locality even after all reasonable care has been taken. People in their daily lives deal with probabilities of the occurrence of future undesirable events and the approach is generally to minimize the risks to a level generally accepted in the community at large. (This is not to say that those who hold divergent views should not try to persuade the public that the general view should be changed.)

Secondly, the type of development must be considered. At Redwood Shores, the residential area is one of mainly single-family dwellings. Even during a maximum earthquake that causes ground failure, lurching and soil compaction on a small scale in such a development, a reasonable design criterion is that the extent of failure impose no greater hazard or damage to the buildings than is accepted in other communities in the San Francisco Bay area. Indeed, in some ways the over-all risks may be less, as risks from fire and other secondary effects, for example, can be minimized by appropriate steps from the start of such a planned project.

The Veterans Administration Hospitals Study, USA

Thirty-four patients and employees were killed when two major buildings of the Veterans Administration hospital collapsed in the San Fernando earthquake of February 9, 1971 (see Section 1.4). These buildings had been constructed before anti-seismic building codes were common. Three other major hospitals, all of modern construction, including Olive View hospital (see Plate 1.6), were so severely damaged in the same earthquake that they had to be evacuated.

Hearings were held by a sub-committee of the Committee on Veterans Affairs of the House of Representatives in Los Angeles on February 18, 1971. In his evidence before the sub-committee, Fred B. Rhodes, Deputy Administrator of the Veterans Administration, spoke of the cooperation of people in the disaster. "I saw policemen and firemen from different jurisdictions, search and rescue teams, doctors from non-VA hospitals, and community volunteers working with our people in the probe of the ruins that continued around the clock for some five days. Giant cranes, bulldozers, and other types of wrecking equipment dipped and probed continuously, but with a tender touch to avoid crushing any survivors."

The Veterans Administration, which administers the largest number of hospitals of any single organization in the world, moved quickly to appoint an Earthquake and Wind Forces Committee to develop requirements for earthquake resistant design of the hospital facilities. Under the advice of the Committee, investigations of the seismic risk at existing and proposed sites of Veterans Administration hospitals was begun in California and progressed into other areas of seismic risk. Sixty-eight of the existing 168 Veterans Administrations hospitals are in areas where moderate or major earthquake damage has occurred. Consultants were retained to study the seismic and geological hazards of each site and, in a pioneering step, strong-motion accelerometers were installed by the Veterans Administration to record future earthquake motions at each site.

The site-evaluation studies indicated that a large percentage of older Veterans Administration hospital buildings in California were unable to withstand earthquakes of even moderate intensity. In the face of these estimates, weaker buildings were vacated and the improvement of others started, while major buildings at the Los Angeles Wadsworth facility were demolished.

New standards for design to withstand seismic forces were developed and incorporated in Veterans Administration Handbook H-08-8. The procedure requires, first, the evaluation of the characteristics of strong ground motion at a site using geological, geophysical and seismological methods. The estimate of the greatest horizontal ground motion which could arise at the site during the lifetime of the hospital is then used as a basis for design of the hospital structures.

Another important requirement is that key hospital facilities will remain operational immediately after the earthquake. In the emergency, utility and access facilities must enable the hospital to continue to care for its own patients as well as to serve the community during the post-earthquake period.

Many problems in operational design against earthquakes have not received the attention that has been given to design of the structures themselves. The situation is a complicated one, involving the maintenance of such life-lines as electric power through emergency generators, continued use of air conditioning systems, water service, water disinfection, sanitary services, as well as key medical services. The Veterans Administration has ruled that normal water service to the hospital will be provided by two independent connections to a public water supply, and in addition, stand-by emergency water will be available which will be capable of delivering water service requirements to the hospital. Water storage facilities are provided close to the hospital buildings in order to minimize the vulnerability to earthquakes of the water service lines.

The Point Arena Nuclear Reactor Site Study, California

In many countries nuclear power plants are now operated to assist in producing electric power. Because of the particular nature of a nuclear reactor, the siting and design of such plants normally come under strict government regulation to ensure that the nuclear power plant can be operated without undue risk to public health and safety.

In the United States, the Atomic Energy Commission has developed seismic and geologic siting criteria in a number of steps over the years. A power company (the "applicant") desiring to add a nuclear power station to its generation system is required to make investigations of the proposed site which will satisfy reactor siting criteria. As well as employing its own professional staff, the Atomic Energy Commission has worked out consulting arrangements with geologists, seismologists, and engineers, so that licensing boards can scrutinize the results of the applicant's investigation. Thus, review proceedings often involve confrontation between groups arriving at different site assessments and there is room for evidence to be given by interested public groups.

The procedure may be illustrated by brief consideration of an application made by the Pacific Gas and Electric Company for a nuclear reactor license for a site on the coast of northern California near Point Arena (see Fig. 1.3). This coastal site, in a relatively uninhabited area, lies 8 km from the main trace of the San Andreas fault, which had been active in this area throughout late Cenozoic time and in the 1906 earthquake had ruptured through the region.

In the United States, critical to the current siting criteria is the establishment of two predicted earthquakes: (i) the *safe shut-down earthquake* and (ii) the *operating-basis earthquake.*

The safe shut-down earthquake (previously referred to by the AEC as the *design-basis earthquake*) is based upon an evaluation of the maximum earthquake potential, considering the regional and local geology and seismology, and specific characteristics of the local sub-surface material. This earthquake would produce the maximum vibratory ground motion for which certain structural systems and components of the reactor plant would be designed to remain functional. These components are regarded as necessary to ensure (i) the integrity of the reactor-core pressure containment walls, (ii) the capability to shut down the reactor and maintain it in a safe shut-down condition, or (iii) the capability to prevent or mitigate the consequences of accidents which could result in exposures to radiation.

The operating-basis earthquake is the less severe case which, considering the regional and local geology and seismology and specific characteristics of local sub-surface material, could reasonably be expected to affect the plant during its operating life. This earthquake would produce vibratory motion for which features of the nuclear power plant necessary for *continued operation* without undue risk to the health and safety of the public would be designed to remain functional.

According to this concept, occurrence of a safe shut-down earthquake might well put the reactor out of normal operation and cause damage but the *safety systems* necessary to prevent disaster would remain functional. On the other hand, the occurrence of the operating-basis earthquake would not necessarily interrupt the functioning of the reactor but might necessitate certain post-earthquake engineering check studies. The operating-basis earthquake is always required to be at least one-half the maximum ground acceleration determined for the safe shut-down earthquake.

Because the Miocene rocks at the proposed site for the Point Arena power plant are folded, fractured and sheared, very detailed field geologic studies of the region were undertaken with the likelihood of future warping, folding, tilting, uplift, subsidence and faulting in the site area in mind. Broadly, the geologic work indicated that during Holocene time faulting in the area had been confined to the San Andreas fault zone. However, extensive marine geophysical work off the coast indicated lineaments and discontinuities on the sea floor which raised questions. These proved difficult to interpret in terms of geologic hazards.

In assessing the severest seismic shaking to be expected at the site, geologic evidence and the seismic record pointed overwhelmingly to a great earthquake originating on the San Andreas fault nearby. A number of independent approaches were taken in estimating the ground shaking that might be expected to follow such an earthquake. These methods included a theoretical modeling of a rupturing fault, an analysis of the attenuation of seismic waves through the type of rock between the site and the San Andreas fault trace, and an assessment of the likely intensity of shaking at the site in the 1906 earthquake.

For example, a conservative MM intensity, based on the measured intensity in San Francisco (6 km from the San Andreas fault) in 1906, was VIII to IX. If this intensity were taken as representative for the operating-basis earthquake, the Neumann curve in Fig. 1.8 indicates that the corresponding peak acceleration would be approximately 0.33 g. This value is in agreement with the curve for peak acceleration 8 km from medium intensity sources (Curve II) in Fig. 1.10.

Before the validity of the seismological assessments was finally decided, the Licensing Division of the Atomic Energy Commission indicated to the Pacific Gas and Electric Company that there were doubts concerning the ability to clarify whether active faults existed under the sea off the site. On June 3, 1972, consequently, the Pacific Gas and Electric Company withdrew its application but directed that the geologic study be completed.

The Alaska Oil Pipeline Study

A much more controversial study of geological risk was undertaken in 1970 to 1974 for the trans-Alaska oil pipeline. The pipeline is not the longest ever constructed, but it is a major project which encounters Arctic phenomena, crosses three mountain ranges, many rivers and streams and active seismic areas. It is intended for the transportation of 2,000,000 barrels of oil per day from the Alaskan north slope along the 1,260 km route to Valdez, an ice-free port located on an arm of Prince William Sound (see Fig. 1.2).

The project was stoutly opposed by a number of environmental organizations on the grounds that it would lead to intolerable damage to wildlife (such as caribou herds), plant life, and to the ecological balance (melting of frozen tundra, oil spill damage, and so on). On the other side, there was some urgency to complete the facility to supply fuel to the United States. (As it turned out, the urgency was far greater than then realized.) Decisions on acceptable risk levels were difficult because geological details were missing and the human population along the route was small.

From a geological point of view, the pipeline storage and pumping facilities would be subject to a variety of seismic conditions, including shaking of the pipeline and displacement of the crust where the line crossed active faults. The pipeline design would also have to take into account slumping of the ground supporting the line, liquefaction of soil, and landslides, as well as melting of surficial permafrost, with accompanying deterioration of foundation characteristics and formation of a zone of water-saturated soil above unmelted permafrost.

The southern terminal at Port Valdez includes storage tanks, docks and transfer equipment to transfer oil to the tankers. The port facilities must be designed so as to be immune from tsunamis, earthquake shaking and fires.

Because detailed geological mapping was not available for the route, special field studies were made using in particular aerial photography. Available seismic records were analyzed and the occurrence rate of earthquakes worked out for different segments of the route. The position and depth of the earthquakes were related to the over-all tectonics of Alaska (see Fig. 1.2).

In its central segment the pipeline has to cross over four major active faults: the Denali fault, the McGinnis Glacier fault, the Donnelly Dome fault, and the Clearwater Lake fault. At each fault crossing the pipeline design must permit accommodation of appropriate offsets across a fault rupture. The three latter faults are less than 80 km long. The great right-lateral, strikeslip Denali fault (1,500 km long) has had many meters of lateral slip on it during Holocene times but evidently little during the last few hundred years. Thus, it is prudent to assume that an offset of about 5 m might occur. This would be associated with a magnitude 8 to 8.5 earthquake (see Table 1.3). For this segment, therefore, specified design ground accelerations need to be greater than for any other segments of the route. A peak acceleration of 0.8 g on firm ground in the fault zone would be an upper design value (see Fig. 1.10).

After the whole pipeline route was divided into segments in each of which the seismic shaking could be treated as constant, appropriate design earthquakes for each section could be estimated. In the 150 km segment, for example, skirting the northern end of the seismic belt that extends from Cook Inlet to Fairbanks (see Fig. 1.2), the largest earthquake known is that of 1937 (magnitude 7.3) with its center situated about 50 km southeast of Fairbanks. The maximum Modified Mercalli intensity was rated as VIII in a region then of sparse population. Only minor damage was reported at Fairbanks. The geological field work located no surface expression of any fault that now seems active within 30 km of the pipeline route in the Fairbanks area.

It is clear that such restricted historical data are anything but statistically optimal and it is difficult to infer from them the future seismic risk. Some weight must be given to the frequency of occurrence, the likely distance of the centers of the earthquakes from the pipeline, and the probable focal depths of such earthquakes. Attention must also be given to the type of structure involved and population (human and wildlife) density. On this basis, for the Fairbanks section, a cautious estimate is that an adequate design earthquake with magnitude no greater than 7.5 might be expected within 100 years centered at a distance as close as 20 km, say, to the pipeline. From Fig. 1.10, we then infer that on firm ground near the pipeline the peak high frequency accelerations from the earthquake shaking would have a 90 per cent chance of lying in the range of 0.3 g to 0.5 g. The mean peak acceleration which should be accommodated would be approximately 0.40 g if it were decided to adopt this high level of caution.

Ordinary Homes

Dwellings of people in seismic areas vary a great deal, from the adobe and torquezal materials of much of Latin America and the Middle East to timber-frame dwellings in New Zealand and California and light wood houses of Japan. The great loss of life associated with historical earthquakes is usually traceable to both poor design and poor construction. As we have noted (Section 1.2), a substantial proportion of the 830,000 deaths in the great Chinese earthquake of 1556 was caused by collapse of the loess caves in which people made their homes.

Even today, home construction prone to earthquake damage may be found in all seismically active countries. In the Dasht-E Bayāz, Iran earthquake of August 31, 1968 where the typical family and commercial house was a one-story adobe mud structure, without footings and with a domed roof, over 12,000 housing units were destroyed and between 7,000 and 12,000 people killed and 60,000 made homeless. In Turkey, where great loss of life has occurred along the Anatolian fault in recent decades, it is common to place large boulders on village roofs to guard against high wind damage. In California, loss of life in the past has resulted particularly from use of unreinforced masonry structures. In the 1906 earthquake, for instance, the collapse of stone and brick masonry buildings in San Jose, Los Banos and Santa Rosa was almost uniform, with 61 identified dead in the latter city. In the following paragraphs we list some recommendations that may be helpful in reducing earthquake hazard in typical single-family houses. Applicability of the items depends on the region and housing type.

i) Basements and open first-floor levels should be adequately braced (see Plate 1.7). Plywood sheathing is valuable for interior bracing of wood frame houses.

ii) Foundations and studs must be watched for dry rot and termite damage.

iii) Roofs and ceilings should be of as light a construction as the climate allows.

iv) Foundation ties should be provided for all types of foundations.

v) At least one fire extinguisher should be provided in the house.

vi) Closets, lighting fixtures and heavy furniture should be fastened to wall-studs.

vii) Gas heaters in basements and elsewhere should be strapped to walls.

viii) Flexible joints should be provided between the utility lines (particularly water) and outside mains.

Of course, special engineering precautions in construction of buildings vary greatly from country to country, depending upon style, available building materials, and size. One UNESCO study estimated that the additional cost to a one-family house in a medium intensity earthquake region is about 4 per cent, while in areas of major earthquakes the seismic resistant design may add 10–15 per cent. Such cost increases may be beyond the purse of poor people.

1.6. References

Adams, R.D., and others: Preliminary Report on the Inangahua Earthquake. New Zealand, May 1968, New Zealand: D.S.I.R., Bulletin **193** (1968).

Algermissen, S.T.: Seismic Risk Studies in the United States. Fourth World Conf. Earthquake Engineering Chile. (1969).

Anon: The Great Alaska Earthquake of 1964, Seismology and Geodesy, Committee on the Alaska Earthquake, Washington, D.C. National Academy of Sciences (1972).

Anon: Earthquake Guidelines for Reactor Siting. International Atomic Energy Agency, Vienna (available in the United States from Unipub. Inc., P.O. Box 433, New York) 1973.

Anon: Earthquake Resistant Design Requirements for Veterans Administration Hospital Facilities. Veterans Administration Handbook H-08-8. 1974.

Berringhausen, W.H.: Tsunami Reported from the West Coast of South America. Bull. Seism. Soc. Am., **52**, 915–921 (1962)

Bolt, B.A.: Duration of Strong Ground Motion. Fifth World Conf. Earthquake Engineering, Rome (1973).

Bolt, B.A.: Nuclear Explosions and Earthquakes: The Parted Veil, San Francisco: W.H. Freeman, 1975.

Cloud, W.K., Perez, V.: Strong Motion Records and Acceleration. Fourth World Conf. Earthquake Engineering, Chile (1969).

Coffman, J.L., Von Hake, C.A. (Eds.): Earthquake History of the United States, Revised Edition (through 1970), Publication **41–1**, National Oceanic and Atmospheric Administration, Washington, D.C.: U.S. Government Printing Office 1973.

Davison, C.: Great Earthquakes. London: T. Murby 1936.

Despeyvous, J. and others: The Skopje Earthquake, 1963, UNESCO (1968).

Gubin, I.E.: Earthquakes and Seismic Zoning. Bull. Int. Institute of Seismology and Earthquake Engineering, **4**, 107–126 (1967).

Gutenberg, B., Richter, C.F.: Seismicity of the Earth and Associated Phenomena. Princeton Univ. Press 1954.

Hershberger, J.: A Comparison of Earthquake Acceleration with Intensity Ratings. Bull. Seism. Soc. Am., **46**, 317 (1956).

Housner, G.W.: Strong Ground Motion. Chapter 4 in Earthquake Engineering, R.L. Wiegel (Ed.), N.J.: Prentice-Hall, 1970.

Idriss, I.M., Seed, H.B.: An Analysis of Ground Motions during the 1957 San Francisco Earthquake. Bull. Seism. Soc. Am., **58**, 2013–2032 (1968).

Jahns, R.H., Bolt, B.A., Clough, R.W., Degenkolb, H.J., Leps, T.M.: Report of Seismic Advisory Board to the City Council of Redwood City. 2 Volumes, April 1972.

Lawson, A.C.: The California Earthquake of April 18, 1906. Report of the State Earthquake Investigation Commission, Washington, D.C. Carnegie Institution (1908).

Medvedev, S.V.: Engineering Seismology. Moscow, English Translation, Israel Program for Scientific Translation, Jerusalem (1965).

Mickey, W.V.: Reservoir Seismic Effects, The Military Engineer, July-August (1972).

Milne, W.G., Davenport, A.G.: Distribution of Earthquake Risk in Canada. Bull. Seism. Soc. Am., **59**, 729–754 (1969).

Newmark, N.M., Rosenblueth, E.: Fundamentals of Earthquake Engineering. N.J.: Prentice-Hall, 1971.

Nuttli, O.W.: The Mississippi Valley Earthquakes of 1811 and 1812: Intensities, Ground Motion and Magnitudes. Bull. Seism. Soc. Am., **63**, 227–248 (1973).

Page, R.A., Boore, D.M., Joyner, W.B., Coulter, H.W.: Ground Motion Values for Use in the Seismic Design of the Trans-Alaska Pipeline System Washington, D.C., Geological Survey Circular **672**, U.S.G.S. (1972).

Richter, C.F.: Elementary Seismology, San Francisco: W.H. Freeman 1958.

Rothé, J.P.: Fill a Lake, Start an Earthquake. New Scientist. **11**, 75–78 (1968).

Schnabel, P.B., Seed, H.B.: Accelerations in Rock for Earthquakes in the Western United States. Bull. Seism. Soc. Am., **63**, 501–576 (1973).

Vitaliano, D.B.: Legends of the Earth. Indiana Univ. Press. 1973.

Whitham, K., Milne, W.G., Smith, W.E.T.: The New Seismic Zoning Map for Canada. 1970 edition. The Canadian Underwriter, June 15 (1970).

Wiegel, R.L. (Ed): Earthquake Engineering. N.J.: Prentice-Hall, 1970.

Chapter 2
Hazards from Volcanoes

2.1. Introduction

Volcanic eruptions are among the most violent, spectacular, and awe-inspiring manifestations of nature. It is hardly surprising, therefore, that from time immemorial they have fascinated man, invoking in him sometimes terror, sometimes religious veneration, sometimes artistic appreciation of beauty, and always respect.

The tremendous forces involved, and the apparent inevitability of their consequences, have commonly led to a feeling of fatalism—the volcano would have its way, and was not to be disputed! Only very rarely have men been bold enough to try any means of altering volcanic behavior, other than by propitiatory offerings or other religious acts. Today, however, we are coming to the realization that there are some more direct things that can be done. Already we can lessen or eliminate some volcanic disasters, and further knowledge and experience will enable us to do still more. Moreover, we are coming also to realize that we *must* learn to do so.

Millions of people live in the shadow of active volcanoes and potential volcanic destruction (Plate 2.1), and will continue to live there. It has been estimated that during the last 500 years some 200,000 persons lost their lives as a result of the more than 500 active terrestrial volcanoes (Appendix C). Of these, some were killed directly by volcanic action, and others died of starvation caused by the destruction of food crops and animals. The figure does not include several tens of thousands killed by "tidal waves" (i.e. tsunamis; see Chapter 3) caused by volcanic eruptions. In 1902, on the Island of Martinique in the Lesser Antilles, in a matter of moments the volcano Mont Pelée almost wholly destroyed the city of St. Pierre, and some 30,000 persons lost their lives. If we are to avoid similar, and probably much worse, disasters in the future we must learn to give adequate warning of coming volcanic activity and to deal with the results of the activity.

Indeed, we must go beyond this, and learn to make direct use of the volcano and its energy. Already we are using some volcanic heat, in the form of natural steam and hot water, to produce electricity and to heat buildings at low cost and with a minimum of pollution of the environment, but this is only a very small proportion of the total volcanic heat that we can learn to use.

Although this chapter deals primarily with volcanic hazards and destruction, it is only fair to state that, in the past, volcanoes have done far more good than harm. They are, on balance, benefactors. Volcanoes have created many thousands of square kms of land surface, both as oceanic islands and on the continents. Volcanic mountains are among the most beautiful of nature's construc-

Plate 2.1. Mayon Volcano, Philippines, with the city of Legaspi in the foreground. Mayon is said to be the most symmetrical volcanic cone in the world, but it is potentially deadly. (Photo by Tommy's Studio, Legaspi)

tions, giving joy and inspiration to many thousands of vacationers. Among the more abstruse benefits derived from them are the Earth's atmosphere and all the waters on the surface of the Earth, without which life itself could not exist.

Not only are volcanic lands extensive, but commonly they are very fertile, especially in the tropics, where leaching removes plant nutrients from the soil very rapidly, and showers of volcanic ash renew the fertility. This very fertility adds to the volcano problem, because it attracts people. Thus, in Indonesia a close correlation exists between the density of the agrarian population and the amount of volcanic activity. Millions of people cluster around the bases of the active volcanoes, although the risk to them is apparent even without the practical demonstration of thousands of deaths and millions of dollars worth of property damage in the last few centuries. The remedy cannot be found in the removal of the people from the zones of high risk, as the food from these rich agricultural areas is already vital, and becomes even more so with the continual increase of world population. The solution must come in learning how to live with the volcanoes—how to warn of the time, type, and place of coming eruptions, how to avoid or alleviate their effects, and how to speed the recovery of devastated lands. These are some of the practical goals of modern volcanology.

One of the very practical problems is expressed by the simple question, "When is a volcano dead?" If it is really *extinct* it presents no further risk, but volcanoes can remain quite—*dormant*—for thousands of years only to return to activity, and often the eruption that marks the return is exceptionally violent. Some of our most destructive eruptions have taken place from volcanoes that had no previous eruptions during all of historic time, and therefore were not regarded as *active*. These include, for example, the eruption of Vesuvius in 79 A.D., and those of Lamington in 1951, Bezymianny in 1956, and Arenál in 1968 (all described on later pages). It is often said that these returns to activity were unheralded, but in most cases it appears that the warnings simply were not recognized. For instance, all of the above-mentioned eruptions were preceded by numerous earthquakes (see Section 1.3), and other signs may also have been present. With proper distribution of instruments, and alert, trained personnel to interpret the records, plus present and increasing knowledge, probably all of these eruptions could have been forecast, at least in a general way.

Volcanic activity has many fundamental similarities the world over, but also there are many differences, resulting in different degrees of danger and destruction. To evaluate the risks and destruction that may occur and the methods that may be used to lessen them, let us first take a look at the general nature of volcanic activity.

2.2. Nature and Kinds of Volcanic Activity

What Is a Volcano?

A volcano is a place where molten rock and/or gas issues from the interior onto the surface of the Earth. Some volcanoes erupt only once, but others erupt repeatedly. The solid products of the eruptions pile up around the vent from which they issue to form a hill or mountain, which also is known as a volcano. Thus the term volcano has two somewhat different meanings, but in any given case the particular meaning is usually obvious.

The proportion of gas to molten rock issuing at the surface varies widely, and the nature of the eruption depends largely on the abundance of gas and how readily it escapes from the molten rock. The gas may be wholly dissolved in the melt or it may be partly present as bubbles. Crystals of various minerals, solidified from the melt, also are commonly suspended in it. The combination of molten rock, dissolved gas, gas bubbles, and suspended crystals is known as *magma*.

The solubility of gas in magma is increased by increase of pressure and at great depth in the Earth most, if not all, of the gas is in solution. Most magma appears to originate between 40 and 300 km below the Earth's surface (see Fig. 1.4). As it rises toward the surface the pressure due to the weight of overlying rocks becomes less and some of the gas starts to come out of solution, forming bubbles, at first small and widely separated. As the magma continues to rise the bubbles grow, increase in number, and begin to coalesce.

The gas in the bubbles is under pressure, but is prevented from expanding partly by the confining pressure of the magma around and over it, and partly by the viscosity of the magma. Eventually the gas pressure may become high enough to burst the enclosing liquid and the gas escapes; and this, of course, is an *explosion*. If the confining pressure is low and the enclosing magma is relatively fluid the gas bubbles escape readily, with a myriad of minute explosions producing a rather mild spattering of magma at the surface. On the other hand, if the gas escapes with difficulty and pressure continues to increase, and if gas is abundant, its eventual escape is violently explosive.

An explosion at the surface may result in a sort of "chain reaction" by which the violent expansion of gas is propagated to deeper and deeper levels in the magma. The surface explosion lifts the uppermost part of the magma into the air, which reduces the weight, and hence the pressure, on the magma

Plate 2.2. Lava fountains and flow along a fissure just outside of Kapoho village, Hawaii, January 14, 1960. The dark cloud rising on the right is ash-laden steam that resulted from the molten lava coming in contact with shallow ground water. Most of the village and about 400 hectares of agricultural land later were destroyed by the spreading lava. (Honolulu Advertiser photo by Gordon Morse)

next below it, allowing the gas in that portion to expand explosively and lift its weight off the next portion, and so on. In this way the explosive eruption may become not a single big "bang", but long-continued violent outrush of gas mixed with magma. Typically, the eruption consists of a series of these blasts, each continuing for a few seconds to several minutes, separated by periods of a few minutes to several hours, or even days; but in some big eruptions individual blasts continue for several hours.

A further complicating factor is that not all of the gas originates within the magma. Some is steam of external origin, generated where rising magma comes in contact with water either at the surface or in near-surface rocks (Plate 2.2), or formed where water has gained access to hot solidified rock within the volcano. Violent steam explosions at Kilauea Volcano, Hawaii, in 1924, contained no magma or magmatic gas, and resulted solely from ground water in surrounding rocks moving into the hot conduits as cracks opened in the volcano. In other instances, even where magma is abundant, the external steam may totally change the character of the eruption. Normally, where fluid basaltic magma reaches the surface, the eruption consists of a relatively gentle fountaining of liquid lava and outpouring of lava streams. In contrast, at Capelinhos Volcano in the Azores, in 1957, and at Surtsey Volcano in Iceland in 1963, similar fluid basalt magma came in contact with ocean water, resulting in violent explosions that threw up great showers of magmatic spray. In both cases, as soon as the volcanic cone around the vent became big enough to shut out the seawater the explosions ended and were replaced by rather gentle lava fountaining.

The explosions tear apart the liquid magma and throw fragments of it into the air, some throwing out only magma fragments, others, like the 1924 explosions of Kilauea, only throwing out fragments of older, solid rock, while many throw out some of each. The fragments, whether ejected as bits of solid rock or as shreds of liquid magma, fall back to the ground (Plate 2.3), or are carried away by the wind to varying distances, sometimes hundreds or even thousands of kms. They are known as *pyroclastic* ("fire-broken") ejecta, or more briefly as *tephra* (a term recently revived from the writings of Aristotle). Sand-size and smaller fragments are called volcanic ash or dust, fragments from about 2 to 60 mm in diameter are *lapilli* (singular, lapillus), and fragments larger than 60 mm are *bombs* or blocks. Bombs are thrown out in a liquid condition; blocks are angular chunks of solid rock.

Some bombs take on rounded or spindle shapes in the air. Strong explosions throw the fragments high enough so that bombs and lapilli have cooled and solidified enough to retain their shapes when they strike the ground; but with weaker explosions they are commonly still sufficiently fluid to flatten out when they strike, or even to stick together, material of this latter sort being known as *spatter*. Many bombs and lapilli are very irregular, and they are commonly so full of gas-bubble holes that they resemble sponges. These irregular sponge-like fragments, whether of bomb or lapillus size, are known as *scoria*, or *cinder*, from their resemblance to cinders formed in a furnace. (It should be emphasized, however, that neither volcanic ash nor cinder is the product of burning.) *Pumice* is cinder that is exceptionally full of bubble holes, and consequently is so light that it will float on water. Both cinder and pumice are solidified magma froth.

Plate 2.3. Region intensely cratered by bombs falling in soft ash, about 4 km west of the summit of Arenál Volcano, Costa Rica, during the eruption of 1968. Leaves and small branches have been stripped from the trees by the falling tephra. (Smithsonian Institution photo by William G. Melson)

In contrast to explosive eruptions are *effusive* eruptions in which there is little explosion, and which produce mostly liquid magma, known as *lava*, which pours downslope away from the vent in streams known as lava *flows*. (The terms lava and lava flows are also applied to the material after it has frozen to solid rock.) Three general types of lava flows have been distinguished, though there are intergradations and many variations.

Pahoehoe is lava that has a relatively smooth or undulating surface, sometimes wrinkled into ropy or drapery-like forms. The flows of pahoehoe are fed largely by movement of liquid lava beneath the crust through lava tubes—natural pipes formed by gradual inward solidification of the slower-moving outer parts of the flow.

Aa is lava that has a surface of very rough, jagged, spiny fragments that resemble granulated fudge or furnace clinker. Beneath the surface layer of clinker is a central layer of massive rock. In the active lava flow the part that later forms the massive layer is flowing pasty magma which bears the load of solid clinker fragments on its top. The aa flow is fed principally by a river of liquid lava open to the sky, and is commonly bordered by levees built by repeated overflows of the river itself. The difference in the feeding mechanisms of pahoehoe and aa is an important factor in attempts to control the flows (discussed in a later section).

The third type of flow is *block lava*. It resembles aa in general structure and mechanism, but the surface fragments are fairly smooth-sided regular polygons, instead of being irregular and spiny.

In general, the magma that forms block lava flows is more viscous than that which forms pahoehoe and aa, and as a result block lava flows tend to be thicker and shorter. Pahoehoe and aa flows are usually between 2 and 7 m thick, and may extend for tens of kilometers. Block lava flows often are 15 to 40, and sometimes several hundred, meters thick, and seldom extend more than 10 km. Pahoehoe and aa flows may advance rapidly. Speeds of advance of 300 m per hour are common in Hawaii, and some flows advance several kms an hour. Block lava flows usually advance only a few meters a day.

Some other types of lava flows, such as pillow lava and hyaloclastite flows, both associated with water, present little or no hazard to people and will not be discussed here.

Lava flows move downhill like any other liquid, and consequently are closely controlled by topography, so that if we know where a lava flow will issue, we can predict its general course. However its viscosity and constant tendency to freeze may result in damming up of channels, in overflows, and even in major changes of channel.

Still more viscous lava, instead of flowing away, tends to pile up over the vent as a steep-sided heap known as a *volcanic dome*. As the dome grows, its outer part usually breaks up into loose blocks that tumble down and form long banks of *crumble breccia* that may nearly bury the more solid part of the dome and give it the shape of a truncated cone. Domes commonly grow in the crater at the summit of the volcano, but they may also form on its flank. Domes generally constitute no direct threat to people, but they are often associated with the deadly glowing avalanches described in the next paragraph, and consequently the presence of an active volcanic dome should serve as a warning. Furthermore, domes are most characteristic of the old age of a volcano, and suggest that the magma body beneath probably has increased in viscosity and gas content. Therefore, even inactive domes should be taken to indicate the likelihood of violent explosions.

Glowing avalanches were originally named *nuées ardentes*, and are sometimes called by the English equivalent, *glowing clouds*. Their most conspicuous feature is a great rapidly-expanding cloud of dust, black in daylight, but glowing a dull red at night. We now know however that the principal part is not the cloud, but an avalanche of incandescent lava blocks, sand, and dust beneath it. Glowing avalanches may move with tremendous speed, and they may travel many kilometers. In several instances speeds have exceeded 150 km an hour. The avalanche portion is largely guided by topography, following valleys, but its great speed and inertia may result in its climbing vertically as much as several hundred meters where it encounters opposing hill slopes, or bends of the valley wall. Such behavior indicates great mobility, which results from fragments in the avalanche and the overlying dust cloud being separated by a cushion of expanding gas, thus largely eliminating friction between the fragments as the mass moves. The gas is partly entrapped air, expanding because it is heated by the incandescent fragments. However, in many instances it is partly, or even largely, gas which is still being given off by the fragments themselves. Glowing avalanches have several different origins, which are discussed in a later section.

Ash flows closely resemble glowing avalanches in their great mobility, and the cause of the mobility appears to be the same. As viscous magma approaches and reaches the Earth's surface the gas in it may separate as a great number of closely-spaced tiny bubbles, and these bubbles expand to transform the magma into a froth—pumice. Further expansion simply causes the froth to disintegrate into the septa that separated the bubbles, each bit of liquid chilling and solidifying in the expanding gas cloud to an angular bit of glass.

The result is an emulsion of bits of very hot glass (ash) and occasional undisintegrated lumps of pumice in the gas (see Plate 2.4). The cloud is heavy, because of its load of solid fragments, and probably also in some instances because of a large proportion of carbon dioxide in the gas. As a result it flows downhill, and spreads out over the ground surface under its own weight. Big eruptions of this sort have produced clouds that traveled as much as 300 km, so rapidly that they reached these great distances still so hot that as the ash particles settled to the ground they stuck together, and in the central part of the mass became densely welded under the weight of the overlying material.

Debris avalanches and landslides (see Chapters 4 and 6) result from the sudden down-slope movement of loose materials on the flank of a volcano. They are driven by gravity, and are often set off by heavy rains and earthquakes. Because abundant loose material on the slope of a volcano is usually of explosive origin, debris avalanches are most common on volcanoes that have explosive eruptions. As the degree of water saturation of the avalanches increases, they grade into mudflows.

Volcanic mudflows (also known as *lahars*) are just what the name implies—slurries of solid fragments in water, some hot, but most of them cold. Like debris avalanches, they are most likely to occur on explosive volcanoes where loose fragmental material is abundant. They flow downhill, under the influence of gravity, and are closely controlled by topography. Depending on the steepness of the slope and the viscosity of the mud, they may attain speeds as great as 90 km an hour, and some have traveled as far as 160 km. Mudflows have various origins, which are discussed in a later section.

As pointed out earlier, the explosiveness of an eruption is determined largely by the viscosity of the magma as it reaches the surface, and by the amount

Plate 2.4. House buried by ash, in the outskirts of Paricutin village, Mexico, in October 1944. Paricutin Volcano is in the right background. (Photo by Tad Nichols, Tucson, Arizona)

of gas it contains. Viscosity also is the principal factor in determining what type of lava flows may form. Both viscosity and gas content are closely related to the chemical composition of the magma. As a general rule, the viscosity increases with increase of the proportion of silica in the magma. Pahoehoe and aa lavas are characteristically of basaltic composition, poor in silica and relatively rich in iron and magnesium, whereas block lava is usually andesite or a still more siliceous type.

The volcanoes of certain regions tend to have more siliceous magmas than those of other regions, hence are apt to erupt more explosively and to produce thick flows of block lava and domes rather than thin flows of pahoehoe or aa. Thus the volcanoes surrounding the Pacific Ocean—those of the Cascade Mountains, the Andes, Japan, the Philippines, and Indonesia, for example—have moderately to highly silicic lavas and tend to erupt explosively, in contrast to the gentle eruptions of Hawaii and other predominantly basaltic areas. Furthermore, the subsurface magma bodies that feed volcanoes typically undergo a change of composition with the passage of time, the eruptible magma becoming richer in silica and gas. These changes are reflected in the formation of shorter and thicker block lava flows and domes, whose presence suggests that the volcano is in a late stage of magmatic evolution, and hence that it has a potential for highly explosive eruptions and the formation of glowing avalanches and mudflows.

Thus both the chemical composition and the physical nature of the products of past eruptions serve as a guide to what can be expected of the volcano in the future. For that reason, if for no other, careful geologic studies should be made of all volcanoes in areas where they could constitute a hazard to humans.

Tephra Falls

In the year 79 A.D. towns around the Bay of Naples were important resorts for the wealthy citizens of Rome. At the western side of the bay (at the site of the present Resina), on the lower slope of a great conical mountain, lay the town of Herculaneum; and farther south, near the southeastern base of the mountain, lay Pompeii. 5 km farther south, at the base of Sorrento Peninsula, lay Stabiae (see Fig. 2.1). The conical mountain is known today as Mount Vesuvius; but its present shape is very different from what it was before the year 79.

Today the "Gran Cone" of Vesuvius is enclosed in a semicircular ridge, known as Monte Somma; but in 78 A.D. and earlier the mountain swept smoothly upward as a single cone, with a rather flat top that contained a steep-sided crater, heavily vegetated. The mountain was vaguely recognized as a volcano, but in all of the long history of the Romans, and the Etruscans before them, there was no record of it having erupted, and if people thought about it at

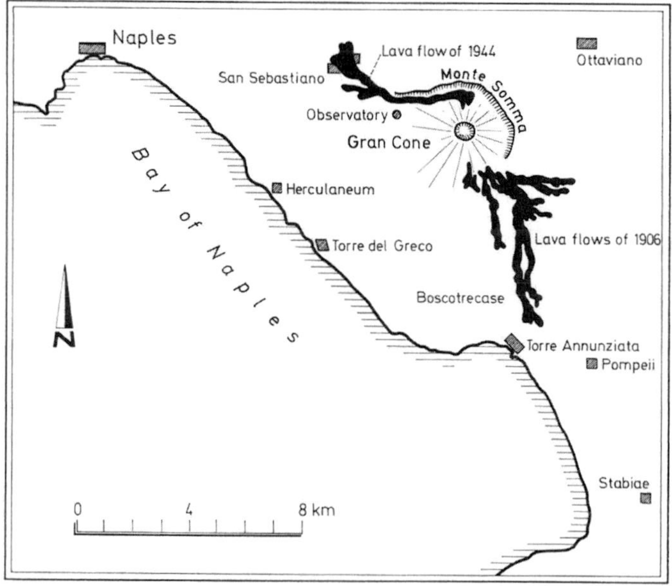

Fig. 2.1. Map of the region around Vesuvius, showing the lava flows of 1906 and 1944, and the location of cities destroyed by the eruption of 79 A.D.

all, they regarded it as extinct. But like many "dead" volcanoes, it was potentially a killer!

Beginning about 63 A.D. many earthquakes (see Section 1.3) were felt in the vicinity of the mountain, and with our present knowledge we might have recognized the signs of a coming eruption; but in the spring of 79 there was no such awareness. Vesuvius' long slumber was suddenly ended on August 24 of that year. The story is told by Pliny the Younger in letters to the historian, Tacitus. Although the letters are primarily an account of the death of his uncle, the famous naturalist, Pliny the Elder, they contain an amazingly clear account of the eruption, and are commonly regarded as the first scientific record of a volcanic eruption—the first document of the science of volcanology. Thus, not surprisingly, the history of volcanology begins with tragedy.

In the year 79 Pliny the Elder was an admiral in command of one of the Roman fleets, with his base at Misenum, at the northwest corner of the Bay of Naples. His sister and nephew were living with him. About 13:00 h on August 24, his sister called his attention to a huge cloud that was rising in the east. The cloud had the shape of a Roman pine tree, with a slender trunk rising nearly vertically to spread out into a broad nearly flat top. (Today we would compare it with the "mushroom" cloud of an atomic blast.)

His scientific curiosity aroused, Pliny immediately ordered a ship made ready to take him to a place where he could observe the happenings more closely. He invited his nephew to accompany him, but the boy appears to have suddenly found a strong interest in his studies—most fortunately so, or the account of the eruption might never have been written! As Pliny was about to start on his expedition he received a message from friends living at the foot of Mt. Vesuvius asking for assistance. Changing his plan, he ordered his war galleys to put to sea to rescue people from the towns along the edge of the bay, and he himself sailed directly toward the volcano. As he drew nearer, hot cinders, pumice stones and "black pieces of burning rock" began to rain down on his ships in ever-increasing amount.

His progress was stopped by a sudden retreat of the sea from the shoreline. (We know now that such retreats of the sea, due to elevation of the shoreline and adjacent sea bottom, commonly accompany eruptions of Vesuvius.) Unable to reach the former shore at the base of Vesuvius, he changed his course and sailed southeastward, toward Stabiae, where his friend, Pomponianus, was staying. The latter had already taken alarm, and had loaded his belongings aboard a ship, ready to flee the area, but the wind was directly on shore and the ship was unable to get to sea. Pomponianus greeted Pliny with "the greatest consternation."

Apparently in an effort to reassure Pomponianus and others, Pliny bathed, had supper, and retired to sleep. Ash and pumice continued to fall, and in the courtyard of the house the accumulation became so deep that soon he would not have been able to get out. So he was awakened, and it was decided to flee across the open fields, with pillows tied on their heads as protection from the falling pumice fragments. Although it was daytime, the darkness was as the darkest night. Reaching the shore, they found the waves still running too high to allow the ships to sail. Pliny, who was a very corpulent man, lay down to rest, but, writes his nephew, ".... the flames, preceded by a strong whiff of sulphur, dispersed the rest of the party, and obliged him to rise. He raised himself up with the assistance of two of his servants, and instantly fell down dead ...." His nephew supposes that he was suffocated, and this does appear to be a possibility in view of the fact that persons suffering from asthma often are very badly affected by sulfur gases; but it seems equally possible that he suffered a fatal heart attack. Not until three days later did it become light enough to locate his body.

At Stabiae the accumulation of ash and pumice fragments was so deep that it nearly filled the courtyards of the houses. Closer to the volcano, at Pompeii, the fragments were somewhat larger and the layer reached a thickness of about 3 m. The town was wholly buried. Herculaneum was also buried, but in a different manner, described later. Of the 20,000 persons living in Pompeii, most escaped. Few remains of horses or chariots have been found, and it appears that many persons fled during early stages of the eruption. The remains of about 2,000 persons have been found, some as skeletons, and some as remarkably detailed casts of the bodies and clothing in the enclosing ash. How did they die? Some were crushed as roofs collapsed under the weight of accumulating ash, others were trapped in buildings and cellars as the ash buried them. Some of these probably suffocated as the limited supply of air they had to breath became depleted in oxygen, others, however, had their hands or pieces of cloth over their faces, as though to shut out noxious fumes. The lower 2.5 m of the tephra deposit consists largely of pumice fragments from about 5 to 15 mm in diameter, and sufficient gas may have been given off by this pumice after its deposition to asphyxiate people. There is little evidence of burning; ash and pumice fragments that have flown any long distance through the air are very seldom hot enough to set fires.

When the eruption had ended, and daylight returned, it was found that the entire top of the mountain had disappeared. In its place was a great hole, 3 km across, that we call a *caldera*. (A caldera, by definition, is a crater more than 1.5 km across.) Such a small amount of debris from the former upper part of the mountain is present in the material thrown out by the eruption that it is clear the mountaintop cannot have been blown off by explosion. Instead, its support was removed by eruption of liquid magma from beneath, and it fell in. Ninety-three years later came another eruption, and that was followed through the centuries by many others, which gradually built the present cone of Vesuvius within the caldera.

Many of the features that have been described at Pompeii have been observed also at other eruptions. In June, 1912, a tremendous eruption took place at and near Mt. Katmai, in Alaska. Although the best-known product of the eruption was an ash flow down the valley now known as the Valley of Ten Thousand Smokes (described later), more than 16 cubic km of ash and pumice were blown into the air and rained down over the surrounding land and ocean. Near the volcano the ash deposit reached a thickness of more than 15 m; and at Kodiak, 160 km to the southeast, it reached about 3 m (see Fig. 2.2).

The eruption began on the afternoon of June 5, and on the morning of June 6 several explosions were heard 240 km away at Seldovia. The most violent

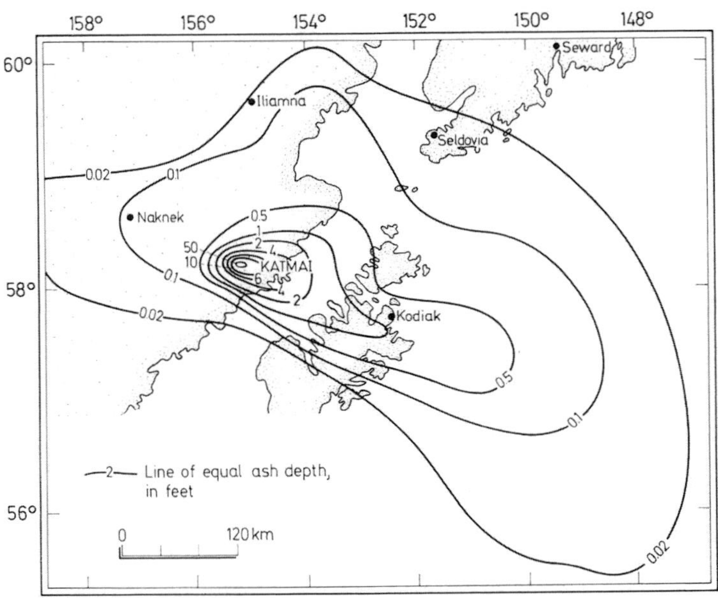

Fig. 2.2. Map showing the distribution of ash from the eruption of Mount Katmai, Alaska, in 1912. (After Wilcox, 1959)

part of the eruption began about 13:00 h on June 6, and continued for 2 1/2 days, with major explosions at about 15:00 h and 23:00 h on June 6 and 23:00 h on June 7, heard at Juneau 1,200 km away.

Fairly strong eruption continued for several weeks, and weaker explosions occurred occasionally for several months. At Kodiak, at 17:00 h on June 6 a dark cloud was seen approaching from the northwest, with unusual thunder and lightning. The dark cloud increased, and by 18:30 h it had blotted out the daylight. Ash fell all night, and by 09:00 h on June 7 had formed a layer 10 cm deep. The radio of the Coast Guard cutter Manning, at the dock at Kodiak, was unusable because of intense static, and the radio station ashore was destroyed by lightning. (Such disruptions to radio communication are common during explosive eruptions.) The captain of the Manning reported that ashore all streams and wells had become choked with ash, and water had to be furnished to the inhabitants of the village by the Manning and another ship.

After a respite of a few hours, ash began to fall again about noon, and by 13:00 h the air was so thick with ash that visibility was reduced to 15 m, and by 14:00 h pitch darkness had set in. Through the rest of the day and the next morning the dust fall was so heavy that a lantern held at arm's length was invisible; sulfurous fumes came at times, and avalanches of ash could be heard sliding on the nearby hills; frequent earthquakes were felt, and the town's populace was in terror. People were cared for aboard the Manning, and in an adjacent warehouse. The ash fall decreased at 14:30 h on June 8, the sky took on a reddish color, and visibility gradually returned; by June 9 people were able to return home.

The 25 cm layer of ash deposited at Kodiak smothered all the small vegetation. Because of the destruction of grass, a herd of cattle had to be shipped away until the pasturage revived, two years later. Larger trees survived. The weight of ash on conifers had little more effect than the normal load of winter snow, and the trees soon shed most of the ash. This commonly is not the case with broad-leafed trees, which are often broken by the weight of the ash.

The contamination of the water supply by falling ash is a common effect of explosive eruptions. During the 1963 eruption of Irazú, in Costa Rica, fine ash suspended in the river water clogged the filters in the water works of the city of San Jose. Ash washed from the streets of the city clogged the storm sewers, resulting in local flooding during the rainy season.

Maps of the distribution of ash and pumice during the 79 eruption of Vesuvius and the 1912 eruption of Katmai both show clearly the effect of direction of the wind (Fig. 2.2). Most of the ash of the Vesuvius eruption was blown southeastward; little fell on the northern slope of the mountain. The Katmai ash moved mostly east-southeastward. About 30 cm fell on Kodiak Island, whereas at the same distance in the opposite direction there was only about one-tenth as much. Clearly, the direction of the wind is of great importance in determining the amount of damage from ash fall at any given point. A similar effect is shown by the eruption of Mt. Spurr, another Alaskan volcano. The eruption began about 05:00 h on July 9, 1953. The outbreak occurred not at the summit of the 3,375-meter mountain, but at a lateral vent at 2,100 m, sending a mushroom-shaped cloud to a height of 15 to 20 km, and scattering ash over the city of Anchorage, 130 km to the east. At Anchorage the ash was 3 to 7 mm thick, but very little fell to the west of the mountain.

Since wind directions commonly differ from season to season, and often are different at high levels than at low levels, the amount of ash fall and the consequent damage at any given locality depend on the time of year in which the eruption occurs, and the strength of the eruption—whether it throws ash to a height of 12 or more km, or only to levels within the lower atmosphere. All of these factors must be considered in any prediction of probable ash damage from eruptions of any particular volcano.

Falling bombs and blocks commonly are restricted to the slopes of the volcano itself and its immediate vicinity, and most of the damage is done by the much more widespread ash fall. Occasionally, however, the larger ejecta do an important amount of damage. An example is furnished by the 1968 eruption of Arenál Volcano, in Costa Rica, when falling blocks broke limbs from trees and crashed through houses more than 3 km from the volcano. Striking in a blanket of soft ash, the blocks formed impact craters, most of them from 3 to 10 m across, but some more than 30 m across. Especially in the larger craters, the blocks commonly disintegrated on impact, so that only small fragments could later be found. The landscape, pock-marked by hundreds of craters, resembled the surface of the moon (Plate 2.3).

The eruptions described above were violent, and threw ash high into the atmosphere to be distributed by wind over broad areas. But ash and cinder falls from less violent eruptions also may do great damage. Two examples are the eruption of Paricutin, in Mexico, from 1943 to 1952, and that of Eldafell,

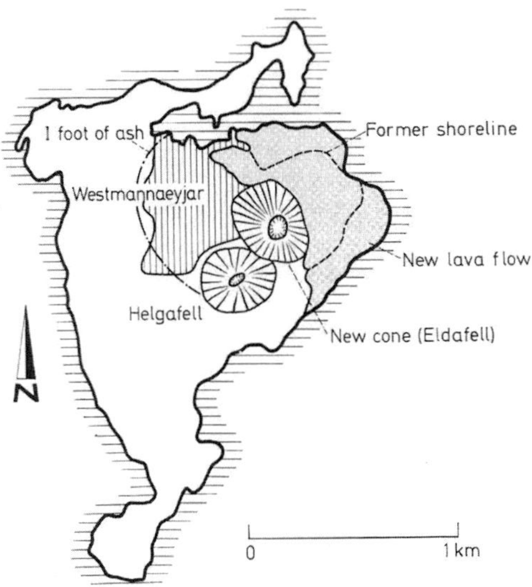

Fig. 2.3. Map of the island of Heimaey, just south of Iceland, showing the cone and lava flow of the 1973 eruption, and the location of the town and harbor of Westmannaeyjar. (After Williams and Moore, 1973)

on the Island of Heimaey, south of Iceland, in 1973. In both instances, as in many other eruptions, destruction was both by ash-and-cinder fall and by lava flows, but we are here concerned with the former. Lava flows are discussed in the next section.

Heimaey, one of the Westmann Islands, is entirely of volcanic origin. In the years 1963 to 1967 submarine eruptions in the ocean 20 km southwest of Heimaey had built the new island of Surtsey and nearby temporary islands, but the last eruption on Heimaey itself had taken place about 5,000 years ago, when the conical mountain known as Helgafell was built. In recent years the town of Westmannaeyjar, at the northern end of Heimaey (Fig. 2.3), had become one of the most important of Iceland's fishing ports. Early in the morning of January 23, 1973, a crack opened in the ground on the lower northern slope of Helgafell and molten lava spurted upward to form a chain of lava fountains more than 1 km long. Soon, however, the eruption became restricted to a short length of the fissure and became more explosive. Ash and bombs were thrown a few thousand meters into the air and piled up around the vent to build a cone that within 3 weeks had grown to a height of more than 200 m and rivaled Helgafell in size. The initial outbreak was at the edge of Westmannaeyjar itself, and much of the ash and cinder was blown westward onto the town (see Plate 2.7). Incandescent bombs set some houses on fire. The deepening blanket of ash completely buried houses in the eastern part of the town, and the weight of ash caused roofs to cave in. Fairly light in itself, the ash became heavy as it became saturated with rain water, and had to be shoveled off the roofs to prevent their collapse.

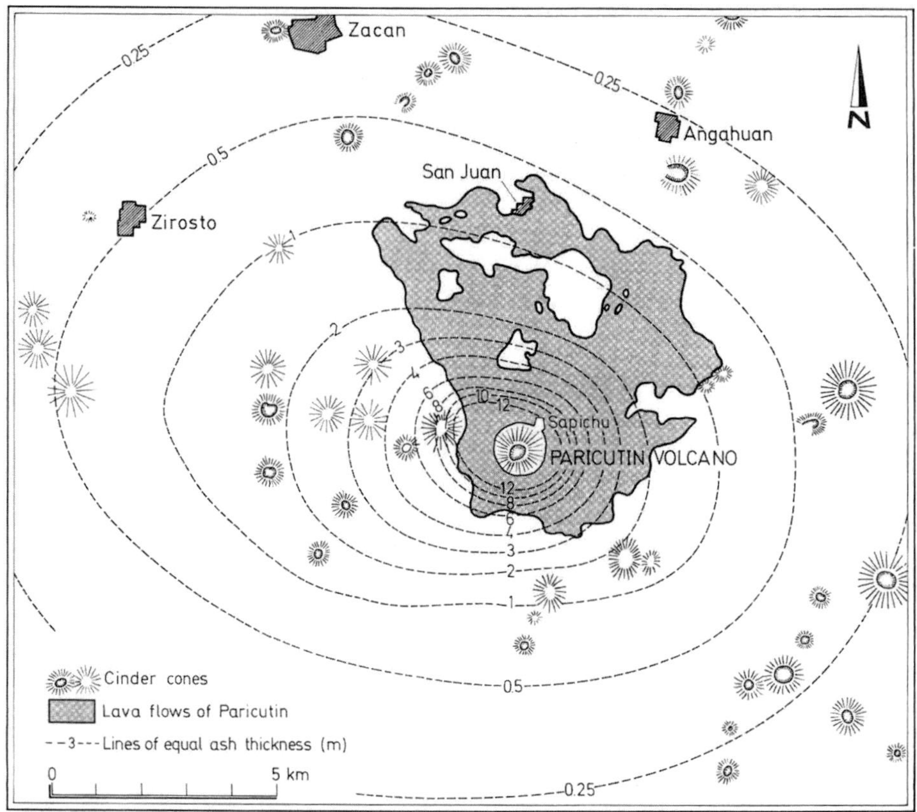

Fig. 2.4. Map showing thickness (in meters) of ash deposits around Paricutin Volcano, Mexico, in October 1946, and the extent of the lava flows in August 1967. The new cone of Paricutin is surrounded by many older cinder cones that mark the vents of former eruptions. (After Williams, 1950)

Most volcanic eruptions last only a few days or weeks, and commonly the wind direction remains fairly constant throughout the eruption so that the ash is deposited largely in one sector extending outward from the volcano. In contrast, the eruption of Paricutin lasted for 9 years, and during that long period many changes in wind direction resulted in a more symmetrical distribution of the ash (Fig. 2.4). The eruption started on February 20, 1943, in a cultivated field near the base of the much older volcano, Tancitaro. By March 6 about 80,000,000 cubic m of ash, lapilli, and bombs had been ejected—an average rate of some 6,000,000 cubic m a day—but thereafter the rate of tephra eruption decreased to an average of about 76,000 cubic m a day. A cinder cone 370 m high was built around the vent, and lava flows spread around its base. Ash buried houses in neighboring villages nearly to their eaves (see Plate 2.4), and caused roofs to collapse, but the most interesting effects were on vegetation. Within 3 to 5 km of the vent, where the ash was more than a meter deep, even the largest trees were killed, while at greater distances large trees survived, but smaller vegetation was smothered. Herds of cattle had to be moved to other areas, but even so several thousand animals died, partly of starvation due to destruction of grazing areas, but partly because their digestive systems became clogged as a result of eating ash-laden vegetation. Another serious effect is the grinding away of tooth surfaces by chewing on ash-covered feed, the teeth eventually becoming so worn down that the animal can no longer eat. Still another result of the Paricutin eruption was the destruction of the sugar-cane crop in the area west of the volcano, not directly by the ash fall, but by an infestation of cane borers that resulted in turn from destruction by the ash of another predatory insect that normally kept the cane borers under control.

Rather rarely the ash contains enough of some poisonous substance to do serious damage, as, for example, the ash thrown out during the eruption of Irazú, Costa Rica, in 1963–1965, which was so acid ($pH > 4$) that it produced acid burns on plants and caused corrosion of metals. In some Icelandic eruptions, such as those of Hekla in 1947 and 1970, thousands of grazing animals in ash-covered areas died of poisoning by fluorine in the ash. The fluorine adheres to the surface of the ash particles, probably as hydrofluoric acid, and the ash in turn adheres to grass and other plants. Experiments by Icelanders have shown that if the fluorine content is as little as 250 parts per million of the weight of the dry grass, it can kill sheep grazing on it within a few days. Fortunately, the fluorine is gradually washed away by rainfall and the range recovers, so that in the case of the 1970 eruption, grazing areas had become safe again in about 6 weeks.

Effects may be even more remote. In certain parts of the North Island of New Zealand sheep developed a "bush sickness" that was eventually found to be poisoning from cobalt derived by plants from a certain cobalt-rich ash deposit. The problem was solved by moving the sheep away from the area underlain by this particular ash.

In 1911 Taal Volcano, in the Philippines, erupted, killing 1,300 persons. A similar, though fortunately less disastrous, eruption took place in 1965. Taal is an island volcano lying in a large lake 60 km south of Manila. The 1965 eruption began about 02:00 h on September 28, with rather mild explosions that threw glowing hot bombs and cinder high into the air and built a cinder cone around the vent. About 03:25 h the eruption suddenly became more violent, as lake water gained access to the red-hot throat of the vent and huge volumes of steam were generated. Great ash-laden steam clouds rose 16 km or more into the air, and wind deposited fine ash as much as 80 km away, the material thrown out being mostly fragments of old rock blasted out of the side of the island. Around the base of the vertical eruption column, ring-shaped expanding clouds formed and rushed outward with hurricane speed over the surface of the island and the surrounding lake, and onto the distant shores. Within a kilometer of the volcano the ash-laden clouds broke off or uprooted all the trees, and in the next half kilometer the sides of the trees toward the volcano were strongly sand-blasted and as much as 15 cm of wood was worn away, although on the other side the bark still remained intact. There was no sign of charring and in the outer part of the area the trees were plastered with mud, showing that the blast was cooler than the boiling temperature of water. As the velocity of the outward-rushing clouds decreased, ash was deposited, forming dunes as much as 2.5 m thick 1 km from the vent and a layer more than 30 mm thick 5 km away. These outward-expanding clouds are like those that form from fall-back around the base of the mushroom cloud of an atomic explosion, and like them are known as *base surges*. They are characteristic of the rather low temperature explosions that result from the involvement of external water, and have been observed during eruptions in the ocean such as those of Surtsey and Capelinhos, as well as at Kilauea in 1924.

What can be done to mitigate damage from tephra falls? Some things have already been indicated. Ash should be shoveled off roofs to prevent their collapse under the increased load, particularly when the ash becomes water-saturated by rain. Dust masks can be improvised by tying a piece of wet cloth over one's nose and mouth. Breaking of the branches of fruit and other trees can be avoided by shaking the ash off the trees, and where ash becomes deep enough to kill large trees some can be saved by digging the ash away from their boles. Surface water supplies may become contaminated with ash, but much of it can be removed by simple cloth filters such as a bag tied over the mouth of the faucet, though as much water as possible should be stored early in the eruption, because later water may become excessively acid due to leaching of the ash. Drains clogged by ash should be reopened to prevent flooding during the heavy rains that often accompany eruptions. Grazing animals should be moved to other areas to prevent ingestion of ash and grinding down of their teeth. Dust filters should be provided for aircraft and other engines.

Base surges have many of the characteristics of the glowing avalanches and ash flows described later, and probably little can be done to lessen losses caused by them except by adequate warning and evacuation of the area before they occur.

Evacuation itself may be very difficult once the eruption has started, because of the greatly reduced visibility. Even with strong lights it may be impossible to see out of doors more than a meter. Under such conditions it is probably best for people to remain indoors in the affected area until visibility returns (usually within a few hours) except for brief trips outside to clear roofs and drains if that is necessary. Because of possible disruption of water supplies, enough drinking water to last several days should be stored indoors as early as possible during the eruption.

Damage to land by ash fall is usually short-lived. On cultivated land a thin ash cover can be plowed into the underlying soil, commonly lightening the soil and adding to its content of readily-available plant nutrients. Natural cementation commonly produces a hard crust on ash deposits, and breaking up this crust by plowing aids plant growth. In some places ash and pumice deposits a few meters thick have been cleared from garden areas, and the soil found to be unharmed.

Climate is a very important factor in the speed of recovery of a devastated area. In the warm wet parts of the tropics plant growth returns very quickly. Where the ash is only a few cms thick plants commonly come up through it within a few months, from seeds and spores buried beneath it. Where thicker ash has killed all smaller vegetation, and has stripped the larger trees of foliage and small branches but has not killed them, the trees generally start to leaf out again within a few months. Even where the ash is so thick that all the vegetation is killed, plant reinvasion is surprisingly rapid. A year after the great eruption of Krakatau in 1883 a little grass already was growing on the deeply ash-covered remnants of the volcano, and two years later 26 species of plants had become established on them; by 1924 they were almost wholly covered with dense forest. In cooler and drier regions recovery is slower, but nevertheless may be quite rapid. During the 1912 eruption of Katmai a meter of ash fell at Russian Anchorage, 40 km east-southeast of the volcano, completely destroying small plants. Alders also appeared dead, but a year later those on the slopes toward the volcano were sending up new shoots from their roots, and those on slopes away from the volcano were green with leaves.

The great eruption from the Laki fissure in Iceland, in 1783, was primarily an eruption of very fluid basalt lava flows and is generally classed as a flood eruption. However, it was much more explosive than most, apparently because the rising lava encountered shallow ground water and surficial lakes. The resulting steam explosions generated about 0.3 cubic km of fine ash, which fell all over Iceland and drifted across the north Atlantic to Great Britain and Norway. The ash caused the death of many grazing animals. A huge volume of sulfur gas, together with the fine ash, caused a bluish haze which also enveloped nearly the whole country, seriously damaging crops and forage plants and resulting in a food shortage for both men and animals. To make things even worse, the haze so reduced visibility that the fishing fleets could not operate. Half of all the cows, three-fourths of all the sheep and horses, and a fifth of the human population of the entire country died as a result.

Lava Flows

During human history lava flows have destroyed great amounts of property, but have seldom taken lives, as their slow advance usually gives ample time for men and animals to get out of their way. In 1950 a ranch foreman, trying to make sure that his cattle had moved out of danger, was temporarily trapped between two lava flows of Mauna Loa in Hawaii, until, making his way to shore, he was taken off to safety by the Coast Guard. If he had not been rescued in this manner, he could still have survived for the few days until the lava flows had cooled enough for him to cross, but under other circumstances, if the flows had remained active and had gradually spread over the intervening area, he might have been killed. Wide active aa flows are difficult or impossible to cross, partly because of their jaggedness and their radiant heat, but even more because heated air above the flow is thin and a person laboring hard to cross the very rough surface may become faint or collapse completely from deficiency of oxygen. Some persons have possibly died in this manner. In 1823, when a very fluid lava flow from vents on the southwestern flank of Kilauea Volcano advanced with great speed over a coastal village, most of the villagers were able to flee to safety, but tradition has it that some old people and very young children could not move fast enough to escape.

Thus, our problem in dealing with lava flows is less the saving of lives than the saving of property, and learning to treat the surface of the flow to make it useable for agriculture in the shortest possible time.

For examples of destruction by lava flows, we can again turn to our classical volcano, Vesuvius. Repeatedly, since the modern cone of Vesuvius started to grow, in 172 A.D., flows have invaded vineyards and villages on the lower slopes of the mountain, one of the biggest of the series being the eruption of 1906. At the beginning of 1905 molten lava stood high in the cone, nearly to the rim of the crater, 1,340 m above the nearby Bay of Naples, and the weight of such a column of molten lava exerts considerable outward thrust that tends to split the cone. Whether for that reason, or because of "tectonic" movements resulting from more fundamental shifting of the Earth's crust, in May the cone cracked open and vents formed on its side 120 m below the crater rim, and for 10 months lava flows poured down the northwestern side of the cone, without doing any serious damage.

On April 4, 1906, new vents opened at the same level on the south-southeast side of the cone. The downward splitting of the cone continued, and about midnight vents opened 530 m below the crater rim and voluminous lava flows started to pour from them. The resultant rapid draining of magma from the central conduit of the cone reduced the pressure on magma at deeper levels, allowing more and more gas to come out of solution, and the eruption at the summit rapidly increased in violence. As the crater walls collapsed, great black clouds rolled out and rained ash over the cone.

On the morning of April 6 still another row of vents opened, 730 m below the summit, and lava poured down the south-southeast flank of the cone, racing with a speed of more than 16 km an hour in a narrow stream close to the vents, then soon spreading out and slowing until the flow fronts were advancing only a few meters an hour. Slowly but inexorably, the aa lava advanced on the vineyards and habitations on the lower slopes of the mountain (see Fig. 2.1). The village of Casa Bianca and much of the Boscotrecase district were totally destroyed, and the lava reached the outskirts of Torre Annunziata, 3 km west of Pompeii. The lava was somewhat more viscous than that which characterizes Kilauea and Mauna Loa, in Hawaii, but nevertheless it was sufficiently fluid to follow even insignificant depressions, dividing and reuniting like a braided river, and leaving islands of undamaged land between its branches. The lava reached a thickness of as much as 7 m, and buried houses to the second story (Plate 2.5), pouring in through windows and doors. Some houses were pushed down, but others were only filled with lava without destroying the relatively weak masonry walls. This small amount of thrust exerted by fluid lava flows on obstacles is important in attempts to divert flows by methods described later.

Another great eruption of Vesuvius came in 1944. While the volcano sided with the Nazis by showering Allied airfields with (volcanic) bombs, a lava flow went northwestward into the Atrio del Cavallo, between the modern active cone and the amphitheatral ridge of Monte Somma (the old caldera rim), and thence to the town of San Sebastiano, which had just been missed by a flow in 1872. The behavior of the flow was much like that of 1906.

Plate 2.5. A lava flow in the streets of Bosco Trecase, during the eruption of Vesuvius in 1906. The thrust of the lava was not sufficient to cave in the masonry walls. (Photo by T.A. Jaggar)

At Mt. Etna, also, lava flows have done great damage. Only two of many examples will be mentioned here. On the eastern slope of Etna the great depression of the Valle del Bove (Fig. 2.5) is the large collapsed crater of an older volcano, partly buried by the younger Etna. Many of the lava flows of Etna are captured by this valley; but in 1928 an eruption on the northeast flank of Etna sent a flow downslope north of the Valle del Bove, destroying vineyards and groves of Sicily's delicious blood oranges, and eventually burying the town of Mascali. Since then a quarry has been opened to obtain building stone from the massive center of the 12-meter-thick aa lava flow, exposing the ruins of buildings beneath it. The lava was more viscous than that of Vesuvius in 1906, and exerted more thrust, so that the weak masonry walls of the houses were very largely pushed over and crushed beneath it.

A still greater eruption of Etna took place on the southern slope of the mountain in 1669. The lava moved southward, destroying agricultural land and several villages, until eventually it came against the wall of the old feudal city of Catania. The wall held for several days, while lava piled up against it and

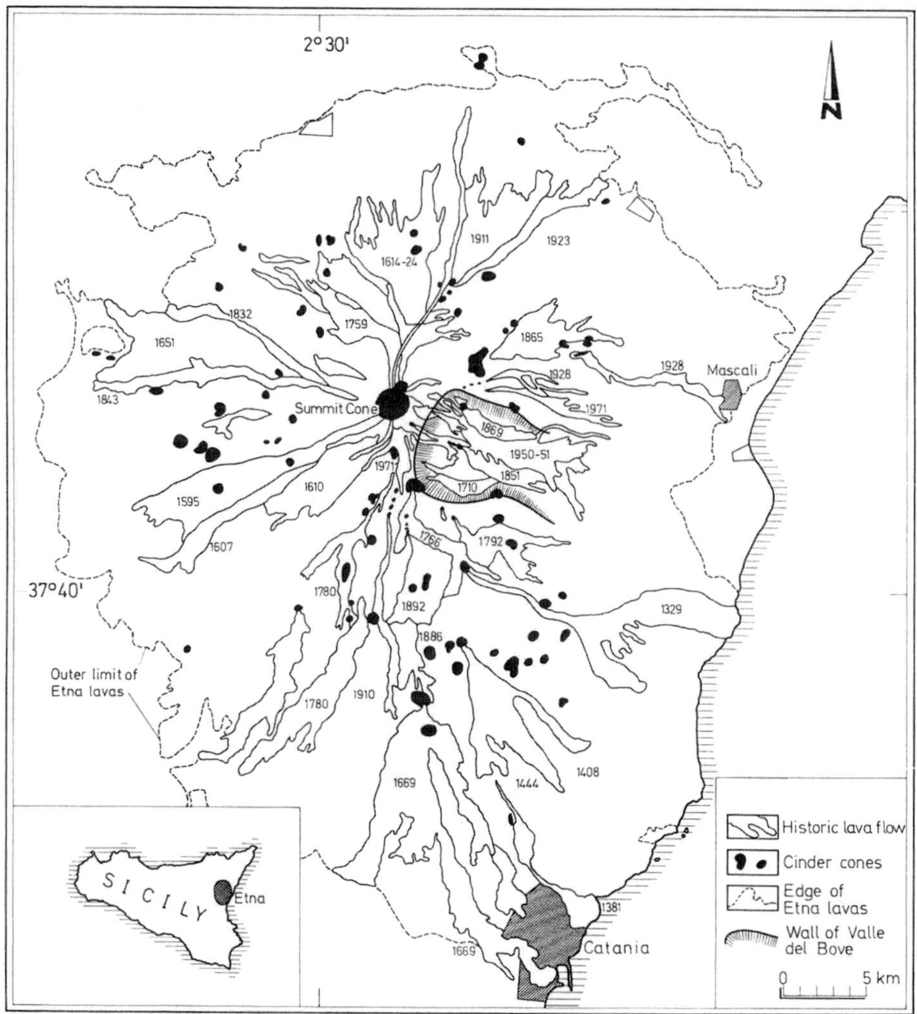

Fig. 2.5. Map of Mt. Etna, showing the location of the Valle del Bove, many lava flows erupted since 1,300 A.D., and the city of Catania and town of Mascali, partly destroyed by the lava flows of 1669 and 1928. (After official Italian government maps, and unpublished mapping by Carmelo Sturiale)

was diverted around it into the Ionian Sea. Eventually, however, a weak part of the wall gave way and lava moved into the city, destroying part of it, pushing over most buildings but burying more resistant ones. The present ground-level entrance to the Orsini Palace, a medieval fortification with thick resistant walls, is at the level of the old second story, but the rooms of the first story remain open and in use. The foundations are buried 9 to 12 m deep by the lava.

During the long eruption of Paricutin, mentioned earlier, lava flows spread around the base of the cone, eventually covering an area of more than 2,400 hectares of forest and cultivated land (see Fig. 2.4) and destroying all of the village

Plate 2.6. Towers of the cathedral of San Juan Parangaricutiro, Mexico, protruding from the rubbly surface of a block lava flow, July 1945. The building has been buried by lava to a level above the top of the nave. Paricutin Volcano, from which the flow issued, is visible in the left background. (Photo by Tad Nichols, Tucson, Arizona)

of Paricutin and most of San Juan Parangaricutiro, where the lava buried the old masonry church to the level of the top of the nave (Plate 2.6). Small amounts of it dribbled in through windows and doors, but the walls were not crushed, and the interior was left largely empty. Once again the small amount of thrust exerted on the walls by the lava is striking!

Since the beginning of written records, in the early 19th Century, lava flows of Mauna Loa, in Hawaii, have covered a total area of more than 650 square km, almost none of this cultivated, but about 100 square km of it ranch land. In 1926 lava buried the fishing village of Hoopuloa, and one of the flows of the 1950 eruption destroyed part of the village of Laupahoehoe, in Kona. The small amount of destruction other than of ranch land is the result of the small proportion of the land that was in use at the time of the eruptions, but land use for other purposes than ranching has increased greatly during recent years, and far greater damage by lava flows must be expected in future eruptions.

Mauna Loa's sister volcano, Kilauea, has covered only about 120 square km with lava in the same period, but has done far more damage because of the greater degree of land use. The lava flows of the 1955 eruption (Fig. 2.6) covered about 440 hectares of cultivated land, buried 10 km of public road, and destroyed 17 houses. The lava flow of 1960 covered about 120 hectares of cultivated land and destroyed a commercial fishpond, a schoolhouse, several beach residences,

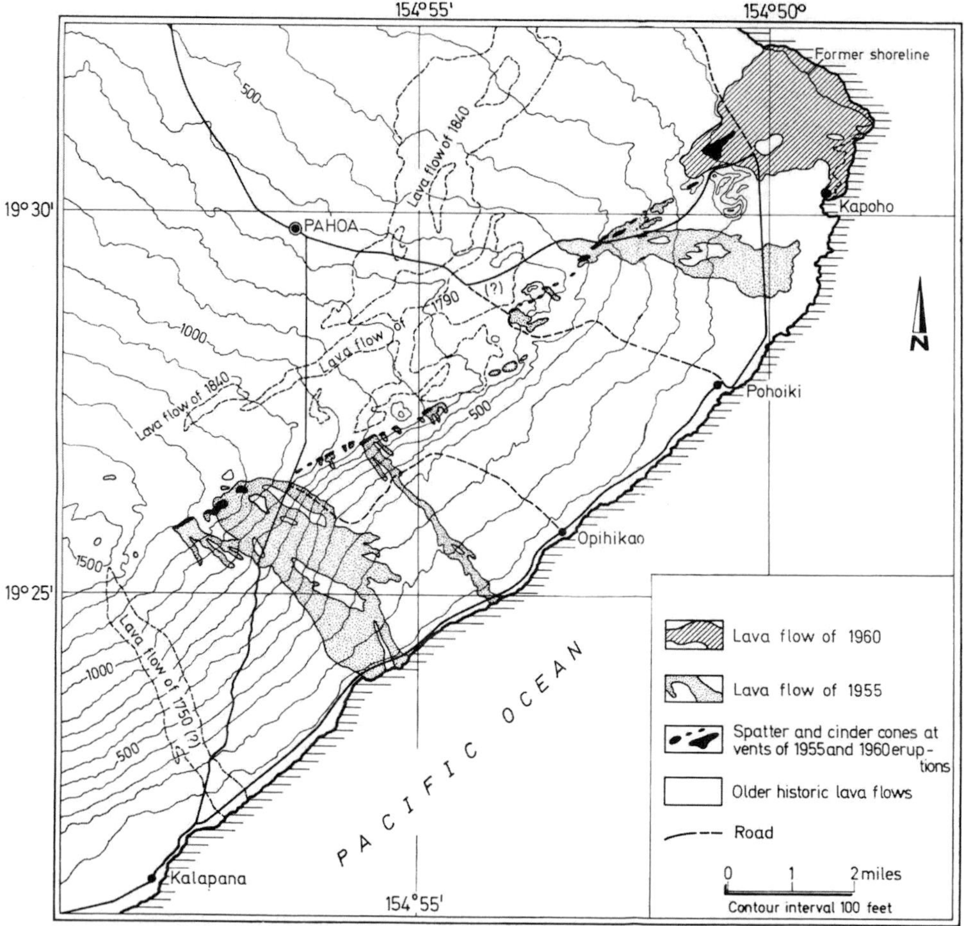

Fig. 2.6. Map of the eastern part of the Puna District on the Island of Hawaii, showing historic lava flows and the location of the village of Kapoho, largely destroyed by the lava flow of 1960. (After Macdonald, 1972)

and most of the village of Kapoho. There were no human casualties in either eruption, and warning of coming events made it possible to save most movable property, including such heavy objects as big store refrigerators. The latest destruction by lava flows has been on the island of Heimaey, where the eastern part of the town of Westmannaeyjar was buried (Plate 2.7).

The greatest lava eruption of historic time was that of 1783 in Iceland, mentioned earlier. The lava of this single eruption covered an area of 560 square km, destroying several farmsteads. But even this is small compared with some great "flood eruptions" of prehistoric times. These flows, which have accumulated to build up great lava plains such as the Columbia Plateau of eastern Washington and Oregon, may extend more than 150 km from their vents. One of the flows in the Columbia Plateau has been traced, on the surface and in bore holes, for more than 300 km, and covers an area of about 52,000 square km. Future flows of this sort could wipe out huge areas of agricultural land and whole cities. We can only hope it will not happen, because none of the methods mentioned in the coming paragraphs holds any promise of controlling lava flows of such magnitude!

Control of Flows

Control of some smaller lava flows does appear possible, however. Three different methods of controlling flows have been suggested, and experimented with to some degree. Two of the methods have been advanced as ways of protecting the city of Hilo, Hawaii, which is in the direct path of probable future lava flows from Mauna Loa (Fig. 2.7). The city is built on prehistoric lavas of Mauna Loa. In 1852, 1855, 1899, 1935 and 1942, lava flows from vents on Mauna Loa's northeast slope advanced toward Hilo, and in 1881 a flow actually entered the outskirts of the present city. The probability of a future flow destroying all or part of the city and harbor is great, and it is certainly desirable that, if possible, steps be taken to avoid such a catastrophe.

One of the suggested methods is by *bombing* the flow high on the flank of the mountain, causing the lava to spread there, where it will do no serious damage, and robbing the supply of lava from the advancing flow front, causing it to stop. Three different situations exist in which bombing might accomplish this result. As pointed out earlier, a mature pahoehoe flow is fed by a stream of lava flowing through a tube that is largely roofed over. If heavy bombs can be dropped directly on such a tube, they may have two results. First, the roof may be shattered locally and the fragments block the tube, causing it to overflow at the point of bombing and diverting the supply of lava from the flow farther down slope; second, the violent stirring of the liquid lava in the tube by bombs striking in it may upset the gas equilibrium in it and cause it to change into aa, which may block the tube because of its greater viscosity. The method was tried in 1935, on a flow advancing toward Hilo (see Fig. 2.7), and although the results were not entirely conclusive, they were encouraging. The main feeding tube was blocked and lava overflowed in a desolate area on the upper slope of the volcano. Molten lava continued to drain from the tube farther down slope, and the flow front continued to advance but most of the supply of lava was cut off, the movement of the flow front slowed almost immediately, and 6 days after the bombing it stopped completely. The eruption also stopped shortly afterward, and unfortunately claims were made that this was the result of the bombing, which certainly was not true, but it does appear that the bombing stopped the advance of the flow front toward Hilo.

The bombing of aa flows depends on a somewhat different principle. Repeated overflows of the open feeding river of an aa flow build up natural levees along the edges of the river, which often hold the surface of the river a few meters above the top of the flow on either side. If the levee can be broken down when the lava stream is at high level the lava will flood out locally, diminishing the amount continuing in the river farther down slope, and thus resulting in the slowing or stopping of the flow front. The effect can be augmented by violent

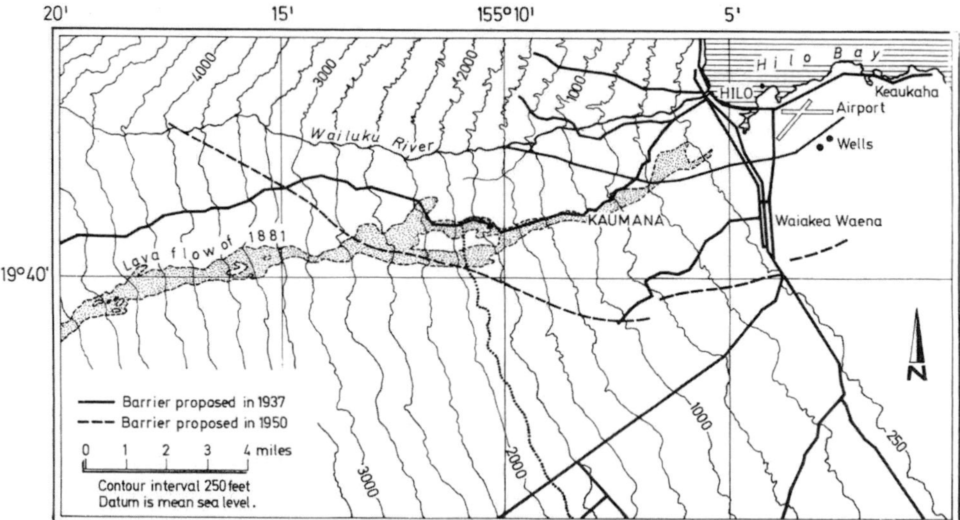

Fig. 2.7. Map showing the location of the city of Hilo, on the Island of Hawaii, the lava flow of 1881, and the course of barriers proposed to divert future lava flows from the city and harbor. (After Macdonald and Abbott, 1972)

stirring from bomb bursts in the river just below the break in the levee, locally increasing the viscosity of the lava and partly or entirely blocking the river channel. Bombing was tried on the aa lava flow of Mauna Loa's 1942 eruption. The best target for effective bombing was chosen during a reconnaissance flight over the area, and although by the time the bombers arrived that target was hidden by clouds, and a less favorable target had to be used, nevertheless, the effort was partly successful. The levee was broken down, and liquid lava spilled out, forming a new flow that continued on downslope along the side of the earlier one and eventually rejoined it a few kilometers below the bombing site. Shortly afterward the advance of the main flow front slowed appreciably, and it appears probable that the temporary reduction in the amount of feeding allowed the viscosity of the lava to increase and slow the flow.

Thus the results in 1942 demonstrate both the possibility of the method succeeding, under favorable circumstances, and some limitations to its use. Visibility is often poor during eruptions, due to clouds and smoke, and it may not be possible to bomb the most favorable targets (though this difficulty is in part removed nowadays by the use of modern aiming devices). Also, successful diversion of the lava depends on favorable topography. The adjacent slope must be such as to lead the diverted lava away from the older flow, or at least not back into it. If a flow is in a well-defined valley which is deeper than the flow is thick, diversion is impossible.

It is interesting to note that the first attempt to divert an aa flow was made on the slopes of Etna in 1669. The flow was advancing toward Catania, and a group of men from that city covered themselves with wet cowhides for protection from the heat and attempted to divert the flow by digging a channel with hand tools through the edge of the flow. To start with, they were successful. Lava flowed out through the artificial channel and formed a new stream that took a new path, reducing the amount flowing toward Catania, but the new stream headed toward the town of Paterno, and several hundred irate citizens of that town armed themselves and drove the men of Catania away. The artificial channel became clogged and the lava continued on to Catania. The attempt points up another consideration: any diversion of lava puts it onto land that it would not normally have covered, and this may result in serious legal problems.

A third possibility is bombing of the walls of the cone at the vent, a method which has not yet been tried. Its success depends on the facts that the cone commonly confines a pool of very fluid lava at a level considerably above the surrounding land surface and somewhat above the head of the active lava flow, and that the cone walls often are thin and relatively weak. If the walls can be broken down by bombing, the fluid lava in the pool will flood out over the adjacent land surface, robbing the flow of much or all of its supply. This took place naturally, without bombing, during the 1942 eruption. One of the present authors (Macdonald) had been watching the vents for hours from a point a couple of hundred meters from the base of the cone. The fountains of lava in the crater of the cone played quite steadily to a height of about 50 m, and lava issued in a series of surges from the pool in the crater into the river that fed the lava flow. Twenty km down slope the front of the flow was advancing slowly toward Hilo, when suddenly the wall of the cone started to crumble and a flood of lava spread outward, forcing a quick retreat to higher ground. As the level of the pool in the crater was lowered, the reduction of load on the magma in the underlying conduit allowed the fountains to grow quickly in height to about 200 m, and at the same time the amount of liquid draining into the river of the flow greatly decreased, the level of the river surface dropped rapidly, and the flow became sluggish. Within a few hours the advance of the flow front had slowed greatly, and a few days afterward it stopped altogether.

Thus bombing does appear to be a means, under favorable circumstances, of diverting relatively thin and fluid lava flows such as those of Mauna Loa and Kilauea. However, whether it can succeed with thicker flows such as the block lava of many continental volcanoes, or the much more voluminous lava floods like those of the Columbia Plateau, is doubtful. Furthermore, it cannot be over-emphasized that successful diversion depends on favorable topography! Bombing would have been useless in attempting to save Kapoho village in 1960, as there was no place to divert the lava. It appears that it probably would have been useless also at Westmannaeyjar in 1973; on the other hand, it might well have saved Hoopuloa in 1926, and it may some day save Hilo.

Another way of diverting lava flows is by means of *artificial barriers*—not dams to impound the lava, but walls to turn the flow from its original course and direct it into some area where it will do less damage, such as were suggested by T.A. Jaggar to protect Hilo. The situation at Hilo is especially favorable for the use of diversion barriers, because the topography is such that flows threatening the city can approach only through a relatively narrow corridor, hence barriers could be built in advance of eruptions with confidence that they would intercept any flows likely to do damage in Hilo. The position of the barriers suggested by Jaggar is shown in Fig. 2.7. The barriers can be built by bulldozers, using heavy rock available locally. The walls suggested by Jaggar varied in height with local conditions, but averaged about 12 m, though more recent experience suggests that a height of 8 or 9 m might be adequate. The channels created by the barriers would accommodate a lava flow about 1 km wide. Admittedly, the volume of lava discharged during the early hours of some of the greatest eruptions of Mauna Loa has been greater than could be contained in a channel of this dimension, but these have been short-lived discharges high on the mountain slope many kilometers from the proposed position of the barriers, and it is very unlikely that lava would reach the barriers in volume even approaching these. All of the historic flows at places anywhere near the barrier sites have had volumes that could easily be accommodated by the proposed channels.

A more serious objection is the possibility that a second tongue of a lava flow might reach the barrier on the up-hill side of an earlier tongue. If the channel behind the barrier had already been filled, the barrier and the earlier lava together would constitute a dam that would impound the later tongue, and, if the latter was of more than small volume, it would overflow the dam and continue its course down hill. Although this is a distinct possibility, the likelihood of two large flow tongues reaching the barrier in this sequence, at distances of several kms from any likely vents, is small. Obviously, barriers do not guarantee protection, but they give considerable promise of success under favorable circumstances such as those at Hilo.

Diversion barriers are still very largely hypothetical, but we have abundant evidence of the small amount of thrust exerted by relatively thin fluid lava flows, as mentioned on earlier pages. Several examples are known of ordinary loose stone walls withstanding the thrust of lava flows, the lava piling up behind the wall and eventually overflowing without pushing over the wall. Walls built for other purposes, and very poorly designed to serve as lava barriers, such as the city wall of Catania, have nevertheless successfully diverted lava flows for considerable periods. Walls were hurriedly thrown up by bulldozers in an effort to divert flows during the 1955 eruption of Kilauea, and although small and poorly designed, they nevertheless were successful to a limited degree. Walls built near Kapoho in 1960 (Plate 2.2) were not designed to divert flows, but were intended rather to raise the level of the ridge south of the flow and confine the lava to the single broad valley, thus preventing its spread southward into other inhabited areas and over the site of a vitally important lighthouse. Like any dam when the liquid gets too deep behind it, the walls eventually were overflowed, but there can be no question that they greatly reduced the amount of lava moving southward, and greatly reduced the amount of destruction. Lava actually advanced onto the concrete apron around the lighthouse, a few meters from the tower itself, and only a little more volume would have brought about its destruction; but it survived, and is still operating today. Furthermore, in the few hours after the lava first came in contact with them, the walls operated successfully as diversion barriers.

The building and observation of the behavior of the walls in 1960 taught us several important lessons, among them that the walls must have a broad base and gentle slopes, and that they must be built of heavy materials. Walls built of cinder, for instance, were a dismal failure. If the material in the wall is less dense than the liquid lava the latter is likely to burrow under or through the wall and float the light material away. If the base of the wall is too narrow the entire structure may be pushed by the lava. A gentle slope on the eruption side results in lava overlapping the base of the wall and pushing down on the structure, diminishing any tendency to push the wall laterally. A gentle back slope results in low velocity of any liquid stream overflowing the wall and reduces or eliminates erosion of the wall by the liquid.

As with diversion by bombing, the local topography is of vital importance; a barrier cannot divert a flow unless the topography is favorable for it.

Few attempts appear to have been made outside Hawaii to use either diversion barriers or dams. Although some years ago a masonry wall was built across a low gap in the rim of the caldera of Oshima, in Japan, to try to prevent lava from spilling through the gap toward a village downslope from it; as yet, lava has not reached the wall.

Whether diversion barriers would work on thick lava flows is uncertain. It would be quite impractical to think of building a barrier 30 m or more high, to equal the thickness of many block lava flows. However, experience has shown that a barrier need not be as high as the flow is thick in order to divert it. As an example, during the 1955 eruption a big aa flow came diagonally against an old railroad embankment, and was diverted by it, probably saving the village of Kapoho from destruction. The embankment was 2 to 3 m high. Although the top of the flow stood as much as 5 m above the top of the embankment, there was no spill-over of lava other than sporadic blocks that tumbled off the edge of the flow, the movement of which was governed by the lower liquid portion, which in turn was guided by the embankment. In addition to being thicker, however, the block lava flows are more viscous. We have had no experience with the behavior of such flows in contact with properly constructed barriers. When the opportunity arises experiments should be carried out, because it is entirely possible that even thick viscous flows can be guided by relatively low diversion barriers.

During the 1960 eruption of Kilauea an interesting experiment was carried out in Kapoho Village, where gradual encroachment of lava was slowly destroying house after house, and of course many fires were set. To prevent the fires spreading to other houses and into surrounding vegetation it was necessary to keep fire engines and crews on hand at all times, but most of the time they were not working. Under these conditions, Fire Chief Eddie Bento decided to try something that had been suggested many times in past decades, though usually, if not always, with tongue in cheek, namely to spray the advancing lava front with water from a fire hose, to see what the effect would be. The idea was greeted with loud derision, and even with condemnation from county officials, but the Chief persisted, and it was found that even comparatively small amounts of water did have an effect. Locally, it was possible to check the advance of the flow margin for as much as several hours, providing time to remove furnishings or other materials from a threatened building, or even to move the building itself. Instead of ridicule, the Chief deserved high praise for his imagination and determination, and for his spirit of inquiry.

Plate 2.7. Lava flow burying the eastern portion of the town of Westmannaeyjar on Heimaey Island, Iceland, in early May, 1973. At the upper left, the lava is advancing into the harbor. At the far right is the prehistoric cinder cone Helgafell, and just to the left of it is the cone built by the eruption of 1973. A thick blanket of cinder and ash covers the part of the town nearer the camera, but much of it has already been removed from the roofs. (U.S. Geological Survey photo by James G. Moore)

The idea bore fruit during the eruption on Heimaey Island in 1973 (Plate 2.7). In a personal letter, Iceland's foremost volcanologist, Sigurdur Thorarinsson, writes: "...the method of pumping water on lava in order to slow up, stop or direct its movement has now, on the initiative and under the leadership of Th. Sigurgeirsson, Professor of Physics at the University of Iceland, been practised on a great scale (900 liters/second) and with remarkable success... The pumping has played an important role in the protection of the town of Westmannaeyjar."

Mud Flows

When we think of mud, it is usually as a nuisance rather than a danger, but during the last few centuries flowing mud has destroyed more property than any other single volcanic process, and has taken thousands of human lives. Second to tephra fall, mudflow is the most prevalent of volcanic risks.

During the eruption of Vesuvius in 79 A.D., that buried Pompeii under ash, Herculaneum also was buried by mudflows generated by heavy rains on thick ash deposits on the upper slopes of the volcano. The debris deposited by mudflows sets almost like concrete. As a result, the ruins of Herculaneum have been much more difficult to excavate than those of Pompeii, being also buried more deeply.

Again, when we think of mud we usually think of fine-grained material. While most mudflows do contain a large proportion of fine debris, most of them also contain many angular blocks of rock, often more than 30 cm, and some of them several m in diameter. Not uncommonly, coarse material predominates over fine, the proportion of different sizes depending on the material that is available. Mudflows tend to sweep up and incorporate everything that is in their paths, and many mudflows contain much organic material, from leaves to tree trunks, and occasionally the bodies of animals or men. Many mudflows are wholly unrelated to volcanic activity, or to volcanic mountains. Only those directly related to volcanic activity are discussed here, non-volcanic mudflows being treated in Chapter 4.

The principal reason mudflows are so frequent on volcanoes is that the slopes of active volcanoes are commonly covered with abundant loose rock fragments that can mingle with flowing water to form mud. Most volcanic mudflows are cold, but some are hot; most are essentially neutral chemically, but some are sufficiently acid to produce serious burns. The motion of mudflows is wholly due to gravity, and their velocity depends largely on the steepness of the slope over which they are moving and the viscosity of the mud, though such things as the channel dimensions and the roughness of the underlying surface also have an effect. The viscosity is largely dependent on the proportion of solid material to water, some mudflows being predominantly water, others containing as much as 95 per cent solids.

On the upper, steeper slopes of volcanic cones the effects of mudflows are largely erosive; but on the gentler slopes on the base of the cone and beyond the moving flow loses its velocity and starts to deposit material, and eventually the entire mass comes to rest as a sheet of debris commonly several meters in thickness. Whole farms and villages have been buried by these masses (Plate 2.8). Because of the high velocity—as much as 100 km an hour—that the flows

Plate 2.8. Church in Cagsaua, Philippines, buried to the top of the nave by a mudflow from Mayon Volcano (background) in 1814. It is reported that the people of the village had gathered in the church for refuge and all were killed

may attain on the steeper slopes, they often continue for distances up to several kms across the gentler slopes adjacent to the volcano; and particularly where they enter streams they have been known to travel as far as 300 km.

Any event that brings a large volume of water onto the debris-laden slope of a volcano is apt to result in a mudflow. Of these the most spectacular is the sudden ejection of a crater lake. Because of the low permeability of the tephra in many volcanic cones, water passes through the cone only slowly, and rain or melt-water from snow often accumulates as a lake in the crater, some of which have volumes of many millions of cubic meters. (Crater Lake, in Oregon, has a volume in excess of 800 cubic km.) An explosion originating beneath the lake may throw a large part of the water out onto the flank of the mountain. Kelut Volcano, on Java, has done this several times, and the resulting mudflows have taken great toll in agricultural land and human lives. Thus, in 1919 the mudflows resulting from explosive ejection of the crater lake destroyed some 200 square km of farmland and took some 5,000 lives. To avoid a repetition of this disaster, Dutch engineers excavated a series of tunnels to lower the lake level, reducing the volume of water from about 65,000,000 to 3,000,000 cubic m (Fig. 2.8). The scheme was so successful that a similar eruption in 1951 formed no large mudflows and killed only 7 persons.

However, the 1951 eruption destroyed the entrances to the tunnels and also deepened the crater about 10 m (Fig. 2.8). The lowest tunnel was repaired, but

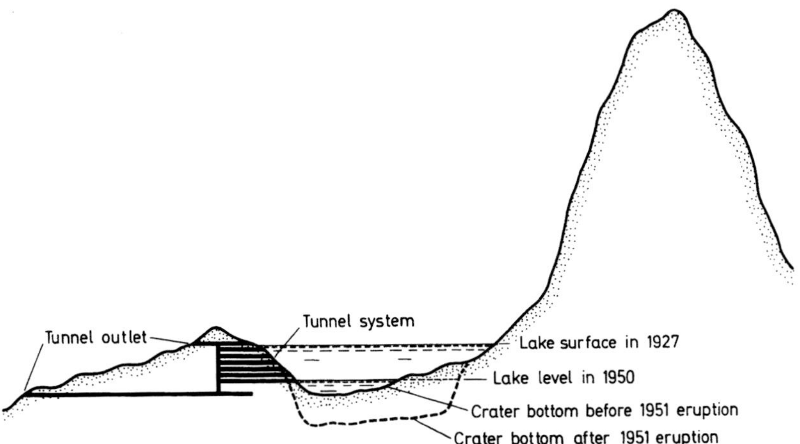

Fig. 2.8. Cross section showing the tunnel system constructed at Kelut Volcano, Java, to lower the level of the water in the crater lake and reduce the danger of destructive mudflows during future eruptions. (After Zen and Hadikusumo, 1965)

by the time the lake refilled to that level it had a volume of 40 million cubic m—enough to again present a serious risk of disastrous mudflows. A lower tunnel was driven by the Indonesian government, but stopped short of the lake, in the hope that seepage from the lake into the tunnel would be enough to lower the lake level. Because of the low permeability of the cone, this did not succeed, and at another eruption in 1966, mudflows again damaged great extents of agricultural land and killed hundreds of people. Belatedly, a new lower-level tunnel was constructed in 1967, and the volume of water in the lake has again been much reduced.

The Kelut mudflows were cool, but in 1822 the Javanese volcano Galunggung ejected the water of its crater lake, forming a steaming-hot mudflow with a volume of 30,000,000 cubic m that traveled 65 km. No figure on the amount of destruction is available.

At White Island, New Zealand, steam explosions in 1914 threw out a mass of landslide debris mixed with mud from the floor of a shallow crater lake and formed a hot mudflow that destroyed part of a sulfur-processing plant and killed 11 men.

In 1817, Kawah Idjen Volcano, also in Java, ejected its crater lake to form a flow of acid mud that did tremendous damage. Years later, Dutch engineers excavated a channelway through the lowest part of the crater rim to keep the lake level low and reduce the risk of recurrence of this disaster.

The sudden release of a crater lake is not necessarily the result of explosive ejection, as it may simply collapse, weakened by explosive undermining, by gas alteration, or by landslides. The collapse may be triggered by earthquakes as exemplified by Kelut Volcano, in 1875.

Rapid melting of ice or snow on a volcano resulting simply from warm rains may release floods of water and generate mudflows. Mudflows of this sort were formed by meltwater from glaciers on the slopes of Mt. Shasta, California, in 1926 and 1931. Air-transported ash cools in the atmosphere and seldom, if ever, reaches the ground warm enough to cause any large amount of melting of ice or snow. However, a covering of ash on ice or snow results in greater absorption of radiant heat from the sun and an increased rate of melting, which may cause floods and mudflows. During the 1956 eruption of Bezymianny Volcano, in Kamchatka, mudflows were formed in this way on the slopes of the nearby Klyuchevskaia and Zimina Volcanoes. Ash flows, glowing avalanches, or lava flows passing over snow or ice or, in the case of lava flows, burrowing under it, may bring about melting that is rapid and voluminous enough to produce mudflows. Along the midline of the glowing avalanche of the Bezymianny eruption (described in Section 2.3), the snow was almost completely melted and the water contributed to a great mudflow that traveled 100 km. In 1915 a small lava flow issued from the crater of Mt. Lassen, California, and caused melting of a snow field that in turn caused mudflows that traveled 50 km. Fortunately they did no serious damage, but with the increase in population in the area, particularly within Lassen National Park, similar flows today might well take many lives. In 1963 a lava flow at the summit of Villarica Volcano, in Chile, melted ice and snow and generated mudflows that destroyed fields and a village at the base of the volcano. In 1877 a similarly-generated mudflow from the upper slope of Cotopaxi, in Ecuador, traveled 300 km with an average speed of 27 km an hour and did serious damage in a village 240 km from its point of origin.

Somewhat different was the disastrous snow melt flood that took place on the flank of Ruapehu Volcano, New Zealand, in December 1953. At the summit of Ruapehu an outer crater rim encloses an inner cone, which in turn contains a crater lake. An eruption in 1945 had nearly drained the crater lake, but had also built the rim of the cone 6 to 8 m higher, and in succeeding years rain and melting snow raised the level of the lake even higher than it had been before. The space between the inner cone and the outer crater rim is filled with ice. When the lake gets high enough it overflows through a low notch in the rim of the inner cone. During recent decades this overflow had been southward, and a tunnel had been melted beneath the ice through which the water drained into the head of a branch of the Whangaehu River. In late 1953, when the water had reached the level of the notch and was overflowing, something, perhaps crevassing of the melting ice, caused movements which in turn caused collapse of part of the rim of the inner cone, releasing a flood of water which further eroded the cone rim and enlarged the tunnel. Rushing down the Whangaehu River, the water formed a wall as much as 6 m high, gathered up loose debris, and became a mudflow. The dense fluid struck a railroad bridge, carrying it partly away, leading to the derailment of the Wellington-Auckland express train, destruction of the engine and several carriages, and the loss of 154 lives.

Studies of Mt. Rainier, in Washington, indicate that prehistoric mudflows there resulted from shattering of lava by steam explosions and rapid generation of meltwater where lava flows moved over ice, snow, or very wet ground. Whether or not their generation was due to glacial melting, some of the prehistoric mudflows of Mt. Rainier were of enormous size. The Electron mudflow, which took place only about 500 years ago, has a volume of about 150,000,000 cubic m; and the older Osceola mudflow has a volume of 1.9×10^9 cubic m. Both reached

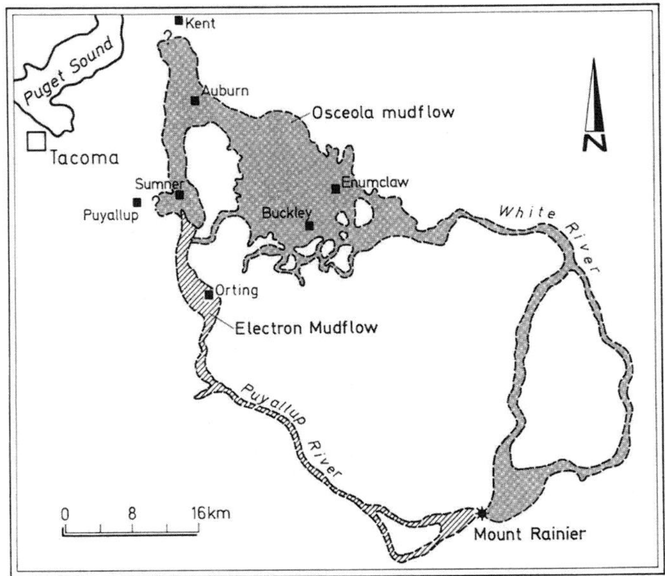

Fig. 2.9. Map showing the Osceola mudflow (5,000 years old) and the Electron mudflow (500 years old) from Mt. Rainier, Washington. (After Crandell and Waldron, 1969)

far out onto the lowland near Puget Sound, and would wreak colossal damage if they were to occur today (Fig. 2.9).

Some mudflows result from glowing avalanches or ash flows entering streams. In 1929, glowing avalanches at Santa Maria Volcano, Guatemala, entered rivers and were transformed into mudflows that traveled 100 km. At Merapi Volcano, in central Java, similar but smaller mudflows formed by entrance of glowing avalanches into streams have done enormous damage to agricultural lands and taken many lives around the base of the volcano.

Avalanches of other origin may also generate mudflows, as in 1888, at Bandai-san Volcano in Japan, when a low-temperature steam explosion blasted away part of the side of the volcanic cone and allowed the part above to collapse. The rock had been partly decomposed into clay by volcanic gases. The collapsing material broke up and formed an avalanche that rushed down hill, mingled with stream water, and formed mudflows, which buried farmlands and villages and killed 400 persons.

Mud may be formed in the air, by contact of ejected ash with rain clouds. Falling to the ground, it may coat vegetation so heavily that limbs are broken from trees, and it may flow in sheets down the sides of the volcano or nearby hillsides. Both were observed, for example, during the eruption of Irazú, Costa Rica, in 1963.

By far the majority of volcanic mudflows owe their origin to heavy rainfall on the debris-covered slopes of explosive volcanoes. Some of the rain may result from condensation of steam in the volcanic gas cloud, but most of it is of ordinary meteoric origin. Particularly in the tropics, torrential rains such as those of the monsoon season often bring about mudflows. The amount of mudflow deposits around the base of many tropical volcanoes is enormous. Indeed, in the skirts of many of them it is difficult to find anything but mudflow deposits and stream deposits formed by the reworking of mudflow material. Much of the material of the mudflows comes from unconsolidated tephra deposits on the mountain slope; but some, as in the 1968 eruption of Mayon, comes from the erosion of glowing-avalanche deposits.

Heavy falls of ash may greatly reduce the permeability of the ground surface and increase the amount of runoff during rains, leading at times to floods and mudflows. Commonly, a thin crust forms very quickly by cementation of the surface of the tephra. This crust has even lower permeability than the uncemented ash, and most of the rain runs off rapidly. Beneath the crust, however, the tephra is still loose, and wherever the crust is broken through, erosion progresses very rapidly. This was well shown at Irazú Volcano in 1963 and 1964, when very rapid erosion resulted from headward growth of gullies by undercutting of the crust. The great amount of debris contributed rapidly to the streams, partly by landsliding of the undercut banks, resulted in mudflows. One torrential rainstorm, on December 9, 1963, created a very fluid mudflow consisting of about 65 per cent water that swept down the valley of the Rio Reventado eleven km, to the outskirts of the city of Cartago. Damage from it was estimated at more than $3,500,000; 300 homes were destroyed, and more than 20 persons were killed.

Tephra does not necessarily have to be young to be transformed into mudflows by heavy rains. Certain island-like areas on the south slope of Mauna Loa, Hawaii, surrounded by recent lava flows, are underlain by volcanic ash hundreds to thousands of years old and as much as 15 m deep. The ash has been partly altered, by weathering, to clay minerals that have the property of absorbing water. The wet material is highly *thixotropic*—so long as it is undisturbed it is quite stable, but if it is jostled it loses its coherence and flows like a liquid (see Chapter 4). In the spring of 1868 the ash had been thoroughly soaked by long-continued rains, and on April 2 came a violent earthquake. The ash turned to liquid mud and flowed rapidly down the mountainside, forming two mudflows, one over a km long and the other three kms long and half a km wide. A village, with about 500 domestic animals and 31 persons, was buried.

In prehistoric times enormous mudflows have occurred in some volcanic districts, such as along the eastern edge of Yellowstone National Park in the Absaroka Mountains, U.S.A., where deposits of such mudflows make up a field 11,000 square km in area, and in places are as much as 2,000 m thick. They were formed 40 or 50 million years ago, long before the nearby lava flows and great collapsed calderas of the Yellowstone Plateau. A little younger are the great mudflow deposits of the western slope of the Sierra Nevada (Mehrten Formation), which once covered an area of 31,000 square km. Only about 3,500,000 years old is the Tuscan Formation, a series of mudflow deposits extending over an area of 5,000 square km in northern California (see Fig. 2.10). No doubt some of the mudflows that formed these great deposits were generated in some of the ways discussed above; but there is considerable evidence that many were formed by slurries of solid debris in water that issued directly from

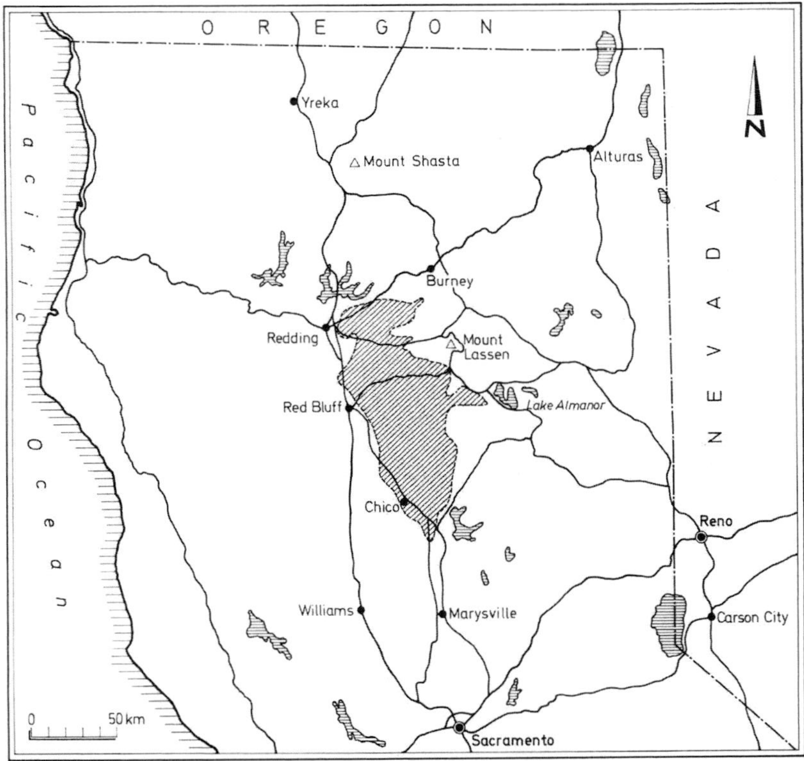

Fig. 2.10 Map showing the approximate area covered by mudflows of the Tuscan Formation in northern California. (After Lydon, 1968)

volcanic vents. This mechanism has never been actually observed in operation. If it is in the future, the resulting flows may have much greater volume than any previously observed.

Among the most intriguing causes of mudflows, because they are least understood, are the floods (see Chapter 7) that accompany the beginning of some eruptions. Those of the 1902 eruption of Mont Pelée are described briefly in the next section. They cannot be explained as the result of heavy rains, or the ejection of a crater lake. During the first 6 hours of the 1947 eruption of Mt. Hekla, in Iceland, somewhere in the vicinity of 3,000,000 cubic m of water poured off the mountain, causing floods in the neighboring streams. G. Kjartansson, who made a careful study of the event, concluded that the amount of water was much too great to have resulted from rainfall and melting snow. Can it have been groundwater driven out of the rocks of the volcano by steam and other volcanic gases as magma rose beneath?

What can be done to lessen the destruction by mudflows? Probably some of them can be diverted from particularly important areas, in the same way as some lava flows, by properly constructed barriers. Others, particularly the very large ones, are probably not amenable to such diversion. This is particularly true because many mudflows follow well-established valleys that are much deeper than the flows are thick. Once again, topography must be favorable or diversion is impossible. Perhaps some small mudflows can be confined behind dams. This was tried at the foot of Kelut Volcano, but the reservoirs quickly filled and overflowed, only temporarily checking the mudflows, which continued down the valley destroying villages and killing their inhabitants. Much larger reservoirs would, of course, confine more material, and some mudflows could be completely confined behind dams such as already exist in parts of the United States. A study of volcanic hazards in the vicinity of some of the volcanoes in the Cascade Range of Washington and Oregon points out another risk, however. Sudden entrance of a sizeable mudflow into a full reservoir would displace an equally large volume of water over the dam, possibly resulting in disastrous floods downstream, and perhaps causing so much erosion at the base of the dam that the structure would fail and bring about an even worse flood. To prevent this, the water level behind the dam should be lowered if any immediate likelihood of a mudflow into the reservoir is foreseen. This in turn, of course, entails prediction of the occurrence of mudflows in the area.

Hills have been built near some Indonesian villages so that the villagers can run up them to safety while a mudflow passes around them, but this also requires warning of the coming mudflow long enough ahead of time so that the people can reach a safe elevation. Again the problem is prediction. One attempt was the installation of thermal sensors on the upper slopes of Merapi, in Java, to warn of the passage of hot mudflows or glowing avalanches, but they appear not to have operated very well, and at the best would give only a very few minutes warning; and of course would give no warning whatever of cool mudflows.

Prediction of the probable paths of many mudflows is relatively easy, since they follow the valleys, but prediction of the time of their occurrence is another matter. Probably the best that can be done is to recognize situations that are conducive to their formation and issue a general warning that their occurrence is likely. A heavy cover of loose tephra that can be mobilized into mud by heavy rains, active domes or lava flows that may melt snow and give rise to floods of meltwater, conditions that may give rise to either hot or cold avalanches into streams, all should be watched for. Other mudflows, such as those that occurred at the beginning of the eruption of Mt. Pelée in 1902, are probably not specifically predictable, although it may be possible ultimately to predict the eruptions of which they are a part, and mudflows should be anticipated as possible occurrences at the beginning, as well as later, in any eruption of an explosive volcano.

Glacier Bursts

Floods and mudflows resulting from melting of glaciers on Villarica and Cotopaxi Volcanoes have already been mentioned, but these are dwarfed by the great floods generated by eruptions beneath some of the glaciers of Iceland. For periods of a few hours these "glacier bursts" may have a volume greater than that of the world's greatest rivers. Some bursts from the Myrdals Glacier, caused by eruptions of the buried volcano Katla, have exceeded 92,000 cubic m per second, and the total volume of water in the burst may exceed 6 cubic km!

Any of the several Icelandic volcanoes buried beneath glaciers may produce these glacier bursts, but most frequent are those from Katla and from Grimsvötn Volcano, which lies beneath the Vatna Glacier. Both are in south-central Iceland. The floods carry and deposit huge amounts of sand and some coarser debris, and have been largely responsible for building the broad "sand plains" that one crosses in traveling across southern Iceland. The floods do great damage to farmsteads, and in past years took many lives. Today toll in lives has been lessened, because Iceland's scientists have learned how to predict the Grimsvötn bursts, though those from Katla remain largely unpredictable. This is because at Katla meltwater remains in a chamber wholly beneath the ice, whereas at Grimsvötn melting extends all the way through the ice and the resulting lake is visible in a hole in the glacier. In both cases the volume of water gradually increases until it is great enough to float the ice slightly, whereupon the water escapes beneath the edge of the glacier. At Grimsvötn, by keeping track of the increase in depth of the lake it is possible to forecast when the water will become deep enough to float the ice and the sudden escape of the water will occur. The level of the Grimsvötn lake rises quite steadily, and most of the melting is caused by hot gas vents beneath the ice, rather than by actual eruptions. It has, in fact, been suggested that the sudden draining of the lake reduces the pressure on the underlying magma body sufficiently to induce the eruptions that commonly accompany the glacier bursts. If at Katla the melting is by actual lava eruption beneath the ice, it may be possible to warn of the bursts by seismic methods that recognize the earthquake pattern of an eruption, but if the melting is primarily by hot gas vents without actual eruption, warning may not be possible.

Glowing Avalanches

1902 was a year of disaster in the Caribbean and surrounding region. On January 18, a strong earthquake did extensive damage near Quetzaltenango and along the Pacific coast of Guatemala, and on April 17 a still stronger quake destroyed Quetzaltenango and a nearby village and did extensive damage throughout the area. About 1,000 persons were killed. About April 20 Mt. Pelée, on the Island of Martinique, began an eruption that was to culminate in catastrophe. On May 7 Soufrière Volcano on the nearby island of St. Vincent erupted, killing more than 2,000 persons. On May 10 Izalco Volcano erupted in El Salvador, taking no human lives, but seriously damaging coffee plantations. In July Masaya Volcano, in Nicaragua, erupted after more than 40 years of inactivity; and on October 24 Santa Maria Volcano in Guatemala exploded, blasting out a pit more than a kilometer wide and 250 m deep on the lower flank of the mountain and laying a thick blanket of ash over surrounding country.

The neighboring volcanoes, Soufrière and Mt. Pelée, both produced deadly glowing avalanches, but in different manners. Brief descriptions will illustrate this and also the characteristic of eruptions that can be expected at these and other similar volcanoes in the future.

Soufrière is a name given to several volcanoes that give off sulfurous gases. The Soufrière of St. Vincent is a cone in the northern end of the island built partly of tephra and partly of lava flows (Fig. 2.11), with slopes of an average inclination of only about 15°. A prehistoric eruption destroyed the summit of the cone and formed a large caldera similar to that formed at Vesuvius in 79 A.D. and, again like Vesuvius, later eruptions built a younger cone within the caldera, at the summit of which is a double crater. The larger and older crater at the south may have been formed during an eruption in 1718, the more northerly by an eruption in 1812. In the beginning of 1902 the southern crater contained a lake more than 150 m deep, with its surface about 300 m below the crater rim.

The last definitely known activity of Soufrière had been in 1812, when an explosive eruption produced a great cloud of tephra and a heavy fall of ash on the flanks of the volcano, killing about 75 people. A small extrusion of lava may have taken place in the depths of the crater lake in 1880. In April 1901 there commenced a series of earthquakes, some of them quite strong, but in July the quakes ceased and all was quiet until April 1902, when quakes began again. Fortunately, the seismic warning was recognized, and by early May most people living on the western flank of the volcano had become sufficiently alarmed to move to the southern part of the island. On the other side of the volcano the people were less alarmed, perhaps partly because the earthquakes were less severe, but partly because it was believed that if an eruption occurred it would resemble that of 1812, and most of the ash and volcanic gas would be blown southwestward by the northeast tradewind.

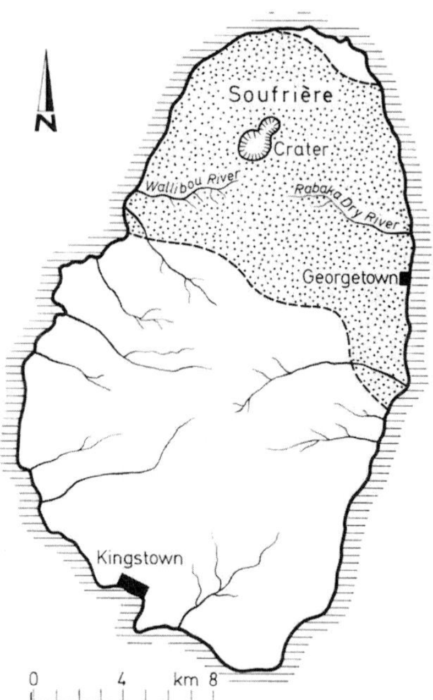

Fig. 2.11. Map of the island of St. Vincent, in the Lesser Antilles, showing the location of Soufrière Volcano and the area devastated by its eruption in 1902. (After Hovey, 1902)

On May 5 it was noticed that the water in the crater lake had changed in color, from greenish blue to brownish yellow, and was being agitated, apparently by rising gas. In the early afternoon of May 6 a small steam explosion occurred in the lake, followed by others, at first at intervals of an hour or two, but increasing in frequency and intensity until by mid-morning on May 7 they had become almost continuous and a cloud of steam and ash was rising 10,000 m into the air. The few people remaining on the western slope could see this, and practically all hurriedly departed. Still, however, most people on the eastern slope were not unduly alarmed, perhaps because the top of the mountain was hidden from them by trade-wind clouds, except for some, who at about 13:00 h, attempting to flee southward across the Rabaka Dry River, found it transformed into a torrent of hot mud 15 m deep, while a similar flood swept down the Wallibu River, on the other side of the volcano. There is little doubt that these floods were caused by explosive ejection of water from the crater lake.

The intensity of the eruption continued to increase, and at about 14:00 h a great cloud of gas and ash, glowing purplish and reddish in the partly obscured daylight, swept down the mountainsides, at the same time boiling upward and outward so that the whole mountain was hidden to viewers at a distance. The cloud descended all sides of the mountain, breaking off and uprooting trees, overturning some walls that lay at right angles to the blast, destroying many houses, and setting some fires. Almost the entire northern third of the island was laid in desolation within a few minutes, and more than 1,500 persons were killed, although a surprisingly large number escaped by taking refuge in cellars

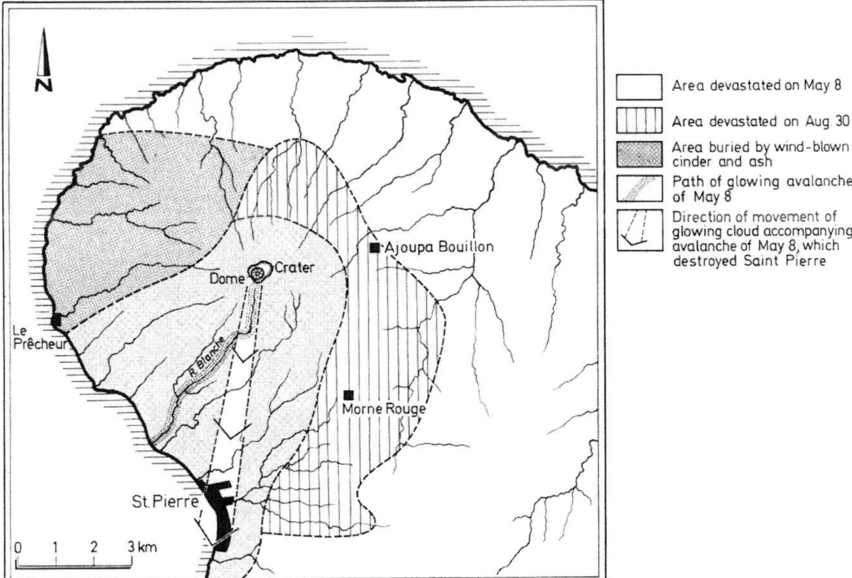

Fig. 2.12. Map of Mt. Pelée, on the northern end of the island of Martinique, showing the paths of the glowing avalanche of May 8, 1902, and the glowing cloud that destroyed the city of St. Pierre, the areas devastated by the outbreaks of May 8 and August 30, and the additional area covered by ash during the eruption. (After Lacroix, 1904)

or resistant stone buildings. These survivors, many of whom suffered no serious injury, described the passage of the cloud as a sudden feeling of intense heat and overpowering suffocation that lasted, fortunately, only a very few minutes. The cloud was an avalanche of incandescent ash and admixed blocks. Tempest Anderson and J.H. Flett, British volcanologists who studied the eruption soon afterward, concluded that the avalanche was formed by abundant ash falling back from the vertically-directed explosion cloud onto the slopes of the volcano, and we believe now that its great mobility was the result of expanding gas between the ash fragments.

Mont Pelée makes up the northern end of the Island of Martinique, 160 km north of St. Vincent. Like that of Soufrière, the cone is composite, made up of layers of both lava and tephra, and has at its summit a double crater (Fig. 2.12). The older of the two craters contains a small lake—the Lac des Palmistes. To the southwest, the younger crater previous to 1902 also contained a small lake which was commonly dry—the Étang Sec. Both craters were contained in a broader bowl-shaped hollow, probably the remains of a collapsed caldera. On three sides the walls of the younger crater rose steeply, but on the southwest side a V-shaped notch led through the crater wall into the head of the valley of the Rivière Blanche, the topography being of great importance in the ensuing events. The valley of the Rivière Blanche descends steeply south-southwestward for a mile from the crater, then bends (of crucial importance) abruptly to the west. Other valleys radiate outward from the summit of the mountain, while 6 km south of the crater lay the city of St. Pierre, one of the gayest and most prosperous of the cities of the Caribbean, one of whose main businesses was the export of rum.

The last eruption of Mt. Pelée had been in 1851, when ash and cinder were rained over the mountain without doing any great damage. The new eruption started about April 20, consisting at first of rather weak ejection of ash and cinders, that gradually built a small cone in the crater of the Étang Sec. During the next week the northeast tradewind drifted ash and sulfurous gases southwestward over St. Pierre, frightening some people and making existence unpleasant for all. As the eruption increased a little, birds started to drop dead of suffocation in the streets; many people fled the city for the safe southern end of the island, but many stayed, and many others came into the city from hamlets closer to the volcano. The population of the city remained at about 30,000. Why did people remain? It appears that the reason was largely political. An important election was to be held, and a person had to be in his home district to vote. The government urged the people to remain, so that they could vote, and a commission appointed by the governor to investigate the eruption concluded that it was taking the same course as that of 1851, and that there was no immediate danger to St. Pierre. The assurance, sincere though it was, was mistaken. To reassure the people, the governor himself took up residence in St. Pierre. He never left.

During the last days of April and the first days of May the Rivière Blanche underwent extreme fluctuations in volume. On April 29 and May 1 it was a roaring torrent; on May 2 it was almost dry; on May 3 it was a torrent again, so far as was observed and recorded, the floods not being caused by abnormally heavy rains. On the morning of May 5 the river was again in flood, when a little after noon a violent explosion occurred in the crater; the muddy water left in the Étang Sec was thrown into the head of the river, and a great flood of hot black mud rushed down the valley, completely burying a sugar mill at the river mouth and killing several men. During the next two days all of the streams around the mountain were in flood, and mudflows formed in several of the valleys. The cause of the floods is obscure; it does not appear to have been entirely rainfall, since the rainfall was not exceptionally heavy. The Lac des Palmistes was not affected. In the early days of the eruption the Étang Sec was a puddle about 200 m across and only a few meters deep. The volume of water in it was surely not great enough to account even for the floods along the Rivière Blanche, let alone those in the other valleys.

Still people lived on in St. Pierre! At 07:50 h on May 8 came a series of violent explosions that shot a towering black cloud high into the stratosphere. At the same moment another cloud shot nearly horizontally southwestward through the notch in the crater wall. Boiling upward as it expanded, it plunged down through the valley of the Rivière Blanche. Its start was witnessed by persons on ships in the harbor of St. Pierre, but they did not have long to watch it, as in less than 2 min it struck the city, having traveled with a velocity of more than 160 km an hour, and within another few minutes nearly all of the city's 30,000 inhabitants were dead (Plate 8.1).

The blast was considerably more violent than that at St. Vincent. The effects have been summarized as follows (Macdonald, 1972, p. 145-6):

"Masonry walls 1 m thick were knocked over and torn apart, big trees were uprooted, 13 cm cannon were torn from their mounts, and a 3-ton statue was carried 4 m from its base. Trees that remained standing were stripped of leaves and branches, and on many the bark was stripped away on the side toward the volcano leaving bare wood that bore the marks of having been sandblasted. Most of the ships in the harbor were upset and sunk or destroyed by fire, only two escaping. Most of the few survivors of the tragedy were on these two ships or had been thrown into the water from the ships that were sunk, and nearly all of them were horribly burned. It is often said that within the city itself there was only one survivor—a prisoner who was in a dungeon with only one tiny window opening on the side away from the volcano. Actually there appear to have been about four survivors, two of them on the very edge of the cloud, and another who by some strange accident escaped even though everyone around him was killed. The survivors had little to tell of the blast other than the sudden pitch darkness, clouds of hot dust that mixed with water to form scalding mud, a short period of intense heat, and a sense of overpowering suffocation. The initial great heat due to the blast itself seems to have been of very short duration—at most only a few minutes.

"The injuries to the dead were grotesque. In many instances the actual cause of death probably was the inhaling of the very hot gas. The bodies were intensely burned, as also were those of the survivors. Many were stripped of clothing by the force of the blast; but others remained clothed and the clothing was not ignited, even though the body beneath it was severely burned. Body tissues were distended, and in many instances skull sutures had been opened up. The injuries were such as would result from sudden heat intense enough to turn water in human tissues into steam, but not high enough or of long enough duration to raise fabrics to kindling temperature.

"The temperature of the blast itself is difficult to appraise. Much of the city burned, and the high temperatures must be attributed in part to the fires, fed by hundreds of thousands of liters of rum stored in warehouses. But the temperature was high enough to ignite the wooden decks of ships in the harbor, where the hot dust accumulated on them, and in parts of the city where there were not extensive fires, the temperature was nevertheless high enough to soften glass objects. It is generally estimated that the temperature of the blast as it left the crater was about 1,000° C, and that the temperature of the cloud that swept over the city was still between 700 and 1,000° C. The thickness of the layer of dust left by the cloud in the city averaged only about 30 cm.

"Destruction of the city was almost total. Little was left save the wrecked masonry walls, twisted sheets of iron roofing, and other metal debris."

The cloud that wrought such havoc in St. Pierre was only the skimmed-off top of a great glowing avalanche that descended the Rivière Blanche. Where it reached the abrupt westward bend in the valley the avalanche turned and followed the stream bed to the sea (Fig. 2.12), but the upper part of the dust cloud that rose from the avalanche overtopped the valley side at the bend and continued straight ahead to St. Pierre.

The eruption continued through the rest of 1902 and on through 1903. Several more glowing avalanches took place, and by then volcanologists were on hand to study them. Some avalanches followed the path of that of May 8, but a few occurred also on other slopes of the mountain. The French volcanologist, Alfred Lacroix, concluded that the great avalanches down the Rivière Blanche originated, not by voluminous fall-back of tephra that had been thrown more or less vertically into the air, but by explosions directed outward at a low angle through the notch in the crater rim. This downward deflection of the explosions was attributed to the presence of an obstruction in the vent—a dome of viscous lava that had begun to pile up early in the eruption. Its presence prevented a vertically-directed explosion, but the junction between it and the underlying crater floor was a path of weakness that directed the explosion laterally (Fig. 2.13 A). Lacroix's conclusion was supported by observations made by Frank Perret during a similar eruption of Mt. Pelée in 1929. Thus we have two mechanisms for the generation of glowing avalanches: a laterally directed explosion, giving rise to what is commonly known as the Pelée type of glowing avalanche; and the voluminous fall-back from a vertically directed explosion, producing an avalanche of the Soufrière type (Fig. 2.13 B).

The glowing avalanches at Soufrière in 1902 differed from those at Mt. Pelée in the lower temperature of the cloud, and the lower velocity, which in turn resulted in a lesser force of the blast and lesser destructiveness. Comparatively few fires were started at Soufrière, many of them by lightning rather than by the heat of the cloud. Many well-built structures survived, and there were many more human survivors within the area of desolation. The avalanches descended all slopes of the mountain simultaneously, instead of being confined to one sector. The relatively low velocity, of only 30 to 50 km an hour, resulted from

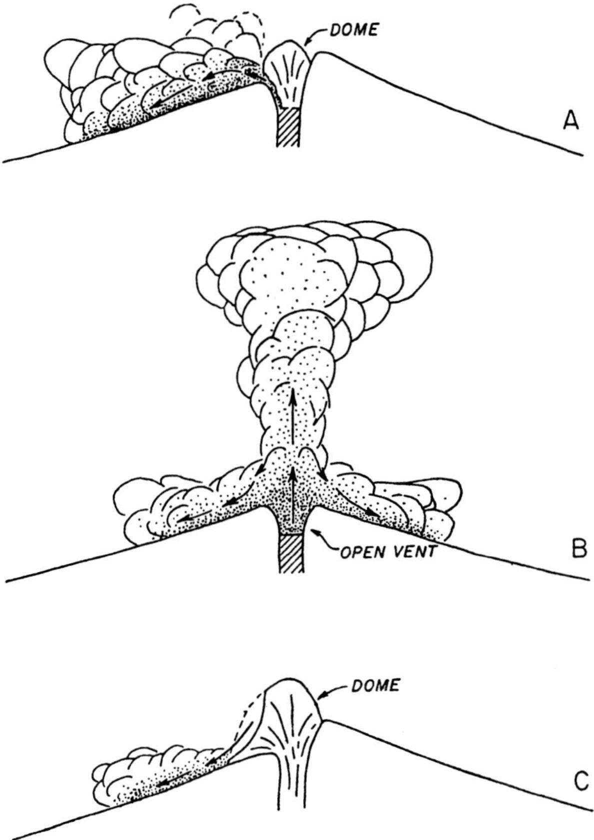

Fig. 2.13. A–C. Diagram illustrating different types of glowing avalanches. (A) Pelée type, (B) Soufrière type, (C) Merapi type. (After Macdonald, 1972)

the fact that the drive was wholly by gravity and the mountain slopes were quite gentle, whereas the avalanches of Mt. Pelée had an added push from the low-angle blast and also from the steep slopes of the upper Rivière Blanche valley.

The glowing avalanches of Hibok-Hibok Volcano, in the Philippines, in 1951 resembled those of Mt. Pelée. Starting in 1949, a dome grew in the crater and eventually extended through a low part of the crater rim onto the upper slope of the mountain. The glowing avalanche of December 4, 1951 (Plate 2.9), which took 500 lives in the outskirts of the town of Mambajao, resulted primarily from a low-angle blast directed outward at the base of the dome. However, another mechanism was also involved, both at Hibok-Hibok and at Mt. Pelée, where the explosions blasted away part of the dome, undermining the upper part, which then collapsed. This mechanism is the same as that which gives rise to the so-called Merapi type of glowing avalanche (Fig. 2.13C).

At Merapi Volcano, in central Java, a series of domes has formed a row across the summit crater of a composite cone, and the latest dome has overlapped

Plate 2.9. Glowing avalanche sweeping past the town of Mambajao (foreground) on the lower slope of Hibok-Hibok Volcano, Philippines, December 4, 1951. More than 500 persons were killed near Mambajao by glowing avalanches during this eruption. (Photo by Hibok-Hibok Studio, Mambajao)

onto the outer slope of the mountain. As the dome grows, its outer part becomes unstable and crumbles, and occasionally large sections collapse and slide downslope, the fragments breaking up into smaller and smaller pieces as they fall. The hot mass gains mobility by liberation of gas from the fragments and/or by the heating of the air between them, so that the result is a highly mobile glowing avalanche (Fig. 2.14). Similar avalanches such as have been reported at Fuego Volcano in Guatemala and Izalco Volcano in El Salvador result from the crumbling of the front of a thick lava flow on a steep slope.

The glowing avalanches of the 1951 eruption of Mt. Lamington in New Guinea, which devastated an area of about 230 square km and took nearly 3,000 lives, were of the St. Vincent type, as were also those of the 1968 eruption of Mayon in the Philippines, and Arenál in Costa Rica. Those of Mayon traveled down valleys at speeds of about 100 km an hour, and formed mudflows where they mingled with stream water. At Arenál the area devastated by the avalanches was only about 13 square km, but two villages were destroyed and at least 78 persons killed. More extensive damage resulted from falling ash and blocks.

Because small glowing avalanches follow valleys, it may be possible to control some of them by means of diversion barriers like those suggested for lava flows, though because of their high speeds the tendency to jump the barrier would be much greater and some spill-over, even of the basal part, appears almost

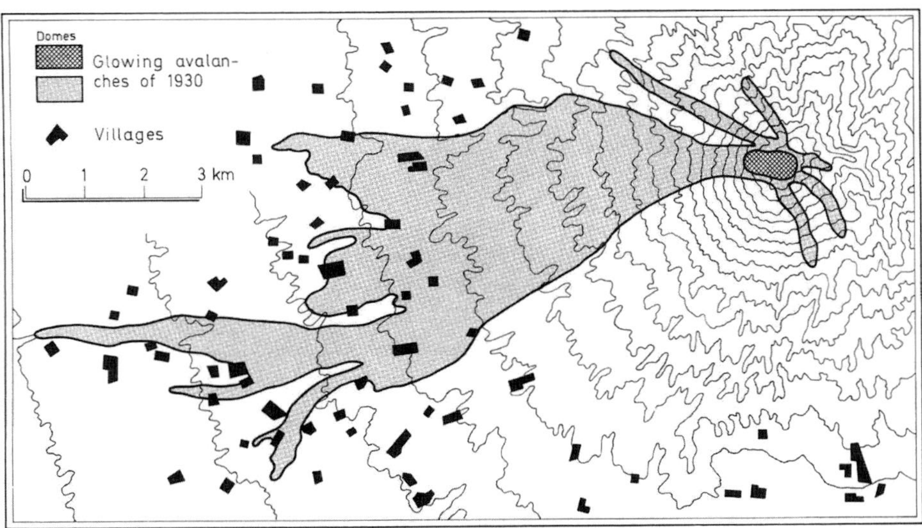

Fig. 2.14. Map of Merapi Volcano, central Java, showing the area covered by glowing avalanches in 1930, and the large number of villages that dot the lower slopes of the volcano. (After Neumann van Padang, 1933 and 1951)

inevitable; no barrier could be built high enough to divert the overlying dust cloud. Much of it would continue, just as the cloud that destroyed St. Pierre passed over the high ridge beyond the Rivière Blanche. Barriers to control glowing avalanches do not appear practical.

Some glowing avalanches might be confined in reservoirs behind dams in river valleys, but as in the case of mudflows the water level in the reservoir should be greatly lowered to avoid spill-over and disastrous floods downstream. The effect on the dam of steam explosions resulting from contact of the hot debris with water is problematical, but should not be ignored. Even in a large reservoir the avalanche probably would not stop at the head of the lake, but would form a mobile slurry that would move partly along the bottom of the lake until it reached the dam. Lowering the water level in reservoirs in preparation for glowing avalanches obviously involves foreseeing their possible occurrence, and again prediction is of vital importance.

Indeed, prediction of possible glowing avalanches and evacuation of people from areas likely to be affected by them appears to be the only defense. Conditions likely to result in avalanches of the Merapi type are apparent where an active dome or thick lava flow is present on the flank of the volcano. Where a dome is present in a crater, Merapi-type avalanches are unlikely, but particularly if one wall of the crater is low, an avalanche of the Pelée type is possible. In both cases, the courses of the avalanches can be anticipated with considerable confidence, being guided by existing valleys. Avalanches of the St. Vincent type appear to be specifically unpredictable, and in their case reliance must be placed on prediction of the eruption in general, plus a knowledge of the past behavior of the volcano, as a volcano that has produced avalanches in the past is likely to produce them again in the future. Probably volcanoes that have experienced a long period of rest should be regarded with the greatest suspicion if they show signs of returning activity, because the first eruption after such a period of dormancy is commonly of great violence, although a long rest is not a necessity for a violent eruption.

Restoration of the surfaces of glowing avalanche deposits for agriculture may be difficult, because of the common abundance of large blocks in them. In avalanches containing relatively few large fragments it should be fairly easy, and should involve essentially the same treatment as with ash-fall deposits. As with ash falls also, the material should be fertile, once any excess of acid is leached out of it, because of the large proportion of easily-decomposed glass that is usually present. Obviously, the speed of revegetation depends greatly on climate, and in the warm humid climate of the Caribbean the deposits of the avalanches of 1902 on the slopes of Soufrière are now heavily vegetated.

Ash Flows

Ash flows resemble glowing avalanches in their mechanics of movement. Indeed there is no clear-cut distinction between ash flows and glowing avalanches of the Soufrière type. Ash flows appear to differ, if at all, largely in their less explosive eruption, and rather than falling back to the ground from a vertically projected cloud, they appear to be caused mainly by overflow from the vent. Unquestionably there is some vertical projection associated with the generation of ash flows, and there may be some simple overflow during some eruptions that generate Soufrière-type glowing avalanches; there may have been some during the 1902 eruption of Soufrière itself.

Eruptions of ash flows generally take place from fissure vents, and the expanding glass-in-gas suspension spreads outward very rapidly. Because the movement of ash flows is governed by gravity, they are guided fundamentally by topography and tend to follow valleys, but because of their high velocity they may climb considerable slopes. Because of their great fluidity they come to rest with nearly horizontal surfaces.

The foregoing accounts of the behavior of ash flows are based almost wholly on deduction, as we have never seen an active ash flow. Only one or two are known to have occurred in historic times, and they were in unpopulated areas; if any near-by observers were present, they did not survive the eruptions. The great speed and deadly character of ash flows make it unlikely that we will ever be able to report on one from close range.

The only unquestionable historic ash flow took place during the 1912 eruption of Mt. Katmai when early in the eruption fissures opened in the head of the valley west of Katmai, and a cloud of ash flowed out of them to pour down the valley for a distance of 20 km, the entire volume of 11 cubic km issuing in less than 20 hours. Toward the end of the eruption, activity at one place built a low cone of ash, and within it rose a dome, later named Novarupta. The deposit of the ash flow filled the valley to a depth of tens of meters, and formed a nearly flat valley floor. For several years afterward no one visited the valley, but when it was finally entered by R.F. Griggs he found the floor spotted with innumerable fumaroles, later shown to be largely the result of ground water coming in contact with the hot inner part of the deposit. He named it the "Valley of Ten Thousand Smokes."

Some ash flows issue from concentric fissures on the upper slopes of big volcanic cones where, commonly, voluminous flows of this type are associated with collapse of the top of the cone to form a large caldera. As an example of this we can cite Aso Volcano, in Japan; but there are many other examples, among them the flows accompanying the collapse of the top of Mt. Mazama to form the caldera of Crater Lake in Oregon. The eruption is described in Section 2.3. Similar flows occurred during the 1956 eruption of Bezymianny in Kamchatka, also described later, and we should perhaps consider the Bezymianny flows as another historic example of true ash flows.

In the past, huge ash flows have taken place in various parts of the world. They make up most of the Rhyolite Plateau on the North Island of New Zealand, huge areas in Armenia, and even larger deposits around Lake Toba in Sumatra. In northern California, the Nomlaki ash flow appears to have traveled at least 100 km, and in southern Nevada and adjacent Utah some flows traveled even farther. Some formed deposits more than 100 m thick, and some individual ash-flow deposits have volumes greater than 1,000 cubic km. Welding of the ash in the flows requires a temperature higher than 535° C, and since even the terminal parts of many flows are welded, they must have cooled little during their movement and thus have flowed with great rapidity. It has been estimated that they probably attained lengths of as much as 300 km in as little as 2 or 3 hours.

The very great speed and volume of many ash flows make it hopeless to try to confine or control them. The only hope of avoiding loss of life from them in the future in inhabited areas is to foretell the eruption and evacuate the area. Prediction will, however, be exceedingly difficult, because we do not know what specific signs to look for, and we will not know until a few have been observed. The most encouraging aspect is that they have been very rare, and no really great ones have occurred in historic time, so it can be hoped that they will be equally rare in the near geologic future.

The surfaces of ash-flow deposits should be nearly as easily reclaimable for agriculture as those of air-fall ash deposits, and more easily reclaimable than those of glowing avalanches, as they contain fewer large fragments, although no experience yet exists in the reclamation of such deposits. One problem may arise from the very loose character of the surface of the deposits, and another from their high surface permeability. Water may not be retained within the reach of plant roots; and the surface may erode very rapidly, choking nearby streams with large volumes of debris.

Volcanic Gases

By far the most abundant gas given off by volcanoes is water vapor, but associated with this in various proportions are other gases, the most abundant of which are carbon dioxide (CO_2), carbon monoxide (CO), sulfur dioxide (SO_2), sulfur trioxide (SO_3), hydrogen sulfide (H_2S), hydrochloric acid (HCl), and hydrofluoric acid (HF), all of which are harmful to plants and animals when they are present in sufficient concentration. For some the concentration that can cause harm is very small. Sulfur dioxide and SO_3 combine with water to give sulfurous and sulfuric acid, respectively. Acid aerosol mists often develop to the lee of fuming vents.

The gases may be given off by the main eruptive vent or vents of the volcano, but often they issue from relatively minor vents that have not erupted either lava or ash. Vents liberating only gas are called *fumaroles*, and liberation of gas without lava or tephra is often referred to as fumarolic activity. Commonly, vents continue in fumarolic activity for weeks, months, or years after the end of lava or tephra eruptions. Fumaroles giving off sulfur gases are called *solfataras*, and low-temperature vents giving off important amounts of CO_2, and sometimes CO, are called *mofettes*. Gases are given off by lava flows and ash flows, either as a general exudation or at well-defined fumaroles.

The acid gases are harmful to both plants and metals. Blowing to leeward of the volcano, they damage foliage and cause fruit to drop, and may bring about complete denudation of the plants, and ultimately their death. Where sulfur gases are the principal offenders the damage to foliage closely resembles that caused by smelter fumes, or by intense city smog.

The Masaya-Nindiri Volcano, in Nicaragua, is a complex double cone with several craters. For several periods of several years during the last century one vent in Santiago Crater has given off a great cloud of water vapor and sulfur gases. The volcano lies in the central basin of Nicaragua, its summit at an altitude of only about 700 m. To the west, many coffee plantations lie on highlands that rise somewhat above the summit of the volcano. The gas cloud blows west-

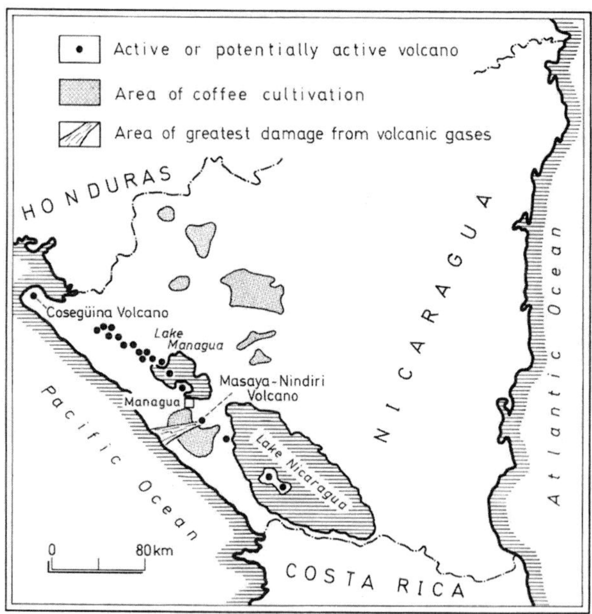

Fig. 2.15. Map of Nicaragua, showing the location of the Masaya-Nindiri Volcano and the area damaged by gases from it. (After Wilcox, 1952)

ward across a strip 5 to 8 km wide (Fig. 2.15), and within that belt some 150 square km of plantations have suffered damage amounting to tens of millions of dollars, while even as far away as the Pacific coast corn and other crops have also been damaged. Fence and telephone wires and metal equipment on the plantations and at a cement plant on the coast, have been damaged by the acids. Very similar damage to coffee and other crops has occurred to the west of Irazú Volcano in Costa Rica.

During the Katmai eruption, in 1912, acid rain fell at Seward and Cordova, 400 and 575 km from the volcano, burning the skin of some persons and damaging both vegetation and metals. Near Cape Spencer, 1,100 km away, fumes tarnished brass. Drifting slowly southward, a whole month later the gases reached Vancouver and acid rains damaged clothes hung outdoors to dry. The poisoning of animals by hydrofluoric acid adhering to ash grains during eruptions of Hekla Volcano has already been mentioned.

Most insidious of the volcanic gases are CO_2 and CO, because they are invisible and odorless. Carbon monoxide causes bleaching of leaves and defoliation of plants, and poisons animals. Carbon dioxide has no serious effects on plants, but because it is heavier than air it sometimes accumulates in pools in depressions, and may drown animals. Where a mofette lies in a small valley, under certain wind conditions the carbon dioxide may accumulate, and animals, or even men, wandering into it may drown. Such "death gulches" are known on some Indonesian volcanoes, and formerly there was one in the Absaroka Range in Wyoming. During the 1947 eruption of Hekla, CO_2 accumulated as shallow pools in hollows, so that sheep entering the hollows were drowned, but men's heads were above the surface of the CO_2, and they walked through the hollows unharmed. In the recent eruption of Eldafell CO_2, and less abundant CO and sulfur gas, accumulated in cellars in Westmannaeyjar, and one man was killed by it. His was the only human death in the eruption.

Innumerable other examples of damage by gas to plants and human casualties could be cited. What can be done to lessen or eliminate this destruction? Various chemical treatments of the affected plants to neutralize the fumes have been suggested, and some have been tried experimentally, most promising of which thus far is spraying with lime, to form a protective coating on the leaves. Whether or not such treatment would be practical is still uncertain. In such areas as the western highlands of Nicaragua the frequent heavy rains, washing the lime from the leaves, would necessitate frequent spraying, and the costs would be high, though perhaps not prohibitive.

General-purpose gas masks such as are used in many industrial plants will usually give adequate protection to persons caught for short periods in clouds of volcanic gas. In most cases there is sufficient air in the cloud for breathing if the harmful gases can be removed or neutralized. Lacking a gas mask, a cloth wet with water, or better with a weak acid such as vinegar or urine, held or tied over one's face, is moderately effective. Where heavy gases have accumulated in hollows or cellars there is insufficient air, and a gas mask is useless unless it has a selfcontained air supply. Knowledge of the possibility of such pools of heavy gas should make it possible, through caution, to avoid many human and animal casualties.

Probably the most interesting attempt to control volcanic gases was made at Masaya-Nindiri Volcano. Starting in 1924, gas liberated in Santiago Crater did heavy damage to the coffee crops, and the growers hired two German engineers to attempt to cap the vent in the bottom of the crater and lead the gases through a pipe to the crater rim, where they would be used to manufacture sulfuric acid. By early 1927 the engineers had installed about a kilometer of pipe, one meter in diameter, and were ready to try to connect it to the gas vent by means of a huge inverted funnel. But the vent was an elongate fissure, and to make it more amenable to capping the engineers proposed to block part of it with rubble obtained by blasting down the crater walls with dynamite. The blasts were duly set off, but the results were not what was expected. The whole crater floor collapsed, much of it dropping about 120 m! Much of the big pipe was destroyed. The vent was largely blocked, and the emission of gas reduced to about a quarter of its former amount. Changing their goal, the engineers blasted down additional portions of the crater wall to complete the blocking of the vent, and the emission of gas was stopped.

The gas-free period lasted for 19 years, but in 1946 a new vent opened and once more fumes started to damage the coffee trees. Again various solutions were suggested, including building a great chimney, 250 m high, to carry the gas high enough into the air so that it would drift harmlessly over the highland. Another suggestion was to drop an atomic bomb in the crater, blocking the vent. The situation was studied by R.E. Wilcox, of the U.S. Geological Survey. Wilcox rejected the idea of the great chimney because of the instability of the foundation on which it would have to be built, and because at any time the location of the gas vent might change, rendering the chimney useless. The atomic bomb he ruled out because of the risks from radioactive material drifting downwind. On the other hand, smaller-scale blasting or conventional bombs dropped in the crater might block the vent and stop the gas discharge, as it did in 1927; and although the vent would almost surely reopen in time, temporary relief might be obtained. In 1953 two medium-sized bombs were dropped in the crater, but without noticeable effect.

The whole problem of control of volcanic gases and reduction of their damaging effects is in need of much further study.

2.3. Cataclysmic Eruptions

The big volcanic cones of continental regions usually consist of alternating layers of tephra and lava flows. They are called *composite volcanoes*. Commonly, toward the end of their history great explosive eruptions throw out many cubic kilometers of glassy ash and pumice, formed by freezing of magma brought up from beneath the volcano. The eruption of this great volume of magma tends to remove the support from below, and frequently results in collapse of the top of the cone to form a caldera.

The Vesuvian eruption of 79 A.D., in which Pliny the Elder died, was an eruption of this sort, and they are often called *Plinian eruptions*. They are the most violent of volcanic manifestations, and consequently are potentially catastrophic. The eruption may consist largely or entirely of vertical projection of ash and pumice, which may fall over areas of many thousands of square kilometers; but usually this is accompanied by glowing avalanches and/or ash flows and mudflows, while lava flows usually are absent or very unimportant.

Because of the great hazard, combining several sorts of volcanic destruction on a large scale, a few of them will be briefly described.

Crater Lake, 5,000 B.C.

Crater Lake, in Oregon, is the result of such a Plinian eruption. It lies in a caldera 8 to 10 km across and 1,130 m deep, formed in the top of a composite volcano about 7,000 years ago. Before the eruption the conical mountain, now known as Mount Mazama, rose to about 3,900 m above sea level. Late in its history concentric cracks formed in the flanks of the cone, and highly viscous magma rose through them to form a series of steep-sided domes and thick lava flows—today's Mt. Hillman, Grouse Hill, Llao Rock, and Rugged Crest. Then came a period of quiet that lasted hundreds or thousands of years. Streams cut deep valleys into the cone, and during the Ice Age glaciers formed near the summit and plowed their way down the valleys, transforming them from

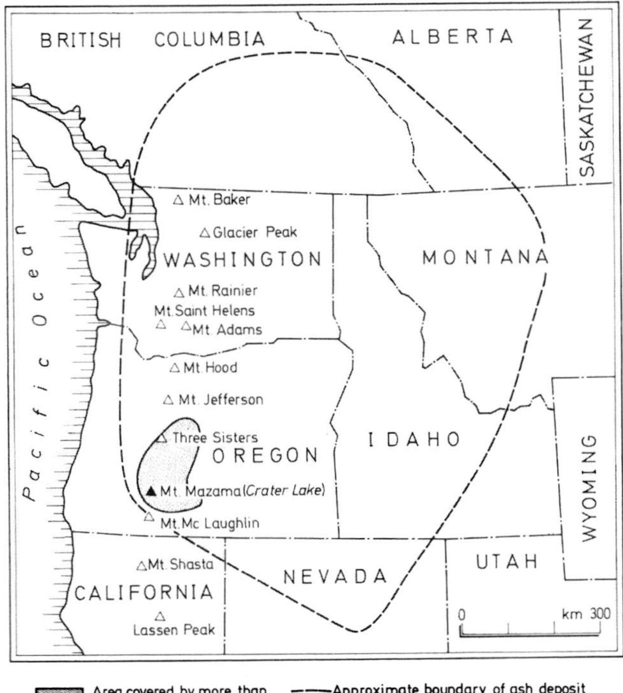

Fig. 2.16. Map showing the location of major volcanoes in the Cascade Range of Washington, Oregon, and California, and the distribution of ash from the great eruption of Mt. Mazama about 5,000 B.C. (After Williams and Goles, 1968)

V-shaped to U-shaped in cross section. The mountain must then have looked much like the present Mt. Rainier, except that the glaciers were more extensive.

Was there any warning that this peaceful-seeming mountain would erupt again? Fumaroles may have developed near the summit, and probably earthquakes shook the surrounding countryside for some days, or even years, before the outbreak, but we do not know. We do know, however, that the quietude was terminated by violence.

The eruption started with rather mild explosive activity that threw out showers of ash and pumice, but its violence rapidly increased, and a great cloud of tephra rose many kilometers into the air, which, blown by the wind, spread out to great distances. Ash fell as far away as Alberta and British Columbia, and closer to the volcano an area of 13,000 square km was buried to a depth of more than 15 cm (Fig. 2.16). As the eruption reached its climax great incandescent avalanches of ash and pumice rushed down the valleys to distances as great as 60 km (Fig. 2.17) and carried chunks of pumice 2 m wide as far as 30 km. Melting ice and snow on the flanks of the mountain must have caused

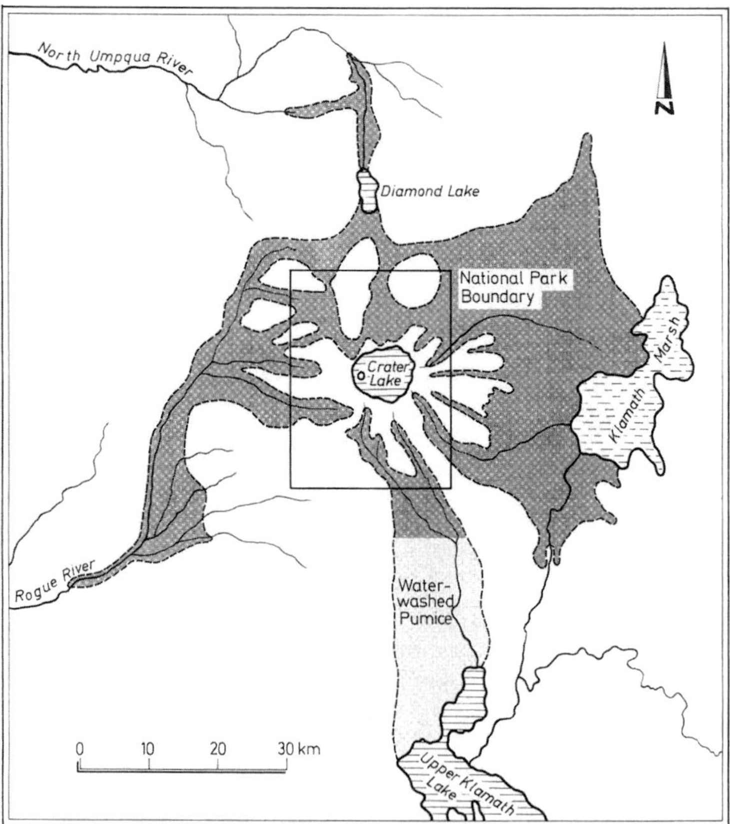

Fig. 2.17. Map of the area around Crater Lake, Oregon, showing the distribution of glowing avalanche deposits from the eruption of Mt. Mazama about 5,000 B.C. (After Williams, 1942)

floods, and the water mixed with ash and pumice to form mudflows. The total volume of ejected material was between 50 and 70 cubic km, and the removal of this great volume of magma resulted in collapse of the mountain top to form the caldera.

By analogy with similar events during historic time, the entire great eruption probably occupied only a few days. Since then water from rain and melting snow has accumulated in the caldera to form the lake, and several relatively gentle eruptions have formed lava flows and the cinder cone of Wizard Island.

Mt. Mazama was just one of a row of great volcanic peaks that extends along the Cascade Range from southern Canada into northern California (Fig. 2.16), and others of these volcanoes could well erupt in a similar manner, bringing about widespread devastation.

Bezymianny, 1956

Bezymianny is one of the Kliuchevskaia group of volcanoes, in central Kamchatka, which, previous to 1956 had no record of eruption and had attracted so little attention that it had not even been given a name. (Bezymianny means "no name.") It was a moderate-sized composite volcano built against the side of an older dome, with a small more recent cone in a crater between them, while to the east a deep notch and gorge led down into the Dry Hapitsa Valley.

On September 29, 1955, earthquakes began to emanate from the old volcano, gradually increasing in number until on October 22 nearly 1,300 of them were recorded at Kliuchi, 24 km from the volcano. On that day a series of weak to moderate explosions began, and by November 9 a layer of ash 6 mm thick had accumulated at Kliuchi. In mid-November the eruption became more intense, ash was thrown 5 km above the mountain and drifted eastward beyond the sea coast. On November 17 it was so dark that house and automobile lights had to be used in Kliuchi, but in late November the intensity decreased greatly, and activity remained at a low level until March, 1956.

On March 30, at 17:11 h, a tremendous explosion occurred. A dense black jet shot eastward from the volcano at an angle of 30° or 40° from the horizontal, quickly reaching a height of more than 30 km, and 15 minutes later another explosion drove the cloud to a height of 43 km. At Kliuchi darkness became impenetrable, and during the next $3^1/_2$ hours 20 mm of ash fell. Ten kilometers from the volcano an observation cabin (fortunately unoccupied) was totally destroyed; 24 km away trees as much as 25 cm in diameter were snapped off at the base; 30 km away bark was blasted completely off the sides of trees toward the volcano, and dead trunks were set on fire. The ash formed a deposit 50 cm thick 10 km east of the volcano, and 25 cm thick at a distance of 30 km.

The explosion cloud was accompanied by flows of ash and pumice fragments which illustrate well the transition between ash flows and glowing avalanches of the Soufrière type. By far the largest of the flows descended through the notch into the Dry Hapitsa Valley, continuing along the valley for 16 km, then mingled with stream water and was transformed into a mudflow that joined many other mudflows formed on the slopes of the adjacent mountains to continue 80 km down valley. The avalanche deposit covered an area of 60 square km to a depth as great as 50 m. When the clouds cleared away, the whole top of the mountain had disappeared.

Diminished activity continued through the rest of 1956. Despite the extreme violence of the Bezymianny eruption there were no reported human casualties, partly due to good luck, but largely because the surrounding country was nearly uninhabited. A similar eruption in a more densely populated area might well cause fearful casualties.

Tambora, 1815

The eruption of Tambora, on the Island of Sumbawa, 400 km east of Java (Fig. 2.18), is generally considered to have been the greatest explosion of recent centuries. As at Crater Lake, the explosion was accompanied by formation of a caldera. Until 1812 the volcano had no record of eruption, and it was believed to be extinct; but in that year mild explosions began in its crater, and in April 1815 tremendous explosions threw ash 20 km into the air. Fragments of pumice as much as 13 cm across fell at a distance of 40 km. At the base of the volcano the tephra deposit reached a thickness of 1.5 m. On Lombok, 150 km to the west, and at Bima, 90 km to the east, it was 50 cm thick. The total volume of ejecta was about 100 cubic km. Crops were almost totally destroyed throughout Sumbawa and Lombok, and to a lesser extent on Flores and Bali. Some 44,000 persons died of starvation and disease on Lombok, and on Sumbawa 48,000 died, partly of starvation and partly directly by volcanic action.

Krakatau, 1883

The great eruption of Krakatau differed from that of Mt. Mazama largely in that the caldera collapse took place beneath the ocean, and was accompanied by great tsunamis (see Chapter 3).

The volcano lies in the strait between Java and Sumatra (Fig. 2.18). Before 1883 the visible part consisted of a group of small islands (Fig. 2.19A), but beneath sea level lay the stub of a big volcano with a caldera 8 km across. Two of the islands, Lang and Verlaten, were projecting portions of the caldera rim, and a third segment of the rim was nearly buried by a younger volcanic cone, Rakata. Two still younger cones, Danan and Perbuwatan, coalesced with Rakata to form the main island of Krakatau.

The last previous eruption had taken place in 1680, and the islands had become heavily covered with vegetation. In May 1883 a few small earthquakes were felt on nearby Java and Sumatra, and on May 20 Perbuwatan burst into eruption, throwing ash 10 km into the air. Activity died away after a few days, and was followed by 3 weeks of quiet, but on June 19 it began again, and was soon joined by another erupting vent at the foot of Danan. Weak eruption continued through July, and by August 11 Rakata also was erupting. Thus, by late August the Krakatau vents had been open and erupting for more than 3 months. Cataclysmic eruptions are not restricted to closed vents!

The cataclysm itself began on August 26, and was essentially over by the 28th. It consisted of a series of giant explosions, the greatest of which, on the

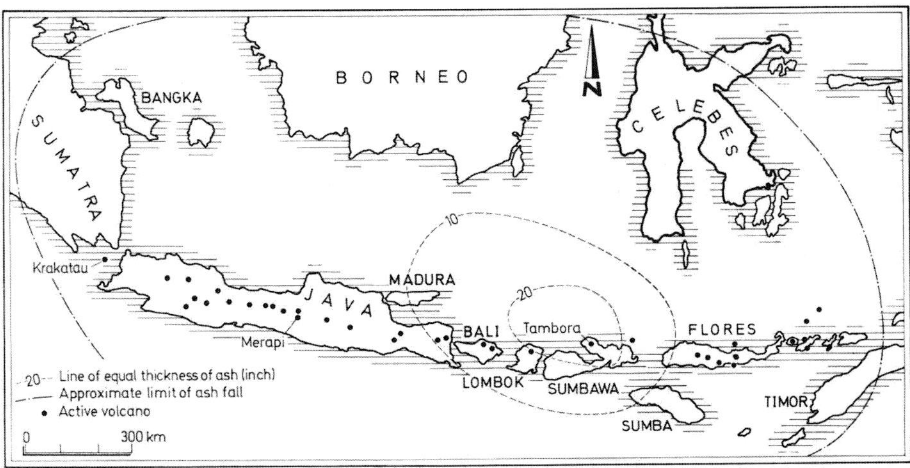

Fig. 2.18. Map of central Indonesia, showing the location of Tambora Volcano and the distribution of ash from its eruption in 1815. (After Petroeschevsky, 1949)

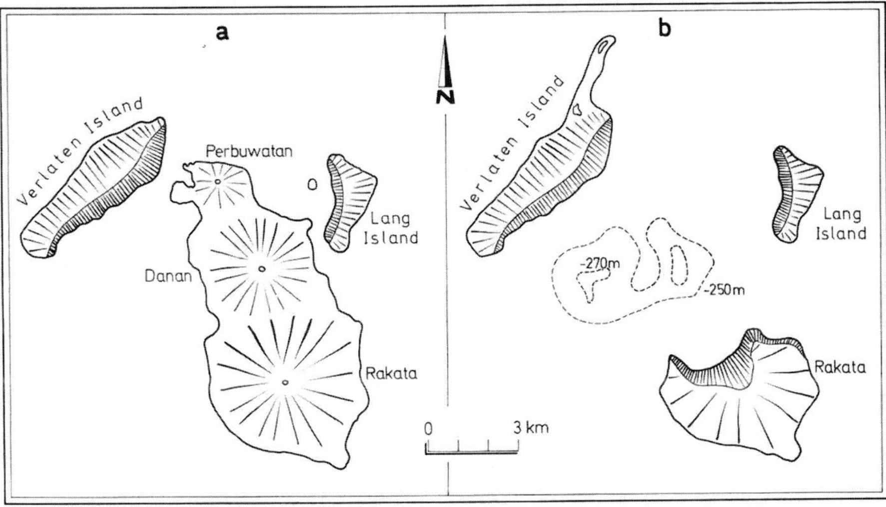

Fig. 2.19 a and b. Maps showing the islands of the Krakatau group, between Java and Sumatra, (a) before the eruption of 1883, and (b) after the eruption. (After Williams, 1941)

morning of August 27, was heard 4,800 km away. Ash was thrown 80 km into the air, and the accompanying air waves shattered windows 150 km away. On Lang and Verlaten Islands 15 m of air-laid ash and pumice is overlain by 60 m of glowing-avalanche deposits. Great rafts of floating pumice blocked the strait. In all, about 16 cubic km of ash and pumice was thrown out, and ash fell to an appreciable thickness over an area of more than 750,000 square km; very fine ash is said to have drifted around the world.

The Krakatau Islands themselves were uninhabited, but on neighboring parts of Java and Sumatra about 36,000 persons died as a result of the eruption, most of them drowned by the tsunamis.

Santorin, 1,500 B.C.

The island volcano Santorin, formerly known as Thera, lies at the southern edge of the Cyclades Islands, 110 km north of Crete (Fig. 2.20). A cataclysmic eruption of the volcano, about 1,500 B.C., which in recent years has been widely blamed for the destruction of the Minoan civilization, resembled in most respects that of Krakatau.

The volcanic structure is complex, composed of several overlapping cones built by eruptions from several different vents. In its center lies a submerged caldera 14 km across, which is believed to have been formed during the eruption of 1,500 B.C. Since then, renewed activity has built a series of domes within the caldera.

The great eruption started relatively gently, as is shown by a thin sheet of air-laid ash. Overlying this are two layers, reaching a maximum thickness of about 60 m, composed largely of ash and lumps of pumice, the structures in which resemble those in the deposits formed by base surges during the Taal eruption of 1965, described earlier.

Destruction is thought to have been the result of the tephra fall and base-surge deposits, and of one or more tsunamis that accompanied collapse of the caldera. On the island of Thera, Minoan cities were buried, and their ruins preserved beneath ash and pumice. Few, if any, people were trapped, however. Forewarned,

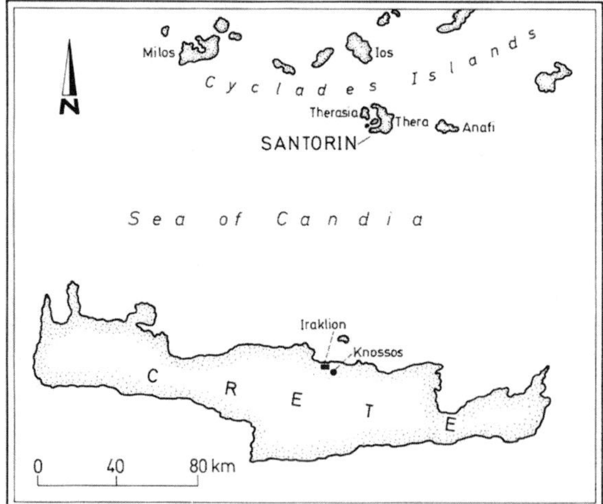

Fig. 2.20. Map showing the location of Santorin Volcano in relation to the island of Crete. The islands of Thera and Therasia are the outer parts of the volcano, and the sea between them occupies the caldera

possibly by earthquakes and the milder early phases of the eruption, they had fled. On the nearer parts of Crete ash and small lumps of pumice falling from the air formed a layer 15 cm thick—enough to affect fruit trees seriously, to kill garden crops and grass, and thereby to cause starvation of grazing animals. There appears to be little question that sizeable tsunamis were generated, though probably they were much smaller than some workers have suggested, but it is likely that they were big enough to create havoc up to a few tens of meters above sea level on the nearby shores of Crete, and probably also brought about flooding of coastal lowlands around much of the eastern Mediterranean. (The legendary floods of Deucalion, the Noah of the Greeks, have been hypothetically identified with these tsunamis.)

Despite the destruction of the cities on Thera and the coastal areas of Crete, and the more extensive damage to agriculture on the latter island, it seems unlikely that the eruption alone could have caused the collapse of the Minoan civilization. More likely the eruption was only one contributing cause, possibly weakening the kingdom to the point where it fell easy prey to invading Mycenaeans or other marauders.

2.4. Evaluation of Volcanic Risk

Appraisal of risk on and near any volcano depends largely on a knowledge of past behavior of the volcano. This, depends on (1) the recorded history of eruptions, and (2) a geological study of the composition and structure of the cone. With due allowance for changes in behavior resulting from evolutionary changes of the magma in the underlying reservoir, a volcano is most apt to do in the future much the same as it has done in the past. In nearly all volcanoes evolution of magma causes an increase in viscosity and gas content, and this in turn leads progressively to shorter and thicker lava flows, formation of domes, and increasingly explosive eruptions. At some volcanoes this sort of evolution can be seen on a relatively short time scale. Thus, at Hekla, in Iceland, a short period of repose is followed by a relatively gentle eruption of fairly fluid lava; whereas a long quiet period is terminated by violent explosion of viscous magma producing abundant pumice and ash, transforming in time into less explosive activity producing more fluid lava. The more siliceous, more viscous, and more gas-rich magma appears to accumulate at the top of the magma chamber like cream on a bottle of milk, and to be tapped off during the early stages of the eruption. Similar behavior was shown by Mt. Fuji, in Japan, during its eruption in 1707.

After the disastrous eruption of Mt. Kelut, in 1919, the Volcanological Survey of the Netherlands East Indies began preparing maps showing areas around some of the most dangerous of the Indonesian volcanoes in which people would be endangered by future eruptions. Areas indicated as dangerous were primarily those that had been affected by historical eruptions, with consideration given to the eruption characteristic of the particular volcano, and the effect of topography on paths of glowing avalanches and mudflows (Fig. 2.21). More recently, Russian volcanologists have prepared similar maps of areas adjacent to the more

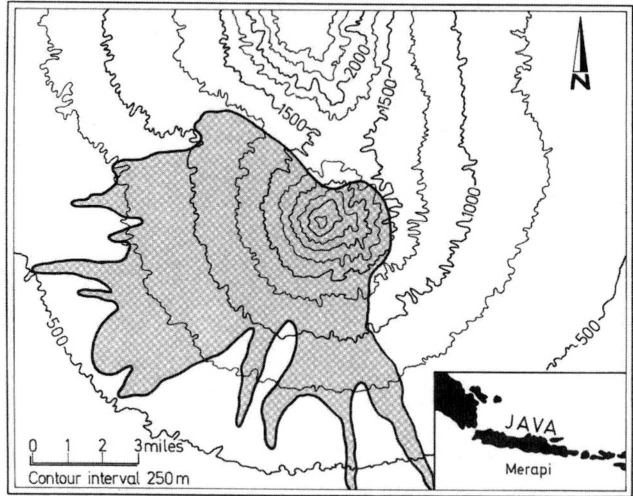

Fig. 2.21. Map prepared by the Netherlands Indies Volcanological Survey, indicating the area most in danger from glowing avalanches at Merapi Volcano, Java. (After Neumann van Padang, 1960)

dangerous of the Kamchatkan volcanoes, showing zones of different degrees of risk. A few other appraisals of volcanic risk, such as that by E.J. Searle for the city of Auckland, New Zealand, have been made.

Currently, the U.S. Geological Survey is studying several of the volcanoes of the Cascade Range and of Hawaii to determine the types of risks to be expected from each, and at least in some instances to indicate on maps the areas that would be likely to be affected by different types of activity. In the Cascades, the greatest risks are considered to be from ashfalls and mudflows. One approach is to assume that any major Cascade volcanoes might have an explosive eruption as violent as that of Mt. Mazama (described earlier), and assuming that meteorological conditions would be like those during that eruption, plot on a map the area that would be covered by ash and pumice thick enough to do serious damage (see Fig. 8.5). An earlier report appraises the risks from ash eruptions of Alaskan volcanoes.

Shortly after the 1960 eruption of Kilauea, one of the present writers (Macdonald) was asked to evaluate the risk from volcanic activity on the Island of Hawaii. The risk is almost wholly from lava flows, the great majority of which are erupted from zones of very numerous fissures (rift zones) 3 to 5 km wide that extend down the flanks of Kilauea and Mauna Loa volcanoes. Some lava flows travel all the way to the ocean; others, of lesser volume, are shorter. The risk of inundation of the land increases, therefore, as the rift zone is approached, and is greatest along the rift zone itself. Damage by falling cinder or ash, or by volcanic gas, is small compared with that from lava flows, but it also is most likely to occur close to the vents, and hence along the rift zones.

That there is risk of damage on the flanks of active volcanoes, and especially along rift zones, is obvious. The problem, however, is to assess the seriousness of the risk – the likelihood that any particular small area will be destroyed or seriously damaged during any given length of time. No one eruption is likely seriously to affect more than a small proportion of the total area. Good records of eruptions in Hawaii extend back less than 2 centuries, and therefore the data are insufficient for a really sound statistical treatment, but nevertheless they do give some indication of the magnitude of the risk. As an example, we can take the corner of the Island of Hawaii east of 155° west longitude—an area of approximately 325 square km which is bisected by the east rift zone of Kilauea, within which, since about 1750, lava has covered approximately 52 square km, or 16 per cent of the total. On this basis, the likelihood of lava covering any given plot of land in any single given year is a little less than one in a thousand. Likelihood of land along or immediately adjacent to the rift zone being covered is probably not more than twice as great as for the area as a whole. In any 25-year period the chance of a given house being destroyed by lava is about 1 in 40.

Within the above-mentioned area the upper parts of big cones are essentially free of risk from lava, because they would stand above the surface of flows of any probable thickness. They are, however, subject to damage from falling tephra and volcanic gas.

Mauna Loa has covered about 5 times as much area with lava during historic time as has Kilauea, but it is a much larger volcano, and a large proportion of the lava-covered land was waste land high on the mountain slope. The risk on the more useful low-level lands, most of which are far from the rift zones, is much less than for the volcano as a whole. Owing to its relative nearness to the rift zone and the relation of the rift zone to the topography, the low-level lands most subject to lava flows are those on the southwestern slope south of 19°22′ N. Of the land within this area below 2,000 m altitude, 14 per cent has been covered by lava during the last 200 years, so that the chance of any given house or small piece of land being destroyed in any one 25-year period is about 1 in 57—somewhat less than on the eastern flank of Kilauea. Below 2,000 m altitude on the southeastern slope of Mauna Loa the chance is only about 1 in 700.

Whether these risks are reasonable or exorbitant is something that must be decided by each individual, on such bases as the usefulness and desirability of the land so long as it is not destroyed, and comparison with actuarial figures on probability of loss of a house to fire in rural areas during a similar period—a risk many people take without hesitation. It should be remembered, however, that insurance against destruction by lava is very expensive, because the insurance companies cannot spread the risk widely (see Section 8.7). On the other hand, the land does not completely lose its value if it is covered by lava. Flow surfaces less than 20 years old are already in use for residential construction, and even for certain types of agriculture. The risk does not appear to be exorbitant, even on the eastern slope of Kilauea; but persons developing land in such areas certainly should be aware that the risk exists, and of its approximate magnitude.

2.5. Prediction of Volcanic Eruptions

It is thus apparent that saving of lives and property from destruction by volcanic action depends heavily on prediction of eruptions. Given sufficient warning, people can be evacuated and much property moved from the threatened area. Needless to say, to be effective, predictions must be accurate. Any large proportion of erroneous warnings will soon lead to all warnings being ignored. The story of the boy who cried "wolf" is very apropos here. Moreover, the effectiveness of the prediction depends greatly on how specific it is as to time, type, and area to be affected. A general prediction may simply state that a volcano is uneasy and may erupt, or that it probably will erupt soon—but since this may mean anytime within the next few days, or months, or even years, it is not of any great help beyond alerting the inhabitants of the area to possibly more specific predictions.

Prediction of the type of eruption depends largely on a knowledge of the past behavior of the volcano. Prediction of the area that will be affected depends on the expected type of eruption, on its strength, and on a study of the topography of the volcano and the surrounding area, which will determine to a large extent the courses of all types of flows. Prediction of the time of the eruption makes use of various lines of evidence discussed below.

Where a reasonably long history of activity exists, and a large enough number of eruptions has occurred to give a useable body of data, it may be possible to recognize a pattern or periodicity of eruptions. Thus, during its 1 1/2 centuries of recorded history Mauna Loa, in Hawaii, has shown a frequent alternation of summit and flank eruptions, with a flank eruption generally following a summit eruption within 3 years. On this basis, a flank eruption can be expected within 3 years after a summit eruption. But the actual interval has ranged from as little as 6 months to as much as 38 months (ignoring early years of very incomplete records), and a summit eruption is sometimes followed by another summit eruption rather than a flank eruption. Obviously, therefore, any prediction on this basis carries considerable uncertainty. At Vesuvius, where we have a record of eruptions over a period of nearly 1,900 years, no pattern or periodicity is recognizable.

The temperatures of fumaroles, hot springs, and crater lakes may rise before eruptions. In June of 1965 the temperature of the crater lake in the inner cone of Taal Volcano began to increase, and by late July it had risen 11° C; on September 28 the volcano erupted. At Aso Volcano, in Japan, the temperature increases to the point that the crater lake may boil completely dry before an eruption. Temperatures of fumaroles may rise several degrees or tens of degrees before eruptions. However, the rise may take place only a few days before the eruption, or several months or even a few years before it. At least in one case, at Aso, the temperature may drop again a few days before the eruption, and in other cases the temperature has returned to normal without any eruption occurring at all. At other volcanoes, notably several of those of Indonesia, no increase of the temperature of fumaroles has preceded eruption. Thus fumarole temperatures are not reliable grounds for prediction of eruptions, but they can serve to alert observers to other signs and help confirm other evidence.

In some instances the composition of fumarole gases changes before an eruption, generally showing an increase in the proportion of HCl and SO_2 to water. Thus, at both Asama and Mihara Volcanoes in Japan, both sulfur and chlorine have been found to increase markedly before at least some eruptions, but in many other instances no change has been observed. Again, although they are not absolute indicators, the compositions of gases may be contributory evidence of coming eruption, and monitoring of gas composition deserves much more attention than it has received in the past. Recent developments in gas chromatography and transmission of chemical data and temperature measurements by wire or radio make practical the continuous monitoring of fumaroles.

Changes in the strength and/or orientation of the Earth's magnetic field have been observed before some eruptions of some volcanoes. Hypothetically, a heating of the volcano should result in a decrease in the strength of its local magnetic attraction and some recent work suggests that this actually occurs, at least in some instances. Thus, a marked change in the intensity of the magnetic field was observed, at a distance of 24 km, 12 hours before the eruption of Piip Crater, Kamchatka, in 1966, while at Mihara Volcano an increase in the westward drift of the orientation of the magnetic field was observed several months before eruption.

Another new method of detecting temperature changes that may indicate coming eruption is periodic infrared photography, either from airplanes or from satellites. It is too new, however, to have yielded any positive results.

Again hypothetically, the presence of a body of hot liquid magma within the solid structure of the volcano should result in an electrical potential, and changes of temperature or form of the magma body should produce changes in the electrical currents within the volcano. Marked rapid changes in these earth currents have indeed been observed several hours before individual explosions of Asama Volcano; as yet, however, very little work has been done on this promising method.

At some volcanoes a marked deformation of the ground surface takes place before eruptions. As an extreme example, in 1943 the ground near the eastern base of Usu Volcano, in Japan, was domed upward 50 m through a period of 5 months before the eruption. At other volcanoes the deformation is far less. At Kilauea Volcano the accumulation of magma in a reservoir only 3 to 5 km below the crater floor results in swelling of the overlying mountain. Leveling by ordinary surveying methods showed that the summit of the mountain was elevated 0.9 m between 1912 and 1921, and then went down 1.1 m as a result of draining of the magma reservoir during the eruption of 1924. The swelling and shrinking of the volcano are usually measured by means of tiltmeters, which measure the changes in inclination of the slopes of the mountain as its top is pushed up or sinks. Swelling of Kilauea indicates that the magma reservoir is being filled and that eruption is possible, and as inflation increases eruption becomes more and more likely. As yet, however, this tumescence of the volcano can be used only as a general indicator of a possible coming eruption, because in the past eruptions have taken place at various degrees of inflation, and at other times marked tumescence may be followed by detumescence without eruption.

Less positive evidence indicates that Mauna Loa also swells before eruptions and shrinks during eruptions. Marked tumescence before eruptions has also been observed at Manam Volcano, just north of New Guinea, and at a few others. Tilt measurements have as yet been made at relatively few volcanoes, but at some of those there appears to have been no tumescence before eruptions. Perhaps detectable tumescence occurs only at volcanoes where the magma reservoirs lie within a few kilometers of the surface. Others such as Etna, where the magmatic hearth appears to lie at a depth of several tens of kilometers, may show no pre-eruption tumescence.

Eruptions are commonly preceded by local earthquakes (see Section 1.3). These may be the result of splitting open of fissures through which the magma rises toward the surface, or of movement of blocks in and around the volcano as it tumefies, or of other causes. Some of the earthquakes are strong enough to be felt over large areas, or even to do serious damage; others are so small that they can be detected only with sensitive seismographs, and some eruptions appear to have no premonitory earthquakes. For instance, none were recorded, even by sensitive seismographs, before the 1950 eruption of Mihara Volcano, even although this was the largest eruption in the history of the volcano.

Some earthquake preludes last less than an hour, but others extend over many years and consist of many thousands of individual earthquakes. That which preceded the 79 A.D. eruption of Vesuvius continued for 16 years. It commenced with a violent quake that did much damage in the vicinity of the volcano, but failed to upset the aplomb of the Emperor Nero, who was just then making his singing debut on a stage at Naples. Despite the shaking, he finished his number. In some instances (as usually at Hawaiian volcanoes) the earthquakes nearly or completely cease when the eruption begins; but in others the quakes continue during, and even after, the eruption. Sometimes the earthquake swarm culminates several days, or even weeks, before the beginning of the eruption, and the intervening period is relatively quiet. Although at Hawaiian volcanoes relatively few quakes occur during most of the eruption, they commonly begin again toward its end and continue for a while afterward as the volcano shrinks due to removal of magma from within it. Whether post-eruption earthquakes at other volcanoes have the same origin is uncertain. On the other hand, increase in seismicity during the eruption in some instances precedes the opening of a new vent or group of vents, as it did during the 1955 eruption of Kilauea.

Volcanic tremor is a rhythmic vibration of the ground that has been shown to accompany the movement of magma. It often occurs just before and during eruptions, but it may also occur without eruption.

There may also be some causal connection between tectonic earthquakes, that result from regional movements of segments of the earth's crust, and some volcanic eruptions. Several workers have claimed that many eruptions, especially in the mountain-building belt along the western edge of the Pacific Ocean, are preceded by strong, deep tectonic earthquakes. Other workers deny any such connection.

At Asama Volcano, T. Minakami has been able to distinguish three classes of earthquakes: A-type quakes, originating at depths of 1 to 10 km; B-type quakes, originating at depths of less than 1 km; and quakes caused by explosions in and just below the crater itself. Using the frequency of B-type quakes, Minakami has derived an empirical formula which he has used with considerable success to predict within a 5-day period individual eruptions of Asama. At Bezymianny Volcano, P.I. Tokarev has found that explosive eruptions are preceded by B-type earthquakes which start 30 to 50 days before the eruption. During the eruption the earthquakes increase in number, but decrease in total energy released. Based on the rate of accumulation of strain ("elastic deformation"), Tokarev has evolved a formula which he has used to predict explosive eruptions approximately a week before their occurrence. No relationship has been found between the frequency of premonitory earthquakes or the amount of energy released by them, and the magnitude of the succeeding eruption.

In spite of limited and local successes, the variability of the patterns of eruption-associated earthquakes makes it impossible in most cases to use them alone as a means of prediction of eruptions. They are most useful in combination with other indicators, especially ground tilting. The latter, when it occurs, makes it possible to distinguish the numerous earthquakes that accompany swelling of the volcano and increasing potential of eruption from the equally numerous ones that accompany shrinking and decreasing potential. Thus, at Kilauea Volcano, in 1950, a swarm of several hundred earthquakes culminated a period of several months of tumescence; but the tilting indicated that the top of the volcano was sinking, and that eruption potential was decreasing.

One of the most successful predictions of an eruption was based on a combination of several indicators. On February 8, 1942, a strong earthquake originated at a depth of 40 to 50 km beneath the northeast flank of Mauna Loa. This was followed, on February 21 and 22, by a swarm of quakes from shallow foci beneath the part of the northeast rift zone of the volcano that lies between 2,700 and 3,050 m altitude, and then by others from foci that shifted gradually along the rift zone to the summit and beyond, reaching a point 8 km down the southwest side of the mountain on March 21. On March 28 another quake came from the northeast rift, while at the same time tilt measurements indicated that the volcano was being inflated with magma. A summit eruption of Mauna Loa had ended in August 1940. Combining the evidence of the earthquakes and tilt with his knowledge of the past behavior of the volcano, R.H. Finch predicted in late March that an eruption would take place within the next few months on the northeast rift zone, probably between 2,700 and 3,050 m altitude. The eruption started on April 28, at 2,800 m altitude.

This excellent prediction was recorded in reports to the National Park Service, but because of wartime security restrictions it could not be published. It was feared that if the enemy learned of the imminent eruption the volcano would be used by them as a navigational beacon. Ironically, in her regular radio broadcast on the morning after the eruption started, Tokyo Rose congratulated the people of Hawaii on their magnificent volcanic eruption.

For reasons that we do not know, animals on and near the volcano often are very uneasy before an eruption. As one example, for 3 or 4 days before the beginning of the flank eruption of Kilauea in 1955 dogs in the vicinity of the later-formed vents were much disturbed, running around excitedly, digging holes in the ground, and snuffling in the holes as though they were in pursuit of some burrowing animal. We could detect no odor of gas in the holes, but seismographs recorded a large number of small earthquakes, which although mostly too small to be felt by humans, may have been felt by the dogs. Other evidence indicates that some animals can feel earthquakes that are not felt by men. Later in the eruption new vents opened 6.5 km to the southwest, and, for a couple of days before the opening, dogs in that area also were much disturbed and excited.

Earthquakes were felt around the base of Arenál Volcano only for 2 days or less before the beginning of the 1968 eruption, but cattle started to move down off the mountain two weeks before that. There were no seismographs to tell us whether there were earthquakes too small for men to feel.

Much more work is needed on methods of predicting volcanic activity. Much of the knowledge gained will be of general applicability, but detailed studies and long records of the behavior of individual volcanoes will be needed, because volcanoes show a great individuality of behavior.

The results that can be achieved by means of proper instruments and scientists familiar with the behavior of the volcano are well illustrated by Kilauea Volcano in late 1973. One of the special invitations to trouble at the Hawaiian volcanoes is the very large number of spectators who come to watch the eruptions. On November 10 more than 500 persons were viewing the activity at Mauna Ulu, a vent on the east flank of Kilauea. At 17:30 h the personnel of the U.S. Geological Survey's Hawaiian Volcano Observatory noticed a strong volcanic tremor on the seismographs and at about the same time the lava began to drain out rapidly into underground conduits from the small lava lake at Mauna Ulu. At 17:45 h the summit of Kilauea started to sink very rapidly, as shown by the tiltmeters. These events closely resembled those which had preceded recent outbreaks on the east flank, and the Observatory staff realized that they probably heralded a new outbreak. They advised the staff of the Hawaii Volcanoes National Park of that fact, and the Park staff moved quickly to evacuate the spectators from Mauna Ulu so that when the outbreak came at 21:47h, cutting the access and emergency escape roads, all of the visitors had been evacuated in time.

As a summary of general recommendations for the surveillance of volcanoes to detect signs of coming eruptions, one cannot do better than quote from a report recently prepared for the United Nations.

"Networks of seismograph stations, to record volcano-related earthquakes and make possible the recognition of seismic preludes to eruptions, exist today at only a few volcanoes. At even fewer are precise geodetic measurements or measurements of ground tilting being made which might reveal pre-eruption tumescence. No volcanic district as a whole is so instrumented; consequently there does not exist anywhere a warning system adequate to cover any volcanic district as a whole.

"It is perhaps unrealistic to hope to have permanent seismic and tilt recording instruments on every potentially eruptive volcano. However, in all (populated) volcanic districts there should be established networks of seismographs capable of detecting unusual seismic activity in any part of the district; and suitable portable seismographs and tilt-measuring instruments with a trained team of operators should be ready for installation on any volcano in any part of the district where the regional network indicates something unusual going on. Trained volcanologists should be available and ready to take to the field, to interpret volcanic events both before and during eruptions, and to give advice on evacuation and other protective measures. During times of volcanic quiet the volcanologists, and other geologists, should be occupied in a complete and detailed study of the geology and eruptive history of all of the volcanoes, to aid in forecasting the types of eruptions to be expected from them.

"Detailed topographic maps should be prepared to aid in identifying the likely courses of different types of flows. Wind patterns at both high and low altitudes should be studied during all seasons to aid in predicting the distribution of ash falls and the course of toxic gases. Probable distribution and thickness maps should be prepared for ash falls from various intensities of eruption for all seasons. Areas of varying degrees of risk from various types and intensities of eruption should be identified and delineated on maps. Emergency procedures, including evacuation procedures, should be established for all communities, including alternate routes and destinations for different types of eruption, and eruptions of different volcanoes, where it is so indicated.

"Periodic temperature measurements and chemical analyses should be made at all fumaroles and hot springs to establish norms and normal seasonal variations for each and make possible recognition of departures from norms which might indicate coming eruptions. Aerial magnetic and infra-red photographic surveys should be made at regular intervals to try to recognize any significant changes in the magnetic field or in ground temperature."

Obviously, the cost of all of this in money would be great, but without it, the toll in human lives and property will probably be greater. Volcanic catastrophes that would dwarf the destruction of St. Pierre are a real geological possibility.

2.6. References

Anderson, T., Flett, J.S.: Report on the eruptions of the Soufrière in St. Vincent, and on a visit to Montagne Pelée in Martinique. Pt. I. Royal Soc. London Phil. Trans., ser. A, **200**, 353–553 (1903).

Bullard, F.M.: Volcanoes: in history, in theory, in eruption, 441 p. Austin: Univ. Texas Press 1962.

Crandell, D.R., Mullineaux, D.R.: Volcanic hazards at Mount Rainier, Washington. U.S. Geological Survey, Bull. **1238**, 26 (1967).

Crandell, D.R., Waldron, H.H.: Volcanic hazards in the Cascade Range. Conference on Geologic Hazards and Public Problems, May 27–28, 1969, Proceedings, pp. 5–18 Office of Emergency Preparedness, U.S. Govt. Printing Office (1969).

Finch, R.H., Macdonald, G.A.: Hawaiian volcanoes during 1950. U.S. Geological Survey, Bull. **996-B**, 27–89 (1953).

Gorshkov, G.S.: Gigantic eruption of the volcano Bezymianny. Bull. volcanologique, ser. 2, **20**, 77–109 (1959).

Healy, J.: Tangiwai railway disaster. Report of Board of Inquiry, pp. 6–8. New Zealand: Government Printer (1954).

Hovey, E.O.: Martinique and St. Vincent, a preliminary report upon the eruptions of 1902. American Museum Nat. Hist. Bull. **16**, 333–372 (1902).

Koto, B.: The great eruption of Sakurajima in 1914. Tokyo Imperial Univ., Jour. College Sci., **38**, art. 3, 237 (1916).

Lacroix, A.: La Montagne Pelée et ses eruptions. 662 p. Paris: Masson et Cie., 1904.

Lydon, P.A.: Geology and lahars of the Tuscan Formation, northern California. Geological Soc. America, Mem. **116**, 441–475 (1968).

Macdonald, G.A.: Barriers to protect Hilo from lava flows. Pacific Science, **12**, 258–277 (1958).

Macdonald, G.A.: Volcanoes. Englewood Cliffs, N.J.: Prentice-Hall, Inc., 1972.

Macdonald, G.A., Abbott, A.T.: Volcanoes in the Sea. The Geology of Hawaii 441 p., Honolulu: Univ. Hawaii Press, 1970.

Mason, A.C., Foster, H.L.: Diversion of lava flows at O Shima, Japan. American Jour. Science, **251**, 249–258 (1953).

McBirney, A.R.: The Nicaraguan volcano Masaya and its caldera. American Geophysical Union Trans., **37**, 83–96 (1956).

Moore, J.G.: Base surge in recent volcanic eruptions. Bull. volcanologique, ser. 2, **30**, 337–363 (1967).

Neumann van Padang, M.: De uitbarsting van den Merapi (Midden Java) in de jaren 1930–1931. Ned. Indies Dienst Mijnbouwk. Vulkan. Seism. Mededel., no. **12**, 135 p. (with English summary) (1933).

Neumann van Padang, M.: Catalog of the active volcanoes of the world including solfatara fields, Part 1, Indonesia, 271 p. International Volcanological Assn., Naples (1951).

Neumann van Padang, M.: Measures taken by the authorities of the volcanological survey to safeguard the population from the consequences of volcanic outbursts. Bull. volcanologique, ser. 2, **23**, 181–192 (1960).

Perret, F.A.: The Vesuvius eruption of 1906. Carnegie Institution of Washington, **339**, 151 p. (1924).

Perret, F.A.: The eruption of Mt. Pelée 1929–1932. Carnegie Institution of Washington, **458**, 125 p. (1935).

Petroeschevsky, W.A.: A contribution to the knowledge of the Gunung Tambora (Sumbawa). Koninklijk Nederlandsch Aardrijkskundig Genootschap, **66**, 688–703 (1949).

Rose, W.I., Stoiber, R.E.: The 1966 eruption of Izalco Volcano, El Salvador. Jour. Geophysical Research, **74**, 3119–3130 (1969).

Searle, E.J.: Volcanic risk in the Auckland metropolitan district. New Zealand Jour. Geology and Geophysics, **7**, 94–100 (1964).

Stehn, C.: The geology and volcanism of the Krakatau Group. Fourth Pacific Science Congress Guidebook, Batavia, pp. 1–55 (1929).

Thorarinsson, S.: Hekla, a notorious volcano. Almenna Bókafélagid, Reykjavik, Iceland, 62 p. 1970a.

Thorarinsson, S.: The Lakagigar eruption of 1783. Bull. volcanologique, ser. 2, **33**, 910–927 (1970b).

United Nations: The role of science and technology in reducing the impact of natural disasters on mankind. Report of the Advisory Committee on the Application of Science and Technology to Development, 36 p. (1972).

Vitaliano, D.B., Vitaliano, C.J.: Plinian eruptions, earthquakes, and Santorin. A review: First International Science Congress on the Volcano of Thera, Acta, 88–108 (1971).

Waldron, H.H.: Debris flow and erosion control problems caused by the ash eruptions of Irazú Volcano, Costa Rica. U.S. Geological Survey, Bull. **1241-I**, 37 p. (1967).

Wilcox, R.E.: The problem of damage by fumes of Santiago Volcano, Nicaragua. U.S. Geological Survey, Report submitted to the Government of Nicaragua (mimeographed), 38 p. (1952).

Wilcox, R.E.: Some effects of recent volcanic ash falls with especial reference to Alaska: U.S. Geological Survey, Bull. **1028-N**, pp. 409–476 (1959).

Williams, H.: Calderas and their origin. Univ. California Pub., Bull. Dept. Geological Sciences, **25**, 239–346 (1941).

Williams, H.: The geology of Crater Lake National Park, Oregon, with a reconnaissance of the Cascade Range southward to Mount Shasta. Carnegie Institution of Washington Pub. **540**, 162 p. (1942).

Williams, H.: Volcanoes of the Paricutin region. U.S. Geological Survey, Bull. **965-B**, 165–279 (1950).
Williams, H.: The great eruption of Cosegüina, Nicaragua, in 1835. Univ. California Pub. in Geological Science, **29**, 21–46 (1952).
Williams, H.: Crater Lake, The Story of its Origin. 98 p. Berkeley: Univ. California Press, 1954.
Williams, H., Goles, G.: Volume of the Mazama ash-fall and the origin of Crater Lake caldera, Oregon. Andesite Conference Guidebook, Oregon Dept. of Geology and Mineral Industry, Bull. **62**, 37–41 (1968).
Williams, R.S., Jr., Moore, J.G.: Iceland chills a lava flow. Geotimes, **18**, no. 8, 14–17 (1973).
Zen, M.T., Hadikusumo, D.: The future danger of Mt. Kelut (eastern Java—Indonesia). Bull. volcanologique, ser. 2, **28**, 275–282 (1965).

Chapter 3
Hazards from Tsunamis

3.1. Causes of Tsunamis

Historical Occurrence

History contains many accounts of great earthquakes near coastlines being accompanied by destructive sea waves that have overwhelmed cities. A famous case occurred in the great Lisbon earthquake on November 1, 1755. A series of high ocean waves washed ashore along the west coast of Portugal, Spain and Morocco, helping to swell the dead in Lisbon (population 235,000) to some 60,000 people. The wave height in Lisbon reportedly reached 5 meters above high tide level. The sea waves swept across the Atlantic and were observed in Holland, England, the Azores and the West Indies. At Kinsale in Ireland, the harbor level rose rapidly about four and a half hours after the earthquake, breaking the cables of two moored ships.

This famous calamity was recalled when on February 28, 1969 a magnitude 8 earthquake occurred in the Eastern Atlantic perhaps close to the same origin on a submarine ridge at 36°0′N, 10°6′W, near Portugal. Casualties and damage were reported from Spain, Portugal and Morocco, and again, a tsunami was generated but in this case it reached a height of only 1.2 m at Casablanca.

Seismic sea waves, or *tsunamis* from the Japanese word, are often called, popularly, "tidal" waves but this is a misnomer, because they are not caused by tidal action of the moon and sun like the regular ocean tides. Rather, they are long water waves generated by sudden displacements under water, the most common cause of significant tsunamis being the impulsive displacement along a submerged fault, associated with an earthquake. Submarine fault rupture, for example, produced the tsunamis generated in the 1960 Chilean earthquake (see Section 3.3) and in the 1964 earthquake in Alaska. A list of great tsunamis is given in Table 3.1. Since 1596, Japan has suffered at least 10 disastrous tsunamis. In 1707, an earthquake in Japan generated huge waves in the Inland Sea which swamped more than 1,000 ships and boats in Osaka Bay.

Less important sources of ocean tsunamis are submarine landslides, as in Sugami Bay, Japan in 1933, and Valdez, Alaska in 1964 and avalanches into bays, as in Lituya Bay, Alaska in 1958. Volcanic eruptions may also be culprits. In the Krakatau caldera collapse in 1883, water waves said to be over 30 m high rolled in upon Java and Sumatra (see Fig. 2.18), killing about 30,000 persons. The waves were so large that the surge was detected as far away as the English Channel! Man-made tsunamis have been generated by the explosion of underwater nuclear bombs at Bikini and elsewhere (but not by the nuclear blasts in U.S. Aleutian tests on Amchitka in 1956, 1969, and 1971 underground).

Table 3.1. Great tsunamis of the world

Date	Source region	Visual run-up height (m)	Report from	Comments
15 B.C.	Santorin eruption		Crete	Devastation of Mediterranean coast
1 Nov. 1755	Eastern Atlantic	5–10	Lisbon, Portugal	Reported from Europe to West Indies
21 Dec. 1812	Santa Barbara Channel, Calif.	Several meters	Santa Barbara, Cal.	Early reports probably exaggerated
7 Nov. 1837	Chile	5	Hilo, Hawaii	
17 May 1841	Kamchatka	< 5	Hilo, Hawaii	
2 April 1868	Hawaiian Islands	< 3	Hilo, Hawaii	
13 Aug. 1868	Peru-Chile	>10	Arica, Peru	Observed New Zealand. Damaging Hawaii
10 May 1877	Peru-Chile	2–6	Japan	Destructive Iquique, Peru
27 Aug. 1883	Krakatau eruption		Java	Over 30,000 drowned
15 June 1896	Honshu	24	Sanriku, Japan	About 26,000 people drowned
3 Febr. 1923	Kamchatka	About 5	Waiakea, Hawaii	
2 March 1933	Honshu	>20	Sanriku, Japan	3,000 deaths from waves
1 April 1946	Aleutians	10	Wainaku, Hawaii	(See text)
4 Nov. 1952	Kamchatka	< 5	Hilo, Hawaii	
9 March 1957	Aleutians	< 5	Hilo, Hawaii	Associated earthquake magnitude 8.3
23 May 1960	Chile	>10	Waiakea, Hawaii	(See text)
28 March 1964	Alaska	6	Crescent City, Cal.	119 deaths and $104,000,000 damage from tsunami

In the open ocean, the length of tsunami water waves dwarfs all the usual sea waves, which are rarely more than 100 m from crest to crest, in that the distance between crests of a tsunami may exceed 100 km. On the other hand, the elevations of the crests of a tsunami are less than 1 m in height and are undetectable to ships in the open sea. The speed of travel decreases with decreasing water depth. Mathematically, the speed is equal to $\sqrt{gd}$, where g is the acceleration of gravity (980 cm/sec^2) and d is the water depth. In the deep oceans, such as the mid-Pacific, where d reaches 5 km, the corresponding tsunami speeds exceed 700 km per hour.

As the tsunami wave reaches shallow water around islands or on a continental shelf its speed decreases sharply. At the same time, the amplitude of the wave increases many times, sometimes to as much as 25 m.

The waves bend as the portion in shallower water moves more slowly than that in deep water. Just as with light waves, such refraction can turn the waves around capes and coastal promontories and bring them inside an otherwise sheltered harbor. Certain coastal regions appear to have a topography which produces a *wave trap,* where the energy of a long stretch of the wave may be focused by the configuration of the sea floor or be reflected from a coastline to concentrate more or less at one place, for instance at Hilo, Hawaii (see Plate 3.1).

With the approach of the tsunami, water level on the coast may first draw down somewhat, leaving reefs dry and denuded beaches covered with fish, At Hilo, in 1923, people unwisely ran out on the beach to pick up fish and were caught by the first onrushing crest. In other tsunamis the first water movement may be a rise. The series of waves in a tsunami may be separated by a few minutes or even an hour or more. The run-up and draw-down vary in elevation considerably from place to place along a coastline, depending on water depth and other factors, and people should not be lulled into thinking the danger

Plate 3.1. Seismic sea wave striking the pier at Hilo in the 1946 tsunami. Note the man in the path of the waves. He was never seen again. Photograph by an unknown seaman on the Brigham Victory. (Courtesy of R.L. Wiegel)

is past after the first great wave. In Hawaii, in 1946, at some places the greatest damage was done by waves as late as the eighth in the series.

The onrush of the tsunami sometimes becomes concentrated in a bay or river mouth to form a wall of water called a *bore,* where the breaking water front may cause great damage, as at Hilo in 1957. Bores may also arise, it should be noted, in constricted regions such as river mouths from the onset of ordinary ocean tides. The bore of the Bay of Fundy is a famous example.

Although most of the problem of geological damage caused by tsunamis relates to ocean shorelines, particularly in the Pacific, tsunamis are a hazard in most other oceans (particularly the Atlantic and Indian) and in many seas, and may even bring calamity when they are on quite a small scale. For example, the explosive eruption of Taal Volcano, in southwest Luzon in the Philippines, in September 1965 led to the generation of a small tsunami in Lake Taal. Water waves capsized boats loaded with residents fleeing from the central island and contributed to the 50 or more deaths, while waves swept the shoreline more than 4.7 m above the lake level.

Although tsunamis in the USSR usually occur on the Pacific shores of Kamchatka and the Kurile Islands, some visit the Black and Caspian Seas, where earthquakes are common. The two strong earthquakes of June 26 and September 12, 1927, in the Crimea region, generated small tsunamis, the peak wave heights recorded on tide gages being approximately 0.5 m.

Tsunamis and Seiches in Lakes and Reservoirs

Tsunamis may be generated by a sudden fall of rock and soil into a reservoir or lake. Such avalanches may be spontaneous, as, for example, the slide (see Chapter 4) which caused overtopping of the Vaiont dam and downstream deaths in Italy in 1965, or be triggered by a substantial earthquake, either natural or unnatural (such as a large explosion).

The most famous example of a landslide-induced tsunami occurred at Lituya Bay, Alaska after a magnitude 7 earthquake on July 9, 1958. The top photograph of the Bay, in Plate 3.2, shows trees growing down to the water's edge in 1954. The main avalanche triggered by the earthquake fell into the northeastern corner of the Bay and produced water waves which ran up the opposite sides to a height of 500 m. The amplitude of the surge was 60 m, stretching completely across the Bay. Boats were carried over trees 25 m high on the peninsula, and water velocities were so great that the vegetation was stripped from the foreshore, as shown in the bottom photograph of Plate 3.2.

This type of geological hazard has not been common but must be considered for the safety of the increasing population in the planning of national parks and marinas around lakes, bays and reservoirs.

Standing waves set up on an enclosed body of water such as a lake or reservoir are called a *seiche* (pronounced sāsh). Seiches are oscillations of the water surface that travel back and forth across it at regular periods determined by the depth and size of the water body. The name was popularized by F.A. Forel, who studied standing water waves on the Lake of Geneva about 1890, where the fundamental period of oscillation is 72 minutes. On a more complicated body

September, 1954

August, 1958

Plate 3.2. Legend see opposite page

of water there may be a number of important standing waves of different periods, as in San Francisco Bay, where the most important seiche motion has a period of 39 minutes.

Seiches are usually caused by unusual tides, winds or currents, but in certain circumstances are produced by earthquake ground motion. The shaking rocks the water back and forwards, setting up seiche waves, which may be destructive to facilities along a shoreline or, inland, may damage sewage and water storage basins.

Slow rhythmic crustal movements of great earthquakes have been known to generate seiches in lakes, rivers and harbors at great distances. The Lisbon earthquake of 1755 caused canals and lakes to go into palpable oscillations as far away as Holland, Switzerland, Scotland and Sweden, while on the Firth of Forth the water rose quickly 20 cm or more soon after the time of the earthquake and boats rocked at their moorings for perhaps 3 or 4 min. The 1964 Alaska earthquake led to the agitation of wells as far away as the coast of the Gulf of Mexico (4,000 km from the epicenter). Seiche surges along the Louisiana and Texas coasts commenced between 30 and 40 min after the earthquake origin time, or about the time surface waves (Love and Rayleigh type) were passing through the area. Minor damage was widespread, with parting of barge moorings in the Mississippi River.

Earthquake-Generated Tsunamis

Because of the great earthquakes that occur around the Pacific, this ocean is particularly prone to seismic sea waves, Dip-slip faulting (see Fig. 1.7) seems to be necessary, whether it occurs at the subduction zones of the island arcs (such as the Aleutians) or on the mid-oceanic ridges (such as between the Azores and Gibraltar in the Atlantic). Extensive Japanese studies, for example, have demonstrated that around the Japanese coast the focal mechanisms of earthquakes that produce tsunamis (*tsunamigenic* earthquakes) are generally dip-slip and, conversely, that strike-slip faulting is almost never accompanied by tsunamis. In agreement with these results, no tsunami was generated by the great 1906 San Francisco earthquake in which the San Andreas fault was horizontally displaced as much as six meters, although the fault runs partly under the sea.

When the ocean floor moves vertically, it forces water before it like a paddle. The submarine trenches off South America, Japan and the Aleutian Islands all face toward the central Pacific and this gives rise to the directional nature of many tsunamis generated in these seismic areas. A glance at a globe shows that the Hawaiian Islands lie close to the great circles at right angles to these generating areas (see Fig. 3.1), which is the main reason why the Hawaiian Islands are subject to considerable tsunami danger.

◁ Plate 3.2. Aerial photos comparing Lituya Bay, Alaska, before (September 1954) and after (August 1958) the sea wave produced by an avalanche. The destruction of the forest around the shoreline by the water wave is marked by the light areas in the later photograph. (After Miller, 1960. Courtesy of U.S. Geological Survey)

Just as an earthquake is allocated a Richter magnitude, so the Japanese have proposed a magnitude scale for tsunamis. The magnitude of a tsunami is defined similarly to an earthquake magnitude in terms of the logarithm of the amplitude of water-wave motion measured at a coast line on a standard tide gage, 10 to 300 km from the tsunami origin. One formula that is used is

$$m = 3.32 \log_{10} h,$$

where h is the maximum height in meters measured at the coast. Such formulae lead to rather variable and crude estimates, as sea waves are much affected by coast lines. A zero magnitude tsunami is a relatively unhazardous one, while a magnitude 3 tsunami would have a run-up as great as 12 m.

Table 3.2. Tsunami magnitude and run-up

Earthquake magnitude	Tsunami magnitude	Maximum run-up (m)
6	Slight	
6.5	−1	0.5–0.75
7	0	1–1.5
7.5	1	2–3
8	2	4–6
8.25	3	8–12

Empirical relations have been worked out between the size of tsunamis and those of tsunamigenic earthquakes. One relationship is tabulated in Table 3.2 and applies to shallow-focus earthquakes. If the focus is deeper, the tsunami size decreases. An empirical limit for the magnitude of tsunamigenic earthquakes and *disastrous* tsunamis (apart from special localized danger) is

$$M = 7.7 + 0.008\,H,$$

where H is the focal depth in kilometers of the earthquake.

The amount of energy in a tsunami ranges up to about 10^{23} ergs for the greatest and is usually from one to 10 per cent of the total energy in the earthquake that caused it.

3.2. Tsunami Risk

Damage and Run-up

From the historical record of tsunami occurrence and heights in a particular region, statistical curves of occurrence can be derived. For example, the Presidio tide gage at the entrance of San Francisco Bay has supplied records of tsunamis dating back to 1868. Nineteen tsunamis have been recorded, ranking in height from a few centimeters to over 1 m in the tsunami from the 1964 Good Friday Alaskan earthquake.

A *risk curve* is constructed by plotting the maximum wave height of a tsunami against the average number of tsunamis greater than a certain height (or magnitude) which occur each year. The curve for the San Francisco Bay entrance indicates that a height equal to or greater than 1 m can be expected once in 25 years, whereas in eastern Honshu, Japan, every 10 years a run-up of about 10 m is to be expected and the same run-up at Hilo would be expected every 100 years. If these statements are put more directly in terms of probable risk we can say for Hilo there is 80 per cent chance in 100 years of getting a wave run-up exceeding 10 m.

When a tsunami enters a confined space, the water level builds up. In the case of an enclosed bay, resonance and interference effects may complicate the level fluctuations as various tsunami waves arrive. Considerable information from Japan and the Hawaiian Islands indicates that, generally, coastal regions that face the area in which the tsunami originated suffer the highest run-up.

In San Francisco Bay, the largest amplitudes have been reported near the Golden Gate. The series of five tide gage records of the tsunami of November 4, 1952 are of critical import, showing that an amplitude of 1.2 m at the outer coast diminishes to 0.6 m at the Golden Gate (Presidio gage). At Benicia in the Carquinez Straits the waves have little amplitude. If we move towards the south Bay, at Hunter's Point and Alameda the amplitudes are less than 30 cm, whereas in the shallow water near Dumbarton Bridge the amplitudes are almost zero. This wave damping sharply reduces the likelihood of over-topping of levees and dikes around developed areas on the south shores of San Francisco Bay (see the Redwood Shores Study, Section 1.5). The evidence indicates that the extreme tsunami height in this area would be no more than 0.5 m above maximum high tide.

The 1964 tsunami which accompanied the Alaska earthquake produced the greatest damage along the California coast in the last 100 years (see Fig. 8.1), with loss totaling almost $10,000,000; of this, Crescent City in the northwest part of the coast (see Fig. 1.3) contributed a loss of $7,000,000.

The population of Crescent City at the time was approximately 2,500. Several waves struck the harbor and the third and fourth damaged the low-lying portion around the southward-facing beach and harbor. The third wave ran inland a distance of over 500 m to a height of about 6 m above mean-low low water. Thirty city blocks were inundated and the one-story wood frame buildings in the area were either destroyed or badly damaged; large logs, carried with velocities as high as 10 m per second through the streets, smashed against storage sheds and other buildings on the waterfront.

A major problem in many recent tsunamis has been the outbreak of fire. Oil storage facilities near the shore caught fire in Kodiak and Crescent City in the 1964 tsunami.

The Pacific Warning System and Hazard Mitigation

The most important scheme to prevent loss of life in the Pacific from tsunamis is the Seismic Sea Wave Warning System (SSWWS).

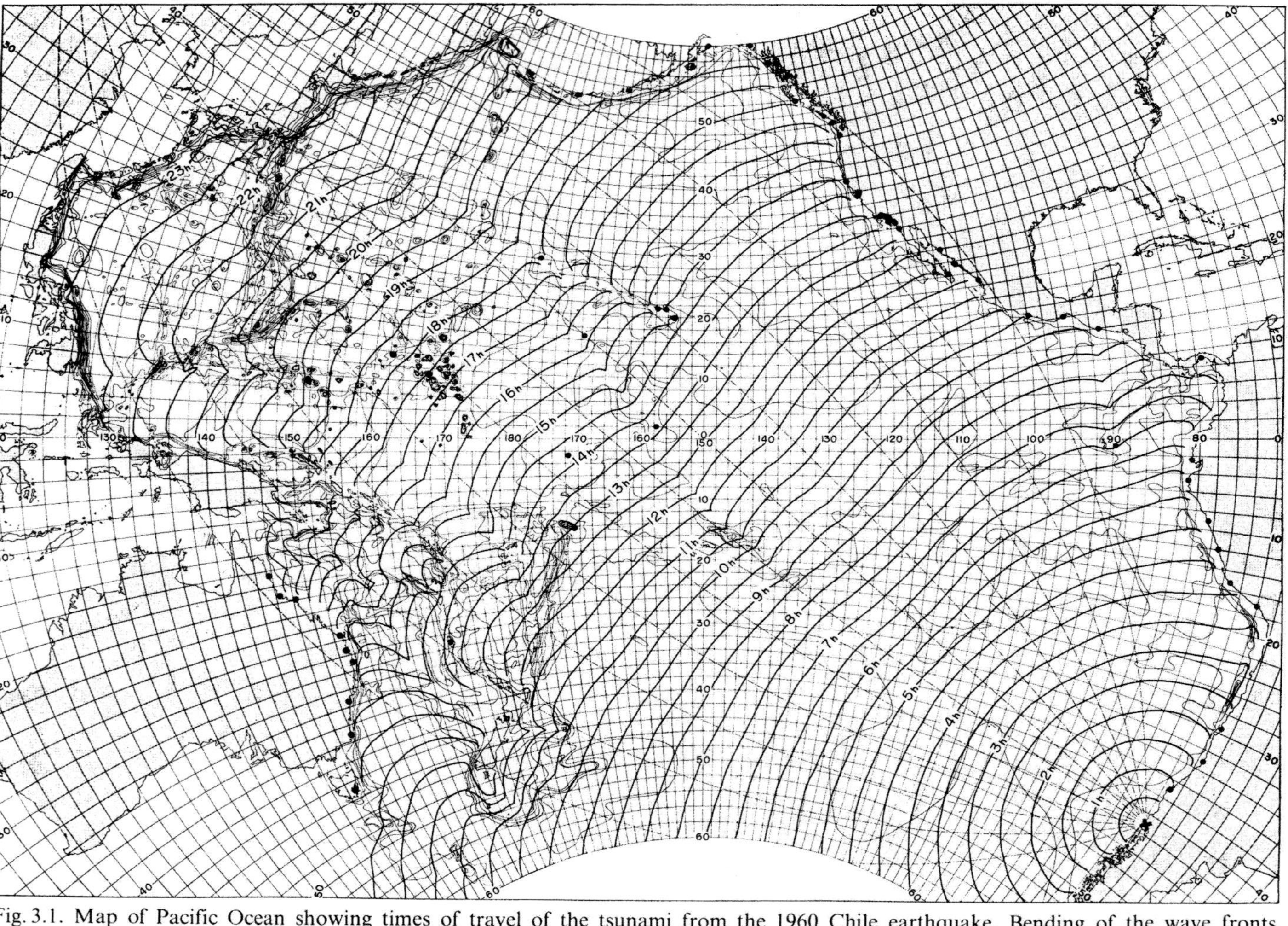

Fig. 3.1. Map of Pacific Ocean showing times of travel of the tsunami from the 1960 Chile earthquake. Bending of the wave fronts shows the effects of refraction from changes in ocean depth. Note the convergence of energy on to Japan. (Courtesy of Japan Meteorological Agency)

After the devastating Aleutian tsunami of April 1, 1946, military and civilian groups criticized the U.S. Coast and Geodetic Survey for the lack of warning in the Hawaiian Islands. The technical steps in the establishment of a warning system were undertaken and a rapid communication system set up through cooperation of the Armed Forces and the Civil Aeronautics Administration (later the Federal Aviation Agency). Another requirement was the preparation of seismic sea wave travel-time charts for different sources around the Pacific (see Fig. 3.1), with lines representing the time of travel from the epicenter of an earthquake to the position on the chart of the first sea wave.

The warning system was made up of a number of seismological observatories including Berkeley, California; Tokyo, Japan; Victoria, Canada and about 30 tide stations around the Pacific (see Fig. 3.1). Warnings were issued for places on the west coast of the United States and Hawaii. Beginning in November 1960, warnings were supplied also to Canada, Alaska and Tahiti. Later, information was supplied on request to Japan, Taiwan, the Philippines, Fiji, Chile, Hong Kong, New Zealand and Samoa.

Seismographic observatories all promised a 24-hour day alert to send information on the position and magnitude of large earthquakes around the Pacific to Hawaii. Tide gage reports could be obtained rapidly and any alert sent out to the cooperating countries by cablegram. The time of travel of a tsunami wave from Chile to the Hawaiian Islands is, for example, about 10 hours and from Chile to Japan about 20 hours. Under this system, therefore, there is ample time for alerts to be followed up by local police action along the coastlines so that people can be evacuated. A full tsunami alert was issued by the Honolulu Warning System in the 1964 earthquake 46 minutes after the earthquake occurred. It should be made clear, however, that the alert issued on the basis of a large submarine earthquake is directed only to civil defense agencies, police departments, and other parts of the warning system. The actual warning to evacuate shorelines is issued only later, on the basis of observation of actual big tsunami waves at stations nearer the source of the earthquake.

The extent and excitement of the system can be gathered from the following warning log entries at the Honolulu Observatory in 1964.

Seismic sea wave warning log Honolulu Observatory

G.M.T.	To	From	Event or remarks
03:44			Alarm sounded. Noted large earthquake. Scaled P Phase 034354. No other phases readable from visible record
03:58			Changing photo records
04:00	HDO	HO (Honolulu)	Advised district officer of earthquake
04:01	Macdonald	HO	Advised Dr. Gordon Macdonald, Acting Tsunami Advisor, of earthquake
04:05	Butchart	HO	Advised Col. Butchart, Vice Director C.D., of earthquake
04:07	HO	CINPAC	Called for information

Seismic sea wave warning log Honolulu Observatory (continued)

G.M.T.	To	From	Event or remarks
04:13	College Sitka Pasadena Tucson Berkeley	HO	Requested immediate readings earthquake of 0336Z. (Z indicates Greenwich Standard time, 10 hours ahead of Hawaian and Alaskan time)
04:16	JMA Tokyo	HO	Requested immediate readings earthquake of 0336Z
04:18	Guam Obs.	HO	Requested immediate readings earthquake of 0336Z
04:19	HO	Manila	P034813
04:23	HO	Hong Kong	P034334 S4755
04:25	HO	Kunia	Called for information
04:30	HO	PTO	Tides officer called for information
04:31	HO	Guam	EP034706
04:32	HO	Fleet WEA Central	Reported earthquake tremors in Kodiak from 0332Z to 0340Z. Kodiak tide gage damaged
04:33	HO	JMA	Matsushiro P034516. Sapporo P034428 S5120. Sendai P034459. Request your readings
04:36	HO	FAA	Reported communication cables to Alaska broken
04:38	HO	Berkeley	P034205
04:48	HO	Tucson	iP034328 ES4932
04:49	JMA	HO	HON iP034554. 1. Preliminary magnitude approx. 8
04:52			Determined epicenter at 61 N, 147.5W, H=033610. Near Seward, Alaska, magnitude 8
04:54	HO	FAA	Reported international airport tower at Anchorage demolished
04:59	HO	Pasadena	iP034350Z
05:01	HO	COMBARPAC	Tidal wave reported heading for Kodiak
05:02	NAVCOMMSTA Hono and FAA Hono for AIG 158	HO	Issued Bulletin No. 1 as follows: This is a tidal wave (Seismic Sea Wave) advisory. A severe earthquake has occurred at Lat 61 N Long 147.5W vicinity of Seward Alaska at 0336Z 28 March. It is not known repeat not known at this time that a sea wave has been generated

The drama continued in the Headquarters of the Hawaii Civil Defense Agency, which in those days had the responsibility of issuing the actual warning to evacuate coastal areas. But before a warning could be issued it was essential to determine that it was really necessary.

The immediate difficulty was that no information could be obtained from Alaska. As in many disasters, all communications with the stricken area had been demolished, so that all that was known for sure was that an earthquake of about magnitude 8 had occurred, probably beneath the ocean, and might have generated a large tsunami. Finally, after hours of tense waiting, contact was established with an amateur radio operator in Alaska who had a mobile transmitter in his car. The news was frightening; Cordoba had been terribly damaged by a great tsunami!

On the basis of this, the evacuation warning was about to be issued; but just then word came from Midway Island that the first waves had passed there and the water had risen less than 0.5 m. Perhaps the tsunami was local or highly directional and Hawaii was in no danger. However, little was known about behavior of tsunamis on tiny oceanic islands, and haunting uncertainty remained. So the evacuation warning was delayed, in hopes of obtaining more information. Then came news of great destruction in Crescent City, California; the wave had taken about four hours to travel from Alaska to California. Severe damage in Hawaii appeared very possible, and as avoidance of loss of life is even more important than avoiding false warnings, no chances could be taken. So the Commanding General decided, on the advice of the scientific advisor, to have the warning sirens sounded, giving ample warning time.

Coastal flooding was extensive in Hawaii; and in a few places, such as Kahului, on Maui, moderately severe damage was done, but at most places damage was minor and there was no loss of life.

In the 1964 earthquake, the city of Kodiak and Kodiak Naval Station were the only places in Alaska which received advanced warning of a tsunami. The U.S. Fleet Weather Central at Kodiak Naval Station, which participates in the Seismic Sea Wave Warning System by maintaining a tide station, provided the local warning, but the earthquake put this tide gage and all communication circuits available (except the telephone) out of action. At 04:10 h (GMT), after report of a 10-m tsunami was received from Cape Chinaak, the commanding officer of the Fleet Weather Central called the Armed Forces Radio Station and had a warning broadcast. The broadcast resulted in a smooth, prompt and complete evacuation of the Naval Station and the Federal Aviation Agency personnel on Woody Island. The evacuation at Kodiak was not as complete or well carried out.

In Canada, the Canadian dissemination agency had withdrawn from the SSWWS in July 1963 so no official warning of the tsunami was provided to Canada from Honolulu. While damage was done along the western Canadian coast, no fatalities occurred.

In California, the first advisory bulletin from Honolulu was received by the California Disaster Office at 05:36 h. The second advisory was received at 06:44 h and disseminated via the State Department of Justice teletype system to all sheriffs, chiefs of police and civil defense directors of coastal counties and cities. At Crescent City, where the bulk of the tsunami damage occurred, the county sheriff contacted the Civil Defense authorities upon receipt of the advisory at 07:08 h. Low-lying areas were warned and evacuation began immediately, but it was not complete. The first two waves caused minor flooding and afterwards many people returned to the area to clean up, a premature return which led to tragedy. The third and fourth waves caused most of the destruction and caught people who had returned and others who had not evacuated. Seven people, including the owner and his wife, returned to the Long Branch Tavern to remove money. Since everything seemed normal, they stopped to have a beer and were trapped by the third wave. Five of the seven were drowned when the boat in which they were attempting to escape was sucked into a creek by the recession, and smashed against the steel grating of a bridge.

There are two problems with a warning system. One is to give too many alerts in which no major wave is obvious to the population. If a warning is given too often, people become blasé and take no proper action. The second danger is that once tsunami alerts are given some of the population will from curiosity go down to the sea shore to watch the wave come in. Newspapers estimated that 10,000 people jammed the beaches at San Francisco to watch the arriving waves in 1964!

Apart from the tsunami warning system, the hazard can be mitigated by using adequate design for wharf, breakwater and other facilities based on techniques of coastal engineering. Often, however, zoning around coastlines is desirable to prevent building in the most low-lying areas where tsunamis are known to overwash the surface level. Sufficient information is nowadays usually available to allow local planners to make prudent decisions (see Chapter 8).

In passing, it might be noted that the recovery of a town can often be spectacular after the occurrence of a natural catastrophe. We may take again as an example Crescent City after 1964. Following the calamity, re-zoning of the town cleaned up the beach, turned the low-lying, poorly developed waterfront area into a public park and relocated businesses together on higher ground. Crescent City is now a safer and more attractive place to live than before the 1964 tsunami hit it. At Hilo, only certain sorts of commercial structures are permitted in the zone subject to tsunami invasion, and an attractive waterfront park has replaced wooden buildings demolished by the 1946 and 1960 tsunamis.

In the USSR, a tsunami warning service was established after the disastrous Kamchatka tsunamis of June 5, 1952. The basis is a network of seismotsunami stations at Petropavlovsk, Kurilsk and Uthno-Schkhalinsk supervised by the Hydrometeorological Service of the USSR. The system has proved valuable though short-comings have arisen from the isolation of the region and the closeness of the tsunamigenic zones. During the years 1958–1964 no cases of tsunami were missed, but five warnings were false.

A long-term plan of improvement of the Pacific Tsunami Warning System has been proposed, whereby the Pacific would be covered by an automated international warning system of 30 seismic stations and 120 tide stations. The vast amount of data generated would be handled by a geostationary satellite relay station. The geostationary satellite would maintain a fixed position 35,000 km above the equator on a selected meridian, receiving and relaying data from seismographs and mareographs independently of weather conditions and damage to submarine cables from earthquakes and ocean-floor mud flows. Whenever large signals were received by a computer in, for example, Honolulu in proper sequence an alarm would be activated.

Quite novel methods of detection of tsunamis may still come to light. After the 1964 Alaska earthquake exceptional very long-period (12 minutes) sound waves through the atmosphere were recorded on microbarographs at the University of California, Berkeley, and some other observatories. These pressure pulses were generated by the vertical displacements of the Earth's crust in the epicentral region. Because, for example, the tsunami did not arrive off San Francisco until about 2.5 hours after the seismic air wave, there is a possibility of using such waves to indicate that earth movements favorable to seismic sea-wave generation have occurred.

3.3. Tsunami Case Histories

Hawaii, April 1, 1946

The 1946 tsunami was the most destructive in the history of the Hawaiian Islands. Over 150 persons were killed, about 90 of them in Hilo, principally by drowning, and many more were injured, while damage to property amounted to $25,000,000.

The breeding ground was the Aleutian trench, where displacements of the sea bottom were associated with an earthquake of magnitude 7.5. The earthquake occurred at 12:29 h Greenwich Time, 3,500 km north of Hawaii. The first rise of the ocean around Hilo (see Fig. 2.7) appears to have started at 06:45 h, giving an approximate speed of the tsunami of 780 km per hour. The wave crests were recorded on tide gages in the Islands (see Fig. 3.2); the mareogram at Honolulu showed that the interval between the first and third wave crests was about 25 minutes, with shorter and less regular intervals between later waves, probably because of the interaction of different waves traveling in different routes around the Islands.

The waves that swept up on the Hawaiian shores varied greatly from place to place. In some places the water rose gently and most of the damage there resulted from the violent run-back of the water to the sea, whereas in most places the waves swept ashore with great turbulence, causing a loud roar and hissing noise. In some places the wavefront resembled a tidal bore, with a steep front and flat wave crest behind it. The energy of the waves was sufficiently great in some areas to tear loose coral fragments up to 1.3 m across and toss them onto the beach five meters above the sea level. The water outflow exposed coastal mud flats for distances of 150 m from the normal strand.

Most observers reported seeing the first tsunami effect as a withdrawal away from the shore. These observations contradict, however, the instrumental records at Honolulu and Waimea which show the first movement as a rise in level (see Fig. 3.2), but as the record indicates, the rise at Honolulu was small and it may have been overlooked at places where there were no recording tide gages. Some of the wave heights advancing across the reef appeared to be as much as 6 m and in some localities the sixth, seventh and eighth waves were said to be the highest.

The reef which protects the northern coast of Oahu reduced the intensity of the waves compared with the unprotected northern coasts of Molokai and

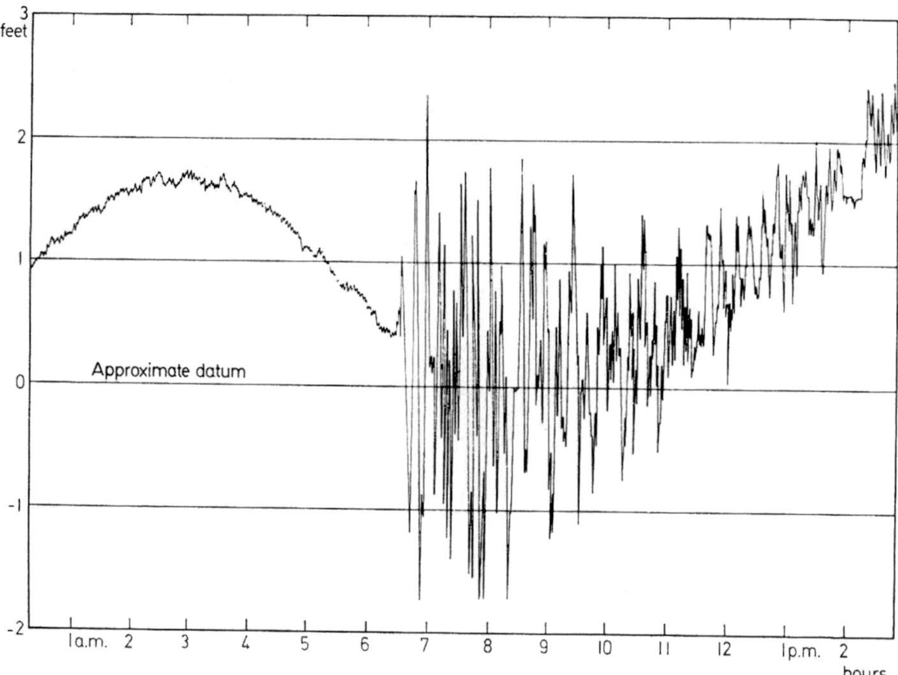

Fig. 3.2. Mareogram of the tsunami recorded by tide gage in Honolulu harbor on April 1, 1946. (Courtesy of USCGS)

Hawaii. There is not much evidence that the wave height was amplified near the heads of V-shaped bays, such as had been noticed previously in Japan and elsewhere. Along several small, steep coves it was found later that the water had risen to greater heights along the valley axis than on the beach at the end.

Structural damage occurred to buildings, roads, railroads, bridges, piers, breakwaters, fish-pond walls and ships; frame buildings along the shorelines suffered extensively, often by being knocked over by the force of the waves and sometimes by the destruction of their foundations. Some houses which were well built and tied together internally were moved considerable distances without suffering severe damage (as in earthquakes). Railroads along the northern coasts of Oahu and Hilo were wrecked, usually by destruction of the road bed and shifting of tracks. Several highway and railway bridges were destroyed, most of them by being lifted entirely from their foundations by the water buoyancy.

Damage was also caused by erosion of sand beaches and foreshores, and water flooding produced considerable loss in house furnishings and personal property.

Chilean Tsunami, May 22, 1960

The earthquake and tsunami arose from a megathrust of the ocean bottom along the coast of central Chile under the South American trench and the Andes.

The tsunami traveled over the Pacific Ocean, hitting the coast of Japan about 22 hours after the shock and spreading around the Pacific Ocean, creating damage at many places. (The path of the waves is shown as a refraction diagram in Fig. 3.1.)

The coastal deformation in this earthquake was enormous, extending from 38° to 43° south along the Chilean coast. Uplift of the ground was observed from 1 to 2 m at Isla Mocha and Isla Guafo at the ends of the fault area, while at Corral and Maullan, in the middle part of the region, the ground subsided about 2 m. The tsunami attacked the coast about 15 minutes after the earthquake with three large waves, which caused a great deal of damage by inundation, particularly in the cities of Saavedra, Mehuin, Corral, Maullan and Ancud. The dead were estimated at 909 and the missing at 834, of which many were due to the tsunami waves. When the tsunami reached the coastal regions of Japan a great deal of damage resulted, about 120 people losing their lives; thousands of houses were washed away or flooded and many hundreds of ships sunk or destroyed.

Since 1952 Japan has had a tsunami warning system, under the Japan Meteorological Agency, planned to warn of tsunamis generated by local earthquakes, using tsunami warning centers with seismographs and tide gauges on the islands of Japan. The aim was to give at least 30 minutes warning before the arrival of the first wave of the tsunami on the coast. However, the system was not designed to issue an alert for tsunamis from earthquakes in distant regions. For this reason, after the disaster of the Chilean tsunami, the Japanese agency entered into keen cooperation with the Pacific Sea Wave Warning System in order to prevent further loss of life from tsunamigenic earthquakes.

3.4. References

Adams, W.M. (Ed.): Tsunamis in the Pacific Ocean. Honolulu: East-West Center Press, 1970.

Anon: The Report on the Tsunami of the Chilean Earthquake, 1960. Japan Meteorological Agency Report **26** (1963).

Bolt, B.A.: Seismic Air Waves from the Great 1964 Alaskan Earthquake. Nature, **202**, 1095–1096 (1964).

Galanopoulos, A.G.: Tsunamis Observed on the Coasts of Greece from Antiquity to the Present Time. Ann. Geofis. **13**, 369–386 (1960).

Iida, K.: Magnitude, Energy and Generation Mechanisms of Tsunamis and a Catalog of Earthquakes Associated with Tsunamis, Proc. Tenth Pacific Science Congress, IUGG Monograph **24**, 7–18 (1963).

Miller, D.J.: Giant Waves in Lituya Bay, Alaska. U.S. Geological Survey, Professional Paper, 354–356 (1960).

Shepard, F.P., Macdonald, G.A., Cox, D.C.: The Tsunami of April 1, 1946. Bull. Scripps Inst. Oceanography, Univ. Calif. **5**, 391–528 (1950).

Spaeth, M.G., Beckman, S.C.: The Tsunami of March 28, 1964 as Recorded at Tide Stations. ESSA Technical Report, C. and G.S. **33** (1967).

Van Dorn, W.G.: Tsunamis, Advances in Hydroscience. **2**, 1–48. New York: Academic Press 1965.

Wiegel, R.L.: Tsunamis, Chapter 11 in Earthquake Engineering. N.J.: Prentice-Hall 1970.

Chapter 4
Hazards from Landslides

Landslides, ground settlement and avalanches interfere widely and persistently with man's activities. The scale ranges from slumps a few meters in dimension, causing only minor inconvenience, to rare gigantic slides or avalanches, kilometers in size, affecting sizeable populations. Because similar soil, rock or snow properties play a part in each process at all scales, they will be described together in this chapter, with the main reference to slides.

Landslides occur on slopes in a variety of geological materials and develop through a variety of mechanisms and causes. They may bring about major disrup-

Plate 4.1. Hillside failure, Hong Kong. (Courtesy of Fugro, Inc., Long Beach)

tions of towns and cities, communication systems, and large structures including dams and bridges (see Plate 4.1). Because of the variety of materials, mechanisms and rates involved, the ordering of aspects of ground movement is a challenge to the naturally-inclined classifier; in consequence, many schemes for classification have been adopted, and a few of their more important aspects are discussed below.

4.1. Classification of Landslides

Material

Perhaps the most obvious difference between various kinds of landslide is the nature of the participating material. Some slides are entirely composed of rock material, others of soils only, a few are mixtures of ice, rock and soil. Snow slides are called *avalanches* and will be dealt with separately in Chapter 6. Thus, it is possible to arrive at a primary classification approach in which the material type and state play the principal part. For instance, the slide consists of rock; the rock is granite, gneiss or sandstone; it was intact or jointed, fresh or considerably weathered, and so forth.

On the other hand, if the slide was formed from rock and mineral fragments, that is to say *soil*, then the soil type may be described. It may consist of very fine-grained materials such as clays, or the slide may be composed of the coarser materials, sands, gravels and so on; the soil mass itself may be dry or saturated, homogeneous or layered. Such a method of classification is not sufficient in itself, since the mechanics of a rock slide or a soil movement are not implicit in a description of the slide-material alone. To understand what has gone on, it is necessary to know something more than the material type. An obvious additional key criterion is velocity.

Velocity

For considerations of effects on people and engineering works, the speed at which a landslide develops and moves is the single most important feature. Few defenses are available against rapid and, therefore, generally unexpected movements that frequently result in damage and injuries. Landslides which move very slowly over periods of months to years seldom cause casualties and may also lend themselves to precautionary measures. Zoning arrangements can preclude the erection of unsuitable structures; highways and utilities can be re-routed.

Additionally, the velocity of the event is usually related to its predictability, for example, precursors in the form of cracks which form and widen over a period of time usually indicate a future landslide. But in the least stable areas, even preliminary cracks may occur so rapidly, or in such inaccessible locations, as to escape notice before violent motion ensues. In the case of slowly-developing land movements, distortion of engineering structures and ground surface features may be increasingly evident before the major motion occurs. In the latter case, structures can be evacuated or detours can be arranged for communication systems in advance of rupture.

However, even when the eventual ground motion is relatively slow, it may, if it occurs on a large enough scale, constitute a major, and possibly insolvable, engineering problem. At the present time, the solvability of most engineering problems is related only to cost and political considerations, and the cost of a field investigation and remedial treatment for a mass movement involving volumes of thousands of cubic meters of soil is large. In the case of the Portuguese Bend landslide of Los Angeles County, California, for example, an initial movement of about 10 m which occurred in 1956, has been followed by continuing creep of 2 or 3 square km of ground surface at a rate of several meters a year since that time. The mechanics of the motion have been investigated in some detail, and measures which might be expected to stop the slide would cost some $10,000,000, which a local government may find unacceptable for the stabilization of a largely residential area. The Portuguese Bend landslide, which is discussed in more detail in Section 4.6, therefore continues to move.

The velocity of a landslide is related to the material behavior and to the mechanism of its generation. For example, earthquakes are usually accompanied by landslides and rockfalls in mountainous terrain. In sufficiently steep or unstable topography (which, through the interconnected relations of tectonics, are usually highly seismic areas), such earthquake-generated landslides may be the principal agent of modification of the land surface. In the 1971 San Fernando, California, earthquake (see Section 1.4), it is estimated that some thousands of landslides and rockfalls occurred in the adjacent San Gabriel mountains. Rockfalls and slides were also a significant feature of the Inangahua earthquake in New Zealand in 1968.

A different generating process is rapid erosion of the toe of a slope by a river or by sea waves, which may also result in the occurrence of rapid land movement (see Plate 4.2). Apart from the immediate cause or trigger, faster movements occur on steeper slopes and are associated generally with rocks rather than with soils because of the more brittle properties of rocks. Under certain circumstances soils can also exhibit this behavior and develop very high velocities. The material properties are examined later in Section 4.2.

It is convenient to classify different landslide velocities in terms of time available for people to shelter, or take remedial action against the slide. This is obviously related to the scale of the event. In general terms, *rapid* landslides or rock avalanches occur in seconds to minutes; *intermediate* rates of movement may be scaled in minutes to hours, and *slow* landslides develop and move in periods ranging from days to years. However, if a very small landslide of a few meters in dimension occurs in a time spanning a few minutes, the rate of movement from a human point of view may be considered slow to intermediate. A landslide with dimensions of hundreds of meters to kilometers is difficult to escape even if it occurs over a period of minutes to hours, and such a large event would therefore require classification as rapid even though the material velocity was relatively slow.

An important component of a landslide and one related to its velocity is the distance that the slide progresses before it comes to rest. This gives rise to another method of classifying slides.

Plate 4.2. Landslide in forested slope, Oregon. (Courtesy of Brann Johnson, Calif. Institute of Technology)

Displacement

A wide range of slide distances has been exhibited in nature. These are naturally related to the magnitude of the event, but depend also on the mass of material and the velocity that it attains. Even relatively small slides may travel a distance of tens to hundreds of meters if there is sufficient water present to transform the sliding mass into a fluid. In arid regions, such landslides are frequently caused by sudden high-intensity rainstorms which saturate the thin layers of soil on steep slopes and generate mud flows. The accumulation of material at the head of undersea canyons can lead eventually to unstable conditions in which a submarine landslide develops. In some circumstances, the movement of the sliding mass becomes violent enough to entrain the ambient water so that the density of the mass becomes reduced. The form of the slide eventually changes to a flow of liquid somewhat more dense than the surrounding water. These slides, called *turbidity currents*, form an important mass-moving mechanism under water, and are thought to have developed during earthquakes in many offshore areas of unstable material. The slides commonly cause breakage of undersea cables following earthquakes.

The high velocity of subaerial soil and rock slides developed by earthquakes in suitable topography frequently results in very large displacements of the slide material. Again the turbulence of the initial motion can cause the entrainment of air or water as a pore fluid in the falling material, thus giving the movement the character of a flow, rather than a slide of solid material. If such a flow develops on a steep mountainside at or near the head of a valley that offers little or no obstacle to the falling mass, it may travel kilometers even when it contains a high proportion of solid material, as in Alaska during the 1964 earthquake and in Peru in 1970 (see Section 1.4).

Even very small amounts of movement in a sliding mass of soil or rock can cause substantial engineering or construction difficulties. In the preparation of the sides of a valley or canyon to form abutments for a dam, the fractures, joints or bedding planes in the rock may be such that movements of a few centimeters occur as the rock or soil adjusts to the new stress conditions. The opening of the joints provides a path for potential water flow when the structure is completed. When close control is exercised in engineering construction, these movements are usually observed and the engineer in charge has to decide what to do, whether to remove the sliding mass and replace it with a prepared fill of soil or concrete, or to fill the cracks with an impervious material of clay or concrete grout. Just such an extremely small movement occurred during preparations for the construction of the Baldwin Hills reservoir in California in 1950; the moving block was not excavated but an attempt was made to seal the open joints. The open fissures which undoubtedly remained have been considered to be a contributory cause to the failure of the reservoir in 1963.

In the preceding discussion, a number of terms have been used such as *flow*, *slide* and *block*, which imply a mechanism of movement. Consequently, another logical method of classification refers to mechanism.

Mechanism

In general, in a landslide, the sliding mass of material can be clearly distinguished from an underlying stationary bedrock or stable soil layer which does not take part in the motion. There is a sliding, or shearing surface across which the displacements occur, but on occasion, when the flow has more of the character of a very viscous liquid, it may be difficult to detect a distinct transition layer. The motion may instead die out gradually with depth. The first of these types of movement is a *slide*, the second a *flow*.

Depending on the character of the material taking part in the slide or flow and the presence in it of joints, cracks, or fissures, the moving mass may take one of a number of geometrical forms; the simplest is where the slide material has fairly large lateral and longitudinal dimensions in comparison with its thickness. Here, the topography and material properties are such that the surface of discontinuity on which sliding or flow takes place is essentially a plane, and in this case, the motion is one of translation of a block or blocks of material downslope. Sometimes only a single moving mass is involved if indeed the underlying slipping surface is flat. In more normal circumstances, the sliding surface is uneven, and the undulations in it cause the slipping mass to break up into a number of blocks separated from one another by fissures or shearing planes. The failure may be initiated at the toe of the slope, perhaps by erosion, so that a block first slides out at the toe of the slope, thus removing its stabilizing influence on uphill blocks, which then may in succession move downslope. This kind of slope failure, which occurs in a variety of forms, is called a *progressive* failure, and may take place either slowly or rapidly.

In relatively homogeneous masses of fine-grained soils in slopes of limited extent a different form of mechanism is common. Here, the sliding surface is roughly cylindrical or spherical in shape and the sliding mass of material rotates during failure. Commonly, as will be seen in Section 4.3, a center can be found about which the block rotates. In soils and rocks containing complex systems of joint planes and fissures, combinations of sliding surfaces may form, including portions of various components of the joint system. Thus, a series of joints lying essentially in a horizontal plane, intersected by joints at a steep angle to the horizontal, may give rise to a sliding mass in which a portion moves out horizontally, whereas an upper volume of material moves downwards along the more vertical joints. Under these conditions, another shearing plane must develop between the two portions of the sliding mass. In materials where less obvious jointing patterns may be present, or in materials which are *anisotropic* (that is, their material properties are different in different directions) very complex sliding surfaces can develop. Some of the features associated with these landslides are shown in idealized form in Fig. 4.1.

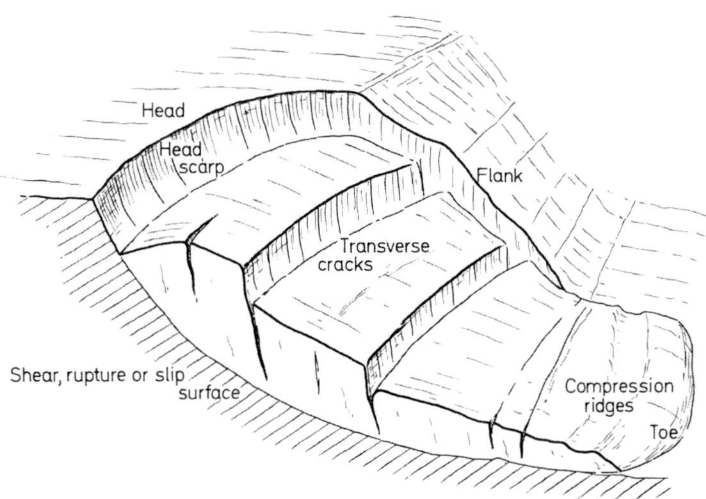

Fig. 4.1. Cross-sectional view of idealised landslide

Where the transition between the sliding mass and the stable underlying material is less clear, the situation is one of flow of material having the characteristics of a fluid flow, in which condition the velocity and displacement diminish gradually from the surface with depth. If the flow is taking place very slowly, the movement is referred to as *creep*, which is an extremely common phenomenon, probably occurring on every hillside in the world. The movement is imperceptible in the majority of these cases unless very precise measurements are performed.

When a jointed brittle rock forms a steep slope, the consequence of rock failure will be a fall rather than a slide; that is, the falling mass of material loses coherent contact with the stable unmoving base, and all or part of the slide may fall freely under gravity. This happens with rocks rather than in soils, (although the collisions between rock blocks reduce the material to a granular mass) and is therefore referred to as a *rockfall.* A number of disasters such as that of Elm in Switzerland in 1881, and Turtle Mountain in Alberta, Canada in 1903 have been occasioned by such rockfalls (see Section 4.6).

Without including too much detail, a simplified classification system is often of use in identifying the causes of a particular landslide, or in assessing the potential of a hillside or rock slope for sliding. Table 4.1 contains elements of the above discussion of various classification systems.

Table 4.1. Classification of Landslides

Movement	Material[a] behavior	Material type	Rate	Names
Fall	Brittle	Rock, Ice, Cemented soils	Rapid	Rockfall, Icefall, Soilfall
Slide	Unstable	Rock Soil Snow	Rapid to slow	Rotational slump Planar block glide Lateral spreading Slab avalanche
Flow	Stable	Rock fragments Sand, Silt Clay Snow	Rapid to slow	Rock flow Sand run Earth flow Mud flow Avalanche

[a] See Fig. 4.5 and accompanying text.

4.2. Mechanics of Landslides

General Considerations

A variety of mechanical causes underly landslide origins, and these must be understood in the assessment of a particular occurrence or hazard. The one common feature in static conditions is, of course, the presence of a slope, since a component of gravity force tangential to the surface is required to generate shearing stresses, and permit work to be done as the slide moves. The lateral accelerations of an earthquake can also supply this force and cause displacements to develop on horizontal ground surfaces on occasion.

A landslide occurs when the downslope component of the forces acting on the earth or rock mass exceeds the strength or shearing resistance of the material. The transition from a stationary hillside to an active slide implies that either the acting force or the soil or rock resistance has changed for some reason. By examining the components making up the acting force and the shearing resistance of the material, a variety of causes and mechanisms for slides can be established.

As a technical definition, a force is the product of the mass of the material and the acceleration to which it is subjected. In the case of a landslide developing under initially static conditions, the acceleration is due to gravity, which acts in a vertical direction and thus has components parallel and perpendicular to the slope. An increase in the force acting to cause failure may therefore be due to an increase either in the mass of the material or in the acceleration.

In the case of landslides not caused by earthquakes, there is, of course, no increase in the acceleration due to to gravity, but there may be an increase in the downslope component of the gravitational acceleration because, for example, the slope has become steeper. The steepening can develop through erosion of material at the bottom of the slope by one of several agencies, addition of material at the top of the slope either by natural or man-made developments, or by local or regional tilting of the Earth's surface. An increase in the mass of material comprising the slope can take place by deposition of soil or rock on the slope surface. The mass density of the material may be increased as a consequence of rainfall or water infiltration from other sources.

On the other hand, a slope failure may develop without any change in the forces acting to cause failure, if the strength of the material supporting the slope decreases for any reason. That is to say, the material forming the slope is held in place by the shearing resistance of the underlying soil or rock layers. This resistance may be altered by chemical processes as a consequence of weathering, or decreased for physical reasons such as the increase of water pressure in cracks or fissures or in the voids of the soil material. In certain types of brittle material, it is possible for the shearing strength to go down as the soil or rock shears. This is unstable behavior. In this case, initial distortions cause a reduction in strength, which leads to further distortions and still lower strength. A progressive failure develops, and the final landslide will generally be preceded by displacements which can be observed with sensitive instrumentation.

Load Changes

An alteration of the downslope component of the weight of a landslide mass can arise from either natural or man-made causes. In nature, processes of slope steepening are common, and develop most often as a consequence of water movements. The most usual form is the removal of material from the base of the slope by wave action or through erosion by rivers. Since, as noted below, the material at the base of the slope helps to stabilize the soil or rock mass, its removal leads to instability. The movements which develop are generally gradual but can, on occasion, take place rapidly during periods of high flood levels or storms. Where slides are generated in this way, it is usual to find that many successive failures of the same region occur. The slide material from one slope failure temporarily stabilizes the area, but is removed by erosion so that a further instability develops, and this debris is, in turn, removed. Areas in which this kind of mechanism has been going on can thus be easily recognized (see Plate 4.9). The scarps where subsurface material is revealed by the removal of the slide debris have a fresh appearance and are discontinuous with the general run of the hillside.

Human activities account for many landslides. The construction of highways, hillside housing developments, dams, reservoirs, drainage and utility structures normally involve the movement of substantial amounts of soil or rock on slopes. If the operation consists of addition of material to the top of the slope or the removal of soil or rock from its base, then the slope is nudged toward failure. In extreme cases, landslides take place during or immediately following the construction process, but the effects of construction can be very subtle in many circumstances. The placement of fill material for a road across a hillside may not, through its own weight, cause the hillside to fail, but may interfere with the natural régime of water flow and drainage through the soil or rock. In this way the weight of the material is perhaps increased or the water pressure is changed in the pores of the soil or in the rock interstices. Either consequence can give rise to a slide which occurs months to years after the completion of construction.

When a dam is built across a valley and water impounded behind the dam to form a reservoir, slope failures along the valley sides can develop, caused by the saturation and consequent weakening of the material at the base of the slopes, or by the erosive action of waves at the toe of the slopes if they are not protected. Material can be removed in this way in minor amounts until a large portion of the valley slope becomes unstable. The presence of the lake will also interfere with the hydrologic regime of rainfall, absorption and runoff (see Section 7.2); the effects of changes such as these on a previously stable terrain are generally unfavorable.

Seismic Effects

During an earthquake the ground shakes in all directions (see Chapter 1) producing accelerations as high as 0.5, or more, times the acceleration of gravity (g) in both horizontal and vertical directions in the area experiencing the strongest ground shaking. When the product of these accelerations and the mass of material in a potential landslide is considered, it will be seen that horizontal transient forces of the same order of magnitude as the weight of the sliding mass are involved. Although these dynamic accelerations last for a short time, they can have an enormous effect in causing sliding on marginally-stable slopes, with

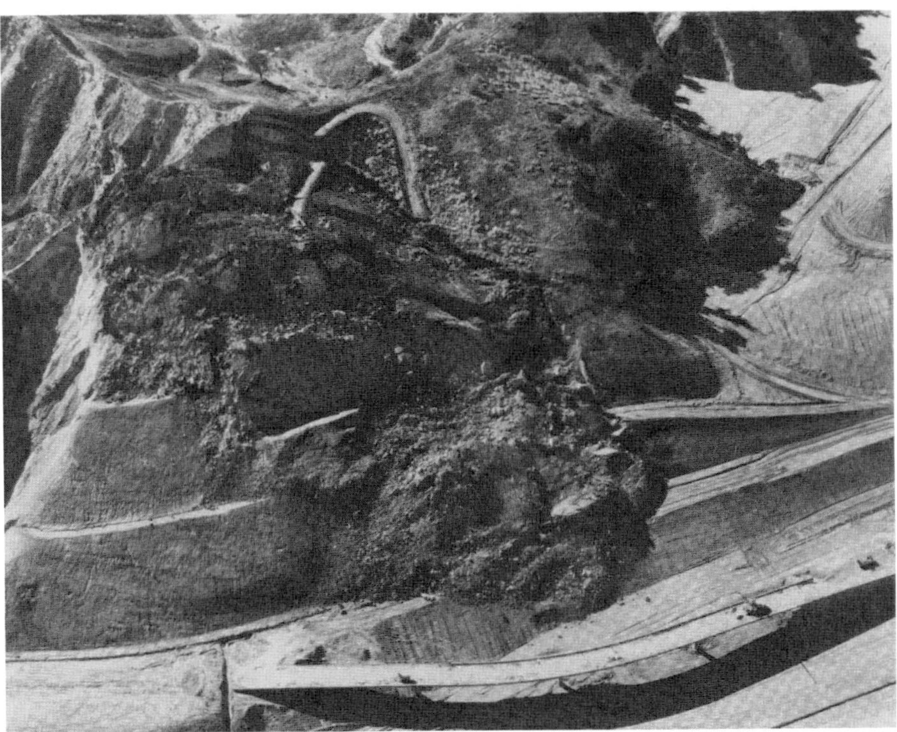

Plate 4.3. Landslide in cut slope during San Fernando earthquake of February 9, 1971. (Courtesy of State of Calif. Dept. of Public Works)

the result that a feature commonly associated with earthquakes is landsliding on a scale depending upon the topography in the epicentral area. During and following the 1971 San Fernando, California earthquake thousands of landslides and rockfalls occurred in the San Gabriel mountains and caused a prominent dust cloud over the strongly shaken area for days (see Plate 4.3).

Besides the direct effect of the earthquake accelerations on the forces acting to cause failure, the vibrations may also play a part in reducing the strength of the soil or rock mass along the surface where resistance to the slide is developed. During strong ground shaking, both normal and shearing stresses along the potential slide surface oscillate in amplitude, and shear-stress reversals may occur. It has been observed in laboratory tests of the shearing strength of soil that under repetitive shear-stress loading certain soils can develop a somewhat lower shear strength than when they are loaded once only to their maximum value; thus, the dynamic shear strength is lower than the static value. Consequently, a potential landslide mass which is stable under static conditions may fail after a number of cycles of vibration caused by an earthquake. In extreme cases, usually involving sands saturated with water, the interaction of the cyclic shearing stresses with the volumetric characteristics of the soil in the presence of the water can result in *liquefaction* of the soil during the earthquake. This is a state of the material in which it behaves for a short time as a dense fluid rather than as a solid mass. Liquefaction will be discussed in a little more detail in the next section.

It is, of course, a characteristic of an earthquake that its onset is sudden and the landslides which it causes occur rapidly. Although certain masses of soil or rock are brought, during an earthquake, to a condition of marginal instability such that failure may be triggered by some other event weeks to months afterwards, the principal effects of an earthquake in terms of landslides are immediate. In terms of Table 4.1, the landslides which develop fall into the classification of rapid to intermediate. Usually, all the major events occur within a few minutes after the termination of the earthquake. The effect of these landslides is intensified by their numbers and by the general disruption caused by the earthquake at the same time, with the consequence that earthquake-generated landslides tend to be more upsetting to human activities than the isolated events which occur under static conditions.

The earthquake constitutes a massive full-scale test of all potentially unstable soil and rock structures over hundreds to thousands of square kilometers. Since it is impossible in any seismically active region thoroughly to investigate all of the potentially unstable areas, many of the effects of an earthquake will always be unexpected.

Shearing Strength and Pore Pressure

The changes in forces acting to cause landslides are relatively easy to understand with a minimal knowledge of applied mechanics, but an understanding of the shearing strength of soils and rocks requires a brief summary of the mechanical behavior of these materials. First, it is necessary to describe the effect of the pressure in the pore fluid on the behavior of a granular material. Fig. 4.2(a) shows a cross-sectional view of an assemblage of soil grains; there are spaces between the grains, referred to as the pores or voids. In considering a rock mass, Fig. 4.2(b) might represent a cross section of a rock specimen in which the pore spaces consist of fissures or cracks. The void space is filled with gas

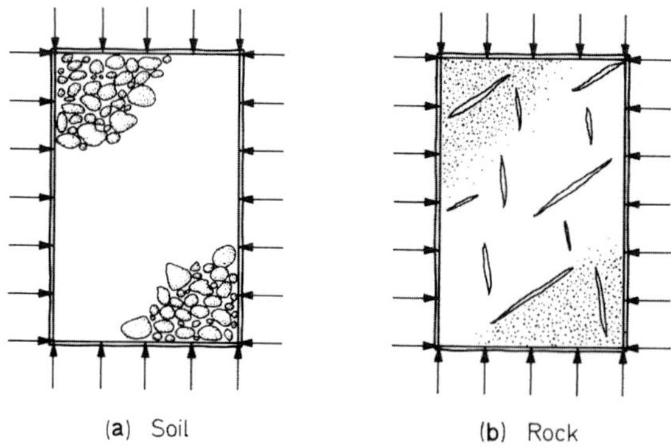

Fig. 4.2. a and b. Material representation, (a) Soil, (b) Rock

(air), liquid (usually water), or both, under some hydrostatic pressure. If the soil or rock specimens of Fig. 4.2 are placed in a flexible container such as a rubber membrane, on the outside of which a pressure can be applied, as shown, the effects of changing this pressure can be examined. In particular, the pore fluid pressure and the total volume of the sample can be measured.

If the pores in the soil or the cracks in the rock are filled with a highly compressible gas (air), then the application of an external pressure to the soil or rock mass will cause a volume change of the solid structure but little alteration in the air pressure. This occurs because of the great disparity between the compressibility of the soil or rock structure and that of the air, which means that the soil grains or rock fragments will be pressed more closely together, and will be stronger under the applied pressure.

However, if the pore space is entirely saturated with water, a change in the external pressure applied to the membrane, will result in a change in the pressure of the water in the pores. The amount of the water pressure increase will depend on the relative compressibility of the solid and liquid components. If a soil structure, such as in Fig. 4.2(a), is saturated with water, then it will usually be the case that the water is relatively incompressible compared to the soil, so that the application of an external hydrostatic pressure will cause the pore water pressure to rise by the same amount. Because little change can develop in the volume of the soil sample due to the incompressibility of the water, its strength in this case is unaffected by the increase in external pressure. On the other hand, in the rock of Fig. 4.2(b), the relatively small amount of pore space present causes the compressibilities of the solid and liquid components to be fairly similar. Therefore the increased external hydrostatic stress is shared and causes some increase in the hydrostatic pressure in the rock, and some increase in the pore pressure in the fissures. The strength of the rock is increased in proportion to the pressure increase felt by the solid component. In either case, the strength and other material properties of the bulk material depend solely on the amount of stress applied to or felt by the solid component. This stress or pressure is referred to as the *effective* stress.

If there is a hole in the membrane, giving access to a region of low water pressures, water will flow out, reducing the pore pressure generated by the application of the external stress and permitting the solid structure to compress. In nature, the function of the hole is supplied by the permeability of the surrounding soil or rock mass. Thus, the fairly rapid application of hydrostatic stress to a saturated soil or rock will normally result in a change both in the pore pressure and in the normal stress in the solid component; the change in the pore pressure will, in the course of time, decrease to zero as the excess pore water drains away from the stressed area through the adjacent material.

An interesting and crucial effect occurs if the saturated soil or rock is stressed by shearing stresses rather than by normal stresses. In this case, for low values of shearing stress, the material distorts without change in the pore pressure. As the shearing stresses are increased and reach values near the failure shearing stress for the material, the solid grains tend to move or slip over one another so that the volume of the structure tries to change, a phenomenon known as *dilatancy*. If the material is soil in a loose condition, volume contraction occurs (negative dilatancy); if, however, the soil grains were tightly packed originally, the application of high shearing stresses tends to increase the soil volume (positive dilatancy). In the case of fractured rocks, the application of a high level of shearing stress tends to open up the fractures so that the rock mass tends to increase in volume like a dense soil. If the void space contains only air, the change in volume of the rock or soil mass on application of the shearing stresses can occur without pore pressure effects. Conversely, when the soil or rock is saturated with water, the shearing-stress-induced volume change is inhibited, and, as a consequence, some pore pressure is generated in the pore fluid. Shearing of a loose saturated soil causes an increase in pore pressure; in a dense soil shearing causes the pore pressure to drop, when no drainage occurs. In the case of rocks, the lack of drainage will also cause a drop in the pore pressure as shearing develops.

Two of the pieces of equipment which are commonly used to carry out tests on soil or rock materials are illustrated in Fig. 4.3. In Fig. 4.3(a) is shown the "direct shear" test. In this, the soil or rock sample is contained within a box which is split lengthwise so that the two halves of the box are not in contact with each other except through the specimen. To this box is applied a normal load N. Then a shearing force T is applied to the ends of the box to displace the two halves of the box and the test material with respect to one another. As the force T is gradually increased, so the test material deforms until, at a critical value of T, failure will take place in the soil, and the two halves of the box will then slide with respect to one another without further increase in T. The material in the box may be free to drain or not according to the rate of loading and its own permeability.

If drainage is permitted to occur by the nature of the test conditions, each application of a normal load N will cause the specimen to decrease slightly in volume to a denser condition. To each of these normal loads will correspond a value of the shearing force required to fail the sample. What is happening is that drainage permits the full load N to be effective in pressing the solid

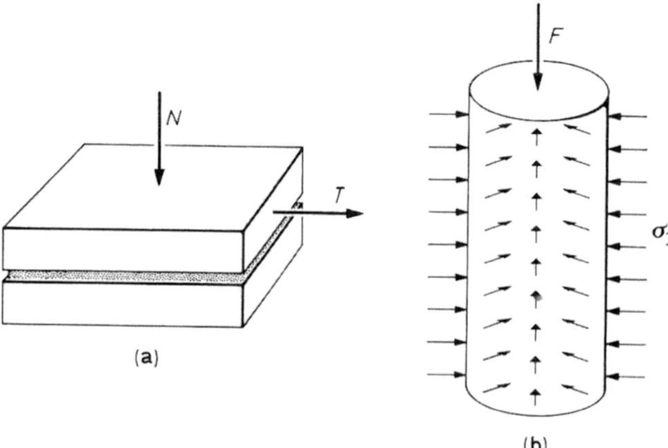

Fig. 4.3.a and b. Test equipment, (a) Direct shear test, (b) Triaxial test

components of the soil or rock together so that the shearing force at failure is proportional to the load N. When the results of actual tests are plotted, as shown on Fig. 4.4, it is found that a straight line (A on the figure) can be drawn approximately through them. This line may or may not have an intercept on the shearing load axis. If such an intercept exists, it is referred to as the *cohesion* of the material; in the absence of such an intercept the material is referred to as *cohesionless*. The angle made by the straight line with respect to the horizontal axis is known as the *angle of internal friction* of the material; the tangent of the angle is the *coefficient* of *friction*.

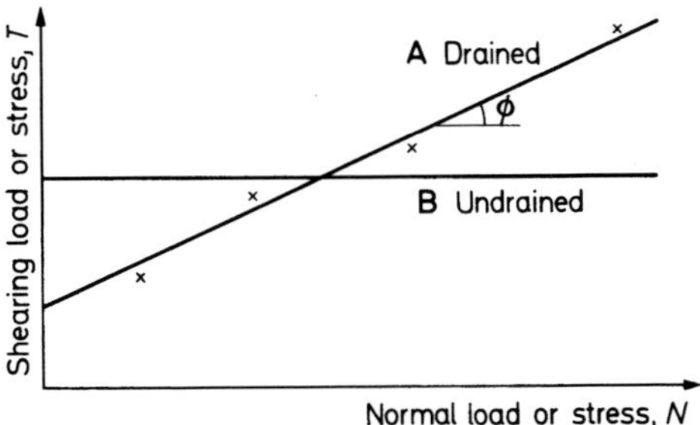

Fig. 4.4. Shearing strength of soil and rock

In the drained condition the full force N is effective in developing the strength of the soil or rock, and it is this load which gives rise to the effective stress described previously. If the material is not permitted to drain under the application of the normal load or if drainage is only partial, then not all of the normal load N will be seen by the solid component as an effective stress, but, instead some pore pressure will be generated. In that case, the shearing force T required to cause failure of the sample will be proportional to the effective force acting and not to the total applied force.

For example, if a sand specimen in the apparatus of Fig. 4.3(a), saturated with water, is permitted to drain under a normal load N_0, this normal load is then an effective load. A tangential load T_0 will be required to fail the sample under this load N_0, where $T_0 = N_0 \tan \phi$, if ϕ is the angle of internal friction of the soil. If now, instead of shearing the soil at a load N_0, an additional load ΔN is superimposed *without drainage being permitted* and next a shearing force is applied, it will be found that failure will occur at the same value of shearing force T_0 as before. This happens because the load increment ΔN resulted in no increase in the effective stress in the soil, and therefore no increase in strength developed. This will be true whatever is the value of the load increment ΔN, providing no drainage occurs. The consequence of this behavior is shown in Fig. 4.4 by the horizontal line B for which the *total load* $N = N_0 + \Delta N$ is shown on the horizontal axis. The material apparently exhibits some cohesion which, as can be seen, is due in this case to the lack of drainage.

A real cohesion is developed in intact rocks because of the bonds between the mineral molecules, and in clays that have been subjected to effective stresses or loads higher than those at which the test is carried out. In the latter case, the higher effective stresses cause bonds to develop between the clay particles which do not break when the effective stress is removed. In effect, the clay "remembers" the highest effective stress to which it has been subjected, and the bonds developed by this stress have to be broken to shear the clay at any lower level of applied stress with or without drainage. Consequently the shearing force or load at failure reflects the highest effective stress applied to the clay. In particular, if clay is sheared under no confining stress at all, its shearing strength is essentially due to cohesion developed by the formerly higher stresses.

The result that the deformation and strength behavior of a soil or porous rock depend on effective stress only is a very important one with broad implications in the fields of soil and rock mechanics, and more recently in studies of earthquake mechanisms (see Section 1.3).

The effective stress in a material is not a measurable quantity in the sense that a gage can be employed to indicate it. However, instrumentation can be used to determine the *total normal stress* acting at a point in a particular direction and other equipment can be employed to measure the pore water pressure. Subtraction of the pore pressure from the total stress then gives the effective stress. It should be noted that the pore pressure is subtracted only from the total normal stress to give an effective normal stress. The *shearing stresses* which are developed in a soil mass are unaffected by the pore pressures, since the pore water cannot take shearing stress. Thus shearing stresses are always effective stresses. Under many real-life conditions in a soil or rock mass the total stress (due to gravity for the most part) may remain unchanged while the pore pressure alters as a consequence of drainage or other effects. In this circumstance, the effective stresses in the mass will also change as the pore water pressure changes and so deformation and strength will alter accordingly. It will be realized that pore pressure changes, unlike total loads, are unseen, and thus a soil or rock mass can be undergoing subtle internal changes without obvious visible effects, unless careful measurements are made. Changes in the pore pressure and thus effective stress are, in consequence, the cause of many sudden and unexpected landslide occurrences.

Another important aspect of material response is its stress-strain or force-displacement behavior before, during and after failure. This can be best illustrated with respect to the other common soil and rock test shown in Fig. 4.3(b). Here, a cylindrical specimen of the soil or rock is jacketed in a rubber or plastic sheath. Loading plates are applied to the top and bottom circular surfaces and the specimen is placed in a chamber containing water or oil through which an ambient or hydrostatic stress can be applied all around the sample. The rubber or plastic jacket prevents the liquid in the chamber entering the specimen. Axial loads can be made to act on the top loading plate. This apparatus is used because more effective control of the drainage conditions can be exercised than in the direct-shear test of Fig. 4.3(a) and the loads are distributed more uniformly throughout the sample. A test is usually initiated by applying an external hydrostatic stress and permitting or preventing drainage depending on the conditions to be simulated. If drainage is permitted, an appropriate time interval must be allowed to let the pore water pressure dissipate; the time depends on the permeability and compressibility of the sample. When the conditions are right, the sample is loaded axially and the resulting axial displacements or strains are measured. Usually either the volume change, if the test is drained, or pore pressure, if it is undrained, is also recorded.

A variety of load-displacement relations is observed depending on the material type. If the sample consists of rock and the ambient stress in the test is low to represent conditions near the surface of the Earth, a *brittle* rock behavior will usually be found corresponding to curve *A* of Fig. 4.5. Here the rock exhibits

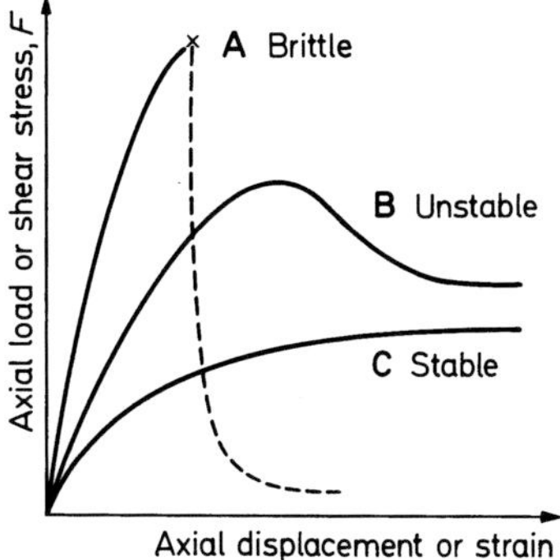

Fig. 4.5. Load-displacement behavior of soil and rock

a deformation gradually increasing up to a peak value of the axial force at which it suddenly splits or breaks. If it is entirely unconfined, it breaks up into fragments. With a small confining pressure applied to the outside of the rubber membrane, the brittle behavior is manifested by a sudden and large drop in the shearing force to some much smaller value. A similar but less extreme type of behavior is also demonstrated by a rock under much higher pressure, by a dense sand or gravel, and by a clay which has been subjected to a large effective load in the past. In these cases, the displacement increases to a peak value of axial force after which the force decreases to some final constant level at which displacement continues steadily. There are two failure values associated with this kind of behavior, (a) the peak value, and (b) the ultimate or residual value finally reached by the distorting material. This curve is shown as *B* in Fig. 4.5. When the specimen is rock at very high confining pressures, or is a loose sand or gravel, or is a clay which has only been subjected to modest past effective loads, the behavior shown as curve *C* in Fig. 4.5 is observed. Here the load and displacement both gradually increase until a peak value of the load is reached at which displacement continues. In this case, the peak load and the ultimate load have the same value. In mechanics usage, curves *A* and *B* of Fig. 4.5 are referred to as *unstable* and curve *C* as *stable* for reasons which are discussed later.

In rocks the higher confining stresses which exist at greater depths in the Earth's crust are accompanied by increases in temperature, so that in a proper test to simulate conditions at depths of a few kilometers, the rock sample should be heated. When this is done, an important variation in behavior is found.

For a certain range of effective pressures and temperatures, the rock deforms essentially as curve *B* of Fig. 4.5, but the fall-off in the peak stress is much more sudden, and, if the axial displacement of the loading plate is continued, the load will build up again to another, lower peak value. This, in turn, is followed by a sudden load decrease, and another build-up. The phenomenon is commonly observed in frictional sliding, and is referred to as *stick-slip* behavior. Since it only occurs in a particular temperature and pressure range, it appears to explain why earthquakes are generated only over a certain depth interval in the Earth's crust (see Section 1.3). At shallow depths, the rock fractures brittlely as shown by curve *A* of Fig. 4.5, and at very great depths, the purely plastic behavior of curve *C* is manifested.

If conditions conducive to a landslide develop in a slope of loose sand or soft-to-medium stiff clay by, for example, a gradual increasing of the load in the potentially sliding mass, a stage will be reached where the shearing stresses along the potential sliding surface are equal to the strength of the material. These materials have the force-displacement characteristic shown as *C* on Fig. 4.5, and thus sliding or displacement of the sliding mass will gradually develop as the force approaches the peak value. In this case it can be seen that the displacements which occur do not result in a decrease of the *shearing strength* of the material. However, when sliding occurs, an essentially automatic reduction of the acting *force* takes place since the downslope motion transfers some of the load to the toe of the slope and relieves the load at the top. Thus, after perhaps a relatively small movement, the slide is stabilized because of the stress-strain relations of the material, and the slight geometrical change caused by the displacement. The behavior is stable, because a small change in load causes only a small displacement.

Now consider that the same slope consists of a soil with the stress-strain characteristics of curve *B* in Fig. 4.5. In this case, when the peak shearing strength of the material in some sliding zone is exceeded, displacements will again occur. However, the curve *B* shows that the *shearing strength* is reduced towards the ultimate value by this displacement. Depending on the degree of unstable behavior evinced by the material, the shearing strength will be reduced more than the load so that sliding progresses and the ensuing displacement along the sliding surface results in another reduction of shearing strength. As a consequence the slide mass undergoes large movements and may achieve velocities large enough to carry it a substantial distance.

In the extreme case of brittle behavior as exemplified by curve *A* of Fig. 4.5, the development of loads big enough to cause the material in the slope to fail results in a complete loss of strength of the material, so that the landslide mass breaks free, in effect, from the surrounding material and can slide with little resistance. Landslides in such materials can attain high velocities and travel considerable distances in appropriate topography. The velocity scale of Table 4.1 is directly related to these material characteristics. The rapid rockfalls or landslides occur in materials with a generally brittle to unstable behavior; slides with intermediate velocities of minutes to hours and generally of a progressive nature, occur in materials which are unstable; and slow or creeping landslides are characteristic of materials with a stable stress-strain relationship.

Pore Pressures

The effect of changes in pore pressures in the development of landslides is important enough to require a separate discussion. Pore pressures are developed in a saturated soil or rock mass by any change in the loading or stressing of the material. If a natural stable slope of a saturated soil exists and a load consisting of fill for a highway embankment, for example, is placed at the top of the slope, pore pressures will be developed in the slope material. Immediately after application of the load the forces tending to cause failure are raised, but effective stresses in the soil and therefore the shearing strength of the soil remains unchanged from the condition before load application. If the load exceeds the resistance of the soil, the slope will fail immediately, but if the load is not sufficient to cause slope failure as soon as it is applied, the excess pore pressures will dissipate in time and the effective stresses will increase. The soil's strength, and the safety of the slope therefore, increase with time also. In general, a slope in such a condition will either fail immediately on application of the load, or if it does not, it is safe.

Let it be supposed, on the other hand, that the load is not applied at the top of the slope, but on a level surface some distance back from the edge. Initially the stability of the slope is unaffected by the load, which however does generate pore pressures in its immediate vicinity. These pore pressures decrease by drainage to the adjacent free surfaces, including the slope. The slope material may therefore experience an increase in pore pressure, without change in the total load, some time after load application. If the slope's stability was initially marginal, the subsequent decrease in effective stresses and strength could cause it to fail. Any mechanism which increases pore pressure represents a hazard. In one case in practice, piles for a bridge structure were driven at a distance from a slope which failed after a lapse of time. The pore pressure increase had resulted not so much from the load on the piles, but from the disturbance and distortion they caused in the underlying clay soil. The clay in this case was of the *quick* type referred to in a case history in Section 4.6.

Pore pressure changes can occur in soil or rock as a consequence of other conditions external to the region under consideration. A slope that has been stable for a long time under existing load and water-table conditions can be rendered unstable by a change in the drainage patterns in the surrounding area. The change can occur naturally as a consequence of seasonal movements of the water table, or may be the consequence of man's activities. The construction and occupation of houses at the top of a hillside area can cause changes in the groundwater conditions through the use of cesspools instead of appropriate sewage or drainage collection systems, by watering of gardens, and through changes in natural runoff and drainage patterns of the hillside because of the presence of paved streets and roofs. In Los Angeles, California, failures of hillsides on which houses have been constructed and occupied has been not uncommon. In many cases, these failures, although difficult to investigate, most probably are due to changes in the amount of water present in the normally arid hillsides.

Another mechanism exists by which pore pressures can be increased, and that is through earthquake vibrations. If a dry granular or sandy soil is subjected to repetitious loading, it will be found that even a relatively dense material will decrease in volume with the number of cycles of load application. Consequently, if the same soil is put in the test configuration of Fig. 4.3(b), drainage is prevented, and axial load applied cyclically to cause alternating compression and extension of the soil sample, the pore pressure will gradually increase. Each cycle of loading leaves a residual increment of pore pressure. The effect is illustrated by a portion of the results of such a test as plotted in Fig. 4.6. The

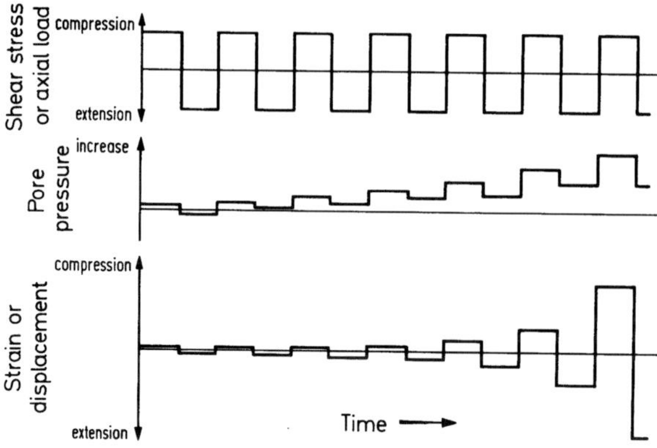

Fig. 4.6. Soil liquefaction under cyclic loading

upper diagram shows the alternating shear stress applied to the sample by cyclic application of the axial load. The second sketch shows the cyclical effects on pore pressure and the tendency of the pore pressure to rise steadily as the test proceeds. Since the externally applied hydrostatic stress or chamber pressure remains constant during the test, the steady increase in pore pressure means that the effective stresses in the soil specimen are decreasing. The consequence of this is that the axial displacement or strain of the sample, shown in the lowest diagram of Fig. 4.6, gradually increases with pore pressure. Eventually, the pore pressure comes very close to the ambient external stress so that the effective stresses are almost zero. When this occurs, the sample is said to have *liquefied* because it now has the characteristics of a dense liquid rather than those of a solid, so that the strains or displacements under the alternating load become very large.

In an earthquake the soil below the ground surface is subjected to alternations of shearing stress, somewhat as shown in Fig. 4.6, except that, of course, the cycles are not regular and the amplitudes are random. Since an earthquake is of relatively short duration, a soil which has fine enough grains, and thus low permeability, will not be able to drain during the time the vibrations go on. Thus, pore pressures may build up as shown in Fig. 4.6. If the earthquake vibrations are sufficiently intense or the duration of the shaking is sufficiently prolonged in relation to the soil properties, the phenomenon of liquefaction in the soil may occur. Because of the requirement of little or no drainage during the period of shaking, and because they tend to be fairly dense, coarse-grained sands and gravels are not generally subject to liquefaction during earthquakes. Instead, the phenomenon is confined to medium- to fine-grained sands in a fairly loose to medium-dense condition. If the material is finer-grained than fine sand or coarse silt, it usually possesses some cohesion. This inhibits the development of the phenomenon since, although pore pressures may build up somewhat the bonding of the grains to one another prevents the loss of contact between them; most silty clays and clays do not therefore undergo a complete loss of shear strength and do not liquefy.

Fine-grained soils in a saturated state are so widespread in their distribution, and earthquake vibrations extend over such large areas, that liquefaction phenomena have been a part of virtually every earthquake that has been closely studied (see Section 1.4). In some earthquakes it has been a major factor in the damage and destruction caused. Liquefaction phenomena played a part in massive soil movements in Alaska in 1964; in Niigata, Japan, also in 1964, they caused major disruption of services and utilities as well as giving rise to substantial building settlements and displacements. In the San Fernando, California earthquake of 1971, liquefaction of material in the San Fernando dam caused a landslide of the upstream portion of the dam structure. Slumping and displacement of other slopes and embankments in this earthquake have also been attributed to liquefaction.

Creep to Failure

All soil and rock materials exhibit the viscous behavior of flowing or *creeping* under sustained shearing stresses. Since shearing stresses are present on hillsides, it is very common to see trees on slopes exhibit a curvature of the base of the trunk, concave upslope; this is characteristic of downslope creep of the soil layer. The velocity or displacement profile with depth in such slowly-flowing materials has a maximum at the surface and diminishes with depth. Thus the trees are continually being rotated downslope, but the natural tendency of the tree to grow vertically causes a curve to develop in the trunk. If the creeping material has the stable force-displacement relation shown in Fig. 4.5, creep proceeds at a relatively uniform rate unless enough displacement takes place to change the geometry of the slope in the direction of greater stability.

On the other hand, if the material has the unstable behavior of Fig. 4.5, shearing surfaces will develop in the region of maximum shearing stresses and strains (usually at the toe or crest of the slope), and there the strength of the material will be reduced to its ultimate value. This zone then acts somewhat like a discontinuity or crack in a piece of metal and causes an increase in the shearing stress in its neighborhood. The material in this region therefore also shears until its strength reaches the ultimate value; in turn the stress increases in the next portion of material upslope. In this way, the progressive development of a shearing surface continues through the potentially unstable area. Where the shearing surface has developed, the material strength is the ultimate or residual value, whereas the material in a region not yet intersected by a shearing surface has its peak strength. Thus under suitable conditions, the average shearing strength along the potential sliding surface decreases until it reaches a value lower than the average shearing stresses imposed by the slide mass. When this occurs, gross movements ensue and a sliding mass moves downslope.

This behavior is demonstrated in Fig. 4.7 which shows the displacement of, for example, a point on the surface of the potential sliding mass as a function of time. The different curves indicate the behavior at different values of shearing stress, which may be caused, for example, by an increasing steepness of slope, or by the application of load. If the load applied is small enough, the rate of displacement will increase after load application, but will slow down until the displacement stops. Should a higher value of load be applied to the top of the slope, the displacement will first increase with time; then usually it will slow down while the slope creeps steadily as the shearing surface, discussed above, develops. Finally, as the shearing surface reaches a critical dimension, the rate of displacement will increase again until complete failure occurs. The application of a still higher load will reduce the time to failure.

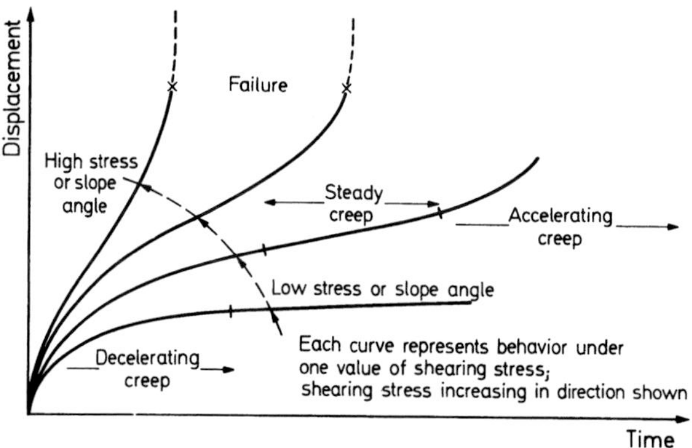

Fig. 4.7. Soil or slope behavior from creep to failure under increasing load

Such a mechanism has been used to explain a number of cases of slope failure occurring many years after the slope has been loaded either by excavation or by filling. In some cases in England, railroad cuts have failed sixty or seventy years after the cut was made. In the investigation of such a failure or in the analysis of a potential slide the shearing strength of the material to be used should be the residual rather than the peak strength. Use of the peak strength will erroneously indicate that a potentially hazardous slope has an adequate factor of safety.

Tectonic Movements

In many parts of the world, particularly at the edges of crustal plates (see Fig. 1.1), uplift or depression of the land surface is going on. In addition, after a major earthquake occurs, areas of substantial uplift or depression up to thousands of square kilometers develop as a consequence of the readjustment of the crust. In other regions, elevation changes are occurring, for example, because the crust is still adjusting to the removal of the ice load of the last ice age. The filling of a large reservoir (see Section 1.3) can also cause measurable displacements to develop in the adjacent land surface. In all of these cases, elevation changes will lead to the steepening or flattening of slopes. The slopes of near-marginal stability undergo, when steepened, a progressive increase in the shearing stress in the soil or rock mass which may result in landslide activity.

4.3. Slope Analysis

A study of landslides falls into one of two classes: the investigation of a landslide which has already occurred, with an accompanying analysis in order to explain its development, or the study and analysis of a particular man-made or natural slope, on which construction is proposed, for the purpose of determining its stability under the changed conditions.

In the analysis of a landslide, actual or potential, an examination is needed of the mechanisms by which failure can develop, and also of the properties of the material involved. Normally the slope stability is estimated in terms of a *factor of safety* defined as the ratio of the forces acting to resist failure, to the forces acting to cause failure. The resisting forces are developed by the shearing resistance of the soil, summed over the potential failure or rupture surface; the acting forces are, as described previously, due to the load of the sliding mass, together with any dynamic forces which may act in an earthquake.

For a particular slope configuration, a variety of mechanisms by which it can fail is usually tried, and a calculation of the factor of safety for each is made. The mechanism with the lowest factor of safety is the one by which failure is most likely to occur, and the actual value of the lowest factor of safety indicates the degree of stability of the slope. A value of around unity indicates, of course, that failure is likely. For construction, the value of the safety factor deemed suitable depends on the economics of the construction procedure and on the consequences of failure. Safety factors of around 1.5 to 2.0 are commonly aimed at. It is, of course, possible, particularly in the case of natural slopes, that the most favorable mechanism for failure may not be found or located in the investigation. For example, a fissure or bedding plane containing water or a material of low shearing strength may go undetected in the usual processes of making boreholes and taking samples, in which case, failure may actually develop before loads reach their design value, or in a slope calculated to be safe. The only precaution is the employment of extreme care in field investigation and analysis.

In the following sections a number of the commonest failure mechanisms are examined.

Infinite Slope

The easiest mechanism to visualize involves a slope long in comparison with the thickness of potentially unstable material. It is also long in the sense that conditions at the top and bottom of the slope are far enough away to have

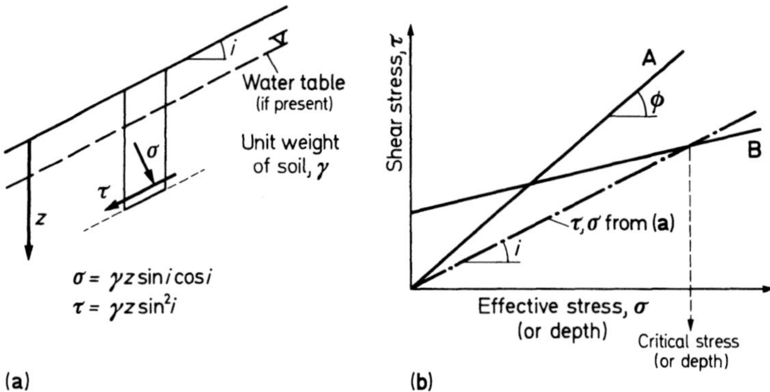

Fig. 4.8a and b. Stresses and strength for an infinite slope. (a) Stresses on soil surface at depth z below slope (b) Relation between stresses in slope and material strength

little effect on the stability of the sloping surface. This situation usually occurs when the potentially unstable layer is relatively thin, and overlies a much stronger material. The conditions may be realized in practice by a layer of soil overlying a sloping bedrock surface or by bedrock itself when jointing or bedding planes run parallel to the surface of the slope.

By considering an elemental column of material in such a slope, the parameters for the analysis can be obtained. They are shown in Fig. 4.8(a) where σ represents the effective stress acting at right angles to the potential sliding surface at depth z below the slope of inclination i, and τ is the shearing stress acting on the same surface. These are the acting stresses. The illustration shown holds for the simple case where the material is uniform between the surface and the depth of interest. If layers of different material and properties are present, then the unit weight γ used in the expressions in the figure must be weighted appropriately. When a water table is present in the slope due to a seepage condition of water flowing through the soil, the appropriate pore pressure can be obtained and subtracted from the total normal stress σ in order to arrive at the effective stress on the potential failure plane.

If the two stresses τ and σ of Fig. 4.8(a) are plotted in Fig. 4.8(b), showing their increase with depth, it is seen that they are related to one another through a line making the same angle as the slope to the horizontal axis. To determine if the slope is stable, it is necessary to plot on Fig. 4.8(b) the shearing strength (which resists failure) versus normal effective stress relationship of Fig. 4.4 for the material forming the slope. It can be readily seen that, if the soil is cohesionless (line *A*), the only requirement for stability is that the angle of internal friction of the material be greater than the slope angle. If, on the other hand, the soil possesses cohesion so that its strength varies according to line *B* in Fig. 4.8(b), then it is possible for this line *B* to intersect the shearing stress line at a particular value of effective normal stress. This corresponds to a certain depth below the ground surface. Below this depth the shearing strength of material is exceeded. If a depth of soil equal to or greater than this depth exists on the slope, the analysis indicates that it will slide off. For the slope to be stable at the angle *i*, the depth of material of property *B* shown must be less than the critical depth. Should the cohesive material also possess a friction angle greater than the slope angle, it will, of course, be stable at all depths.

These considerations also apply to snow slopes, and will be returned to later, in Chapter 6.

Finite Slopes

The situation encountered more frequently is the limited slope where the depth of material which is capable of failing is comparable to the slope dimensions. In this case, although a plane surface of failure is still possible, it is not parallel to the slope surface and more complicated surfaces have also to be considered. The failure surface most often used is a circular arc as represented in the two-dimensional view of Fig. 4.9. In the simplest case the slope is cut to height *H* and angle *i* rapidly enough in a homogeneous clay that no pore pressure dissipation can occur during the excavation. In this event, it may be assumed that the shearing strength of the soil along the proposed circular arc remains approximately constant at its value before the slope was excavated.

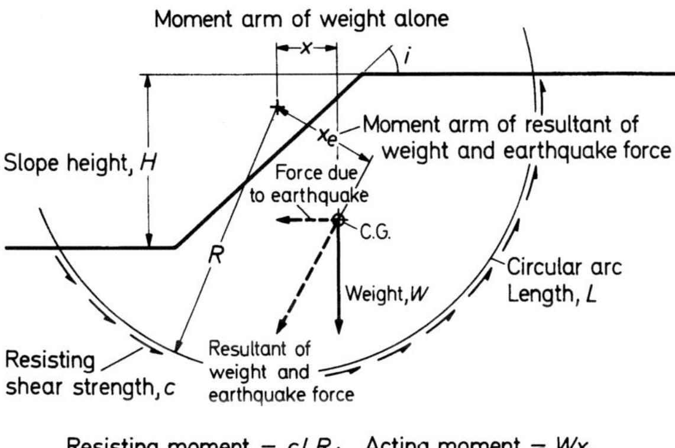

Fig. 4.9. Stability of slope calculated using circular arc failure surface

A circular arc is selected with some center and radius *R*; from this the length *L* of the arc can be measured or calculated and the weight of the potential sliding region determined. The position of the center of gravity (CG) can also be obtained. From these quantities and the soil's strength, the factor of safety for this circular surface can be calculated as shown in Fig. 4.9. However this is not necessarily the circular arc on which failure is most likely, so another circular surface must be selected and the same calculation followed through, and so forth. Failure is most probable on the arc with the smallest factor of safety.

In practice, of course, few problems are as simple as this; usually the soil properties vary with depth, pore pressures are present, and the stability of the slope must be estimated under both drained and undrained conditions. For such circumstances, the usual technique is to divide the slope into a series of vertical slices (parallel lines in Fig. 4.9), the weight of each of which is calculated, and the components of the weight acting normal and tangential to the shearing surface through its base can be estimated. From the normal force acting across the shearing surface, the force due to pore pressure can be subtracted to give the effective component of this force, and on this component the shearing strength depends. The shearing forces and soil-shearing resistance can then be summed for all the slices, and the factor of safety calculated for a particular circle. Usually six or seven slices are sufficient to give adequate accuracy in the calculation. Once again a number of such circles must be employed before the one most likely to fail can be arrived at. Various methods exist to give assistance in the selection of the most likely circles, and these are described in detail in books listed in the References (Section 4.7).

More complicated failure surfaces can be employed when the material properties suggest them, and this so-called *method of slices* can then be used to give a factor of safety for these cases. For example, if the material exhibits strongly horizontal layered characteristics, it is unlikely that a failure surface would slice through the layers as shown in Fig. 4.9; instead, a failure surface would probably follow a horizontal plane of shearing for some portion of its distance and emerge at ground surface through a more steeply sloping portion at each end. Failure surfaces also tend to follow jointing patterns in rocks and the failure method appropriate to these orientations must be selected. A simple illustration of such a failure plane in a jointed rock with bedding planes at an angle less steep than the slope angle is shown in Fig. 4.10(a). In many cases, of course, the arrangement of joint planes is such that a three-dimensional faceted sliding block must develop, as shown in Fig. 4.10(b); analysis methods are available to handle this situation.

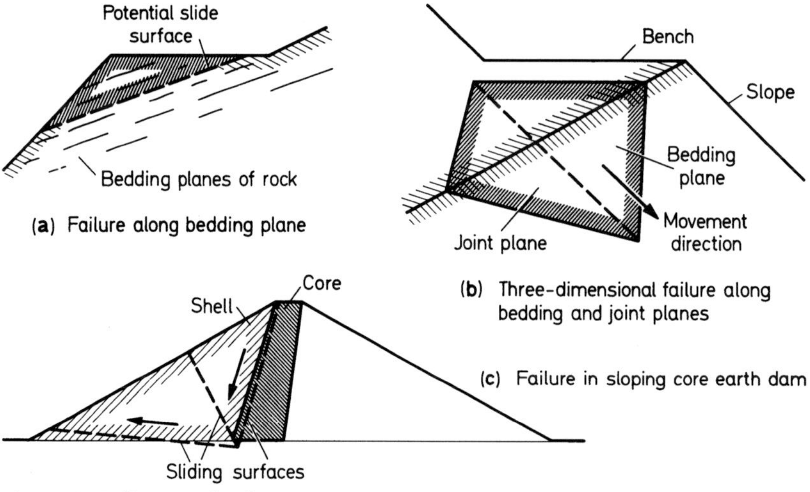

Fig. 4.10. Failure mechanisms

A technique employed in the construction of earth dams is to use a relatively coarse granular material to support an impervious central core which may, as shown in Fig. 4.10(c), be sloping. The core is constructed of fine-grained material, such as silt or clay, which is usually weaker than the supporting soil. In this case, failure, should it occur, would most likely take place on a shear plane through the core, but this requires a mechanism such as shown in Fig. 4.10(c) to develop. Here the sliding mass may consist of two blocks, one of which moves out on a more or less horizontal shear surface, possibly lying in the underlying natural materials at the damsite, and the other moving downwards along the failure surface through the core. The two sliding blocks have an interface along which they slide with respect to one another.

Seismic Considerations

In the analyses discussed so far, only vertical forces due to gravity have been taken into account. When an earthquake occurs, as discussed in Chapter 1, lateral and vertical accelerations of a dynamic nature are generated. Through the material's mass, these change the forces acting to cause failure in the slope, while at the same time the soil's strength property can be affected through, for example, the generation of higher pore pressures during the shaking period.

A first level of analysis can be undertaken by ignoring the dynamic character of the earthquake and the dynamic nature of the response of the soil or rock structure to it. In this approach the effect of the earthquake is considered to be represented by an *equivalent static force* caused by the earthquake accelerations and applied in addition to the weight of the potentially sliding mass. The equivalent static force can be applied through horizontal and vertical components. It is common practice to neglect the vertical component of earthquake accelerations in such an analysis. If the peak earthquake lateral acceleration is taken to be some proportion of the acceleration of gravity such as 0.1 or 0.2, (called the *static seismic coefficient*), then the earthquake-generated force can be represented on a diagram such as Fig. 4.9 by a horizontal force vector acting through the center of gravity of the sliding mass in the least favorable direction, that is, outwards. The size of the force is taken to be the seismic coefficient times the weight of the sliding mass. The sum of the two force components, vertical and horizontal, is then a resultant acting at some angle to the direction of the weight vector, as shown in Fig. 4.9. This resultant force can be calculated for an assumed earthquake effect, and its lever arm (shown in the figure as x_e) can be measured or calculated. Then the factor of safety of this slope during the assumed earthquake acceleration can be calculated. It is seen from Fig. 4.9 that the earthquake obviously reduces the safety of the slope in consideration.

This method of analysis, which neglects the dynamic nature of the phenomena involved in earthquakes, can be applied to relatively small slopes or earth dams of a height usually less than 30 m. For larger structures, more complicated methods of analysis are required. A technique known as *finite element analysis* which involves the use of a large digital computer is frequently employed. With such a method, both the dynamic stresses acting to cause failure in the earth or rock mass can be calculated and also some estimation can be made of the effect of these dynamically varying stresses on the soil properties (see Section 4.7 for References).

4.4. Site Investigations for Landslides

Normally, investigations of natural hillsides are not carried out unless some form of engineering construction is proposed, in which event the construction will interfere with the natural régime in some way which must be determined as part of the investigation. The natural slope may be cut into or added to in order to form a highway or bench for supporting a structure, or drainage structures may be installed to divert water from the area. A prudent approach indicates the necessity of determining the stability of the slope under the changed conditions.

Plate 4.4. Blackhawk landslide, Mojave desert, California. (From R.L. Shreve thesis, Calif. Institute of Technology)

It is usual to begin such an investigation by an examination of aerial photographs of the region; indeed, such photographs may guide the highway alignment or play a dicisive role in the location of other structures prior to any work on the ground. A geologist or soil engineer experienced in aerial photo-interpretation can readily determine the existence of many past landslides on aerial photographs. In many parts of the world, landslides are much more common than would normally be supposed, and it may be difficult to avoid landslide areas entirely, particularly for extended construction such as highways, pipelines or cables. Sometimes on an aerial photograph a landslide is distinct especially when it has occurred recently, the scarp at the head of the sliding mass and the crumpled compressed ground at the toe immediately striking the eye (see Plate 4.4). Although slides that took place many years ago have usually been sufficiently altered by erosion to make them somewhat more difficult to detect, they are usually characterized by rumpled topography caused by the movement of the sliding mass over the irregular underlying surface. In arid terrains the ground surface shows evidence of the joint-and-fracture pattern of the underlying geological structures and this too can give clues to potential landslide hazards.

Once a site has been chosen for construction, a detailed investigation is usually undertaken, including a preliminary geological evaluation of the rock types in the area, with the distribution and orientation of bedding planes, cracks, fissures, and faults. The hydrology of the region is also examined to determine the position of the ground water table, the quantity of seepage taking place below the ground surface, and the natural drainage patterns of the area. Potentially hazardous features such as bedding planes aligned more or less parallel to the slope of the hillside, seepage underground or on the face or toe of the hillside are studied in particular detail. The geological study will usually include an evaluation of the mechanical characteristics of the rock, to the extent to which they depend on the presence of thin beds of finer-grained materials such as clay or shale, as slides frequently develop upon such planes of weakness in the base material. Any slides that have developed on similar slopes or in similar rocks in areas adjacent to the construction site are also noted.

The next stage of the site investigation consists of an engineering study of the detailed soil and rock profile by means of bore-holes, and the extraction of soil samples for laboratory testing. The water table height, if one exists, and the shearing strength properties of the various materials present, are also obtained. Both drained and undrained shearing tests of the type discussed earlier will normally be carried out on the soil material. The study proceeds further to an analysis of the stability of the proposed construction involving the slope geometry and the material characteristics which have been determined. A number of potential sliding surfaces are postulated, and the factor of safety of the slope calculated for each potential surface of failure. The failure mechanism with the least factor of safety is used for hazard assessment. Normally, it is required that a minimum factor of safety of 1.5 or greater apply to the as-constructed condition to ensure confidence in the future stability of the slope.

In seismic areas, a static stability calculation is usually performed first, then a lateral seismic coefficient assessed on the basis of the earthquake acceleration considered likely to occur at the site within the lifetime of the structure (see

Section 1.5) is employed in a further analysis, and a new factor of safety evaluated. Because the failure of the slope during an earthquake is associated with the evaluation of the probability of the earthquake's occurrence, it is usual to accept a lower factor of safety for the design under earthquake vibrations. Frequently, it may be specified that the acceptable factor of safety with the assumed design earthquake is about 1.1 or 1.2, but on occasion it may only be required that the earthquake does not reduce the factor of safety below 1.0.

The factor of safety which is considered acceptable for a particular construction site depends on the cost, size and importance of the structure concerned as well as on the consequences of its collapse. A lower factor of safety is acceptable for a structure whose repair costs are slight or which endangers no lives than for an important structure whose collapse could cause casualties.

The assessment of the equivalent seismic lateral acceleration to be employed in design involves the selection of the earthquake or earthquakes which could cause the strongest ground motion at the site (see Section 1.5).

If the structure is important enough to warrant a detailed dynamic analysis of the slope stability and site response, an acceleration-time record must be constructed to represent the estimated ground motions at the site during the design earthquake. This acceleration history is then used in the computer calculations of the soil and structural behavior. Frequently, more than one such record will be used in analysis because of the statistical nature of the accelerations assumed.

4.5. Detection and Control

Many, if not all, landslides are preceded by a period of creep which gradually increases until failure occurs. Under some conditions creep may go on for years before the climax; in other conditions the creep period may be only a few seconds long. However, many landslides could be detected in advance were suitable instrumentation installed on the slope and the readings monitored at regular intervals. Only in a few cases have slopes been instrumented, sometimes to give warning of any problems in a slope whose safety factor has deliberately been lowered during a period of construction. When the construction is finished, in these circumstances, the slope safety will be re-established. In the absence of measurements, most slope failures take place without warning except on rare occasions in inhabited areas when local residents notice new cracks in garage floors, driveways, house walls or sidewalks. Sometimes hillside movement is indicated by cracks or breaks in sewer or utility lines before failure occurs. By the time such effects are noticed, it is usually too late to avoid the final slide.

Two kinds of measurements can be made: surveys of the movement of points located on the slope surface, and observations of the changing inclination of tubes installed through the soil or rock mass considered to be slide material. The first of these requires detailed regular surveying if the smallest movements are to be detected or, alternatively, the connection of points on the potential sliding zone by wires to continuously recording equipment established on stable ground. The second technique consists of installing a special type of plastic tube in boreholes drilled for the field investigation. Periodically, an inclinometer is lowered down the plastic tube in order to make measurements of the tube inclination. If a slide mass is moving downslope, the inclination of the tube will reflect this. Integration of the records with height above the base or with depth from the top of the tube enables the absolute displacement of various layers in the ground to be measured. The position of the top of the tube is obtained from surface surveys.

Even if such measurements are taken, it is not easy to determine at what stage in the movement the mass may accelerate to a catastrophic level. The slopes around the Vaiont reservoir in Italy, filled about 1960, were considered to be unstable some time before disastrous slide movements occurred and measurements were made at regular intervals of points on the slope surfaces. These indicated that the slope was moving, but the final failure was not predicted (see Section 4.6). In Japan in 1969 a test hillside was being saturated with water in a study of the mechanism of slope failure, when a slide occurred killing several of the research workers.

Should the motion of a hillside be detected and analysis indicate that a landslide might ensue, a number of means of prevention or control are available. When small volumes of material are involved, earth-moving equipment can, in a very short time, relieve the load at the head of the slope or add material to the toe to increase the stability and stop the movement (though sometimes the weight and vibration of the equipment may also itself precipitate trouble). Choice of what to do depends upon the availability of equipment and material, the topography of the site, access conditions and space in which to carry out the remedial operations.

If it is suspected that pore-water pressures, owing to a rise in the ground water table or water flow into cracks, are the underlying cause of the potential movement, these may be reduced directly by drilling drainage holes. It is not always possible to accomplish this because of the low permeability of many materials of low shearing strength. In such cases, *electro-osmotic stabilization* has sometimes been successful. This method involves installing electrodes in the ground suitably located with respect to the potential slide and applying a voltage potential difference across them. In response to the potential difference, the water in the soil tends to flow away from the anode and towards the cathode, thereby draining the soil. There is, in addition, a further stabilizing effect on the soil through the pore water pressure gradients which develop. The technique is used as a temporary stabilization method for steep walls of excavations in fine-grained materials when the excavation will eventually be backfilled.

When high pore pressures in fine-grained material have been suspected as being the cause of a ground movement, another stabilization method involves boring horizontal holes into the fine-grained layer. The pore water pressure is reduced by blowing hot air through the holes to dry out the soil. The drying process increases the shearing strength of the ground.

Where the slopes are man-made, city and county ordinances usually provide certain requirements to be followed during their construction. Slopes made of fill, for example, may be required to be no steeper than an angle of one vertical on one and one-half horizontal, and the compaction of the soil forming the slope must be carefully controlled to exceed certain minimum unit weight and therefore shearing strength requirements. The drainage of water onto and away from such slopes is usually carefully specified. Where cut slopes are involved in hillside materials, the relation of the slopes to the bedding planes must be considered, and certain city codes supply minimum standard soil property values which must be used in the stability analysis.

4.6. Landslide Case Histories

Usoy (Pamir), USSR, 1911

The largest landslide to occur in recent history was at Usoy in the Pamir mountains of the Soviet Union (then Russia) in 1911. This landslide has an interesting history. There was an earthquake at the same time as the landslide, but because the area was remote and sparsely inhabited, it was two weeks after the event before the Russian Government had word of it, and a further two months before they knew about the landslide. The area of the landslide remained uninvestigated for two and one-half years until 1913 when a military expedition visited and reported on the region. At this time, the origin of earthquakes was still a subject for debate among seismologists, and when the report on the earthquake and landslide became available, some seismologists concluded that the events were in fact simultaneous, and that it was the impact of the landslide's fall which had caused the waves recorded as an earthquake. The exact times of the two events were, of course, not known.

Prince B. B. Galitzin, one of the early experimenters in seismology, undertook a calculation of the energy radiated by the earthquake, as indicated by the seismograms obtained at the recording station at Pulkovo. He computed the amplitude of elastic waves generated in the Earth's crust by the energy released at the source, and by comparing the results of his calculations with the displacements recorded at Pulkovo was able to arrive at an estimate of the energy release at the source. He concluded that the energy was so great that the earthquake must have been the primary event, not the landslide. There is historic interest in the event and in the calculation, since Galitzin was the first to attempt an analysis of this kind. His work was discussed extensively in the seismological literature and led eventually to techniques for establishing from instrumental records the magnitude and energy release in an earthquake. Subsequent analyses, however, have indicated that Galitzin may have underestimated the source energy by as much as a factor of 100.

Subsequently, in 1915 a seismological expedition was sent to investigate the landslide; it came back with a report including a survey of its extent and volume. The amount of material involved was estimated to be 2.5 billion cubic m, that is to say, 2.5 cubic km. The expedition surveyed the landslide in enough detail for the original location of the mass and its final resting place to be known. Thus the energy dissipated in the slide could be better estimated, although the efficiency with which it was converted to elastic waves was probably small. The landslide energy, although large, was substantially smaller than that released in the earthquake.

The Pamir landslide, of soil and shattered rock, overwhelmed the village of Usoy of 54 inhabitants, blocked the valley and dammed the river Murgab causing a large lake to form. Eventually the lake level rose until it submerged another hamlet, Sarez. The new lake, Sarez, was complete when the water cut a new channel through the landslide material to equalise the inflow and outflow of the lake. As reported by the first military expedition, the landslide dam was 301 m high, the maximum lake depth 284 m, and its length 53 km. By comparison, the Nurek (USSR) man-made earth dam has a planned height of 310 m, and a volume of 58 million cubic m.

Although no information is available on the material properties, it would appear that this was a landslide in relatively brittle material (the underlying rock) on a hillside of originally marginal stability, triggered into movement by the vibrations of the earthquake. Since the slide seems to have taken place in rock, it appears unlikely that the generation of pore water pressures by the earthquake vibrations would have been a significant factor. The magnitude of the earthquake has been approximately estimated at 7.0 on the Richter scale. Events such as the Pamir landslide are common during earthquakes, but fortunately seldom on such a large scale. A similar slide developed in the Madison Canyon in Montana during an earthquake in August, 1959. This slide, of about 27,000,000 cubic m in volume, also blocked a valley, causing a lake to develop behind it. The slide descended upon a camp site and killed approximately 26 persons. It occurred in a bend in the canyon where the rock wall had been eroded by the Madison river to angles as steep as 45°. The canyon walls consisted of highly fractured and altered metamorphic rock.

Because other structures were affected by the rising lake level, the U.S. Army Engineer Corps worked rapidly to excavate a channel through the landslide mass to let the lake waters drain before further damage occurred. They were successful in completing a channel in time to control the release of the lake waters. In the case of both the Pamir and Montana landslides, the falling debris was caught by a U-shaped valley and was brought to rest in a mass, blocking the valley. Distances traveled by the rock and soil masses were not great. A different slide development with the same kind of triggering has occurred in other places and will be described next.

Sherman Glacier Slide, Alaska, March 1964

The great March, 1964 earthquake in Alaska (see Section 1.1) caused a number of landslides to develop, two of which will be discussed in this chapter. The first of these is the Sherman Glacier slide, illustrated in Plate 4.5. At a distance of approximately 140 km from the epicenter there is a region of high mountains and glacier-filled valleys. One of the mountains, now called Shatter Peak, was shaken so violently that a portion of rock near the crest of the mountain broke off. As the material fell down the mountainside, it disintegrated and the mass attained substantial velocities. In this case, there was no narrow canyon to contain

Plate 4.5. Sherman Glacier landslide, Alaska. (Courtesy of Austin Post, USGS)

the material arriving at the base of the mountain; instead a glacier extended several kilometers across the valley. The landslide arrived at and traversed the glacier, spreading as it went, to form eventually a flat, lobe-shaped deposit several kilometers in lateral dimensions and only a few meters in thickness. The volume of material involved was about 23,000,000 cubic m.

A number of curious features are associated with this landslide. Two of these are longitudinal flow-lines and a raised margin around the edge of the stationary mass. There was no sorting of the dèbris by size with distance along the slide axis. The slope of the glacier in the direction of furthest extent of the landslide is only a few degrees. The question therefore arises: how could an incoherent mass of soil and rock fragments travel such a distance?

The following mechanism has been suggested: as the broken-up slide material fell, it encountered a ridge running transverse to its direction of flow across the base of the mountain and above the level of the glacier. The rock fragments were launched into the air by the ridge, and formed a relatively flat sheet of flying debris as they encountered the snow-covered surface of the glacier. It is hypothesized that the sheet of debris trapped air between it and the glacier's surface to form an *air cushion* on which the landslide slid freely. At the edges of the flow, the air was able to escape and so the moving sheet was retarded at its periphery and subsided onto the glacier surface. This helped to trap the air better beneath the moving portion and accounts for the bulge around the edge. Eventually the supply of material from the mountain ceased, the air escaped from the moving mass, and the slide came to rest on the glacier surface. It appears that the final stages of the movement must have been relatively gentle, since a number of large rock masses in their final position are broken up into fragments, but the fragments lie closely adjacent to one another on the slide surface. To the present, only a qualitative analysis of this mechanism has been carried out. Under the conditions involved, it is difficult to arrive at quantitative estimates of the material properties, velocity, air pressure, and thickness of the air cushion.

The air-cushion mechanism has also been suggested for the Blackhawk landslide in the Mojave desert of southern California, which is thought to have occurred many thousands of years ago. In this case, a volume of about 320,000,000 cubic m of crystalline limestone was involved, and it flowed a distance of 8 km across the alluvial desert floor at an average slope of $2\ ^1/_2\,^\circ$ (see Plate 4.4). Without the air pressure-lubrication hypothesis, it is even more difficult in this case to see how the slide material could go so far across such slopes. This slide overlies an even older layer of similar topography, interpreted to be a previous slide.

On an entirely different planetary body, on the other side of the moon there is a feature which has been named Tsiolkovsky, a small mare basin ringed with mountains. Behind one section of the mountain ring wall is a large landslide-like area which exhibits all of the features of the Sherman Glacier and Blackhawk landslides except for the scale; the lunar landslide has a lateral dimension of about 100 km and includes a volume estimated at 3,000 cubic km if its thickness bears the same proportion to its lateral dimension as seen in the Sherman or Blackhawk slides.

There is, of course, no information on the material properties in this case nor the events that caused such a landslide, if indeed it is. It does not appear to have been associated with the formation of the mare itself, since the landslide must be a younger event than the mountains from which it originated. The slide material itself is substantially younger than the plains on which it lies, according to a comparison of the number of craters on its surface and on the adjacent lunar surface. The extent of this mass movement may be related to the low value of lunar gravity (one sixth of the Earth's, approximately). The absence of air on the moon causes difficulties with the airlubrication idea.

If it is supposed that the material properties of terrestrial and lunar rocks are identical, similarity considerations indicate that such a lunar landslide would be equivalent to a feature on Earth of one-sixth of the scale. The volume would be reduced by a factor of about 216. With these considerations, the lunar event (with a reduced volume of about 15 cubic km) becomes comparable to the 1911 slide in the Pamirs, and to the Saidmarreh landslide of Iran (described later) although, even with the reduction in scale it is still an enormously large feature.

The Sherman Glacier and, presumably, the Blackhawk and Tsiolkovsky slides occurred in unpopulated areas and thus caused no human property damage or casualties. This was not the case in the next landslide to be discussed.

Yungay, Peru, May 1970

As described in Section 1.4 the earthquake in Peru, in May, 1970, caused a fall of ice, snow and rock from Mt. Huascaran in the Andes which over a short distance became a water-saturated granular mass of immense proportions. The slide reached a velocity of approximately 150 to 200 km per hour as it traveled down a valley, overwhelming the town of Yungay and overlapping the town of Ranrahirca.

This slide was almost a duplicate of one which occurred in January 1962 on the same mountain (see Section 6.3), without an earthquake. Both events may be classified either as avalanches or landslides, since they began as ice and snow avalanches which gouged out rock and soil in their travel.

In this case it does not appear that there was any lubrication mechanism acting to give the landslide its great speed. Rather the mixing process in the early stages of landslide development caused it to assume the character of a viscous flow which attained great velocities because of the steep slopes involved.

Turtle Mountain, Alberta; Elm, Switzerland

In 1881 in Elm, Switzerland, and in 1903 in Alberta, Canada, two rapid rockfalls occurred which were not associated with earthquake events, and proved disastrous for the inhabitants living below the rock masses that broke free. At Elm the rock mass was located on steeply dipping layered materials in which failure occurred rapidly. The rock had been undermined during excavation of a quarry for slate, and in September 1881 first one, then a second mass of rock fell down over the quarry and spread laterally on the valley floor. The second slide killed about 20 people. Between them, the two rockfalls had left a greater mass unsupported and in a few minutes it also fell, in a slide that became a high-velocity flow with accompanying high winds. The rock mass flowed across the valley floor and up the far side of the valley to a height of 100 m.

The town of Frank, in the southwest corner of Alberta, is dominated by Turtle Mountain which consists of massive beds of limestone lying on much weaker shales and sandstones. Coal was mined in coal beds in the shale and sandstone layers. Although the limestone beds dip away from the valley in which the town lies, joints and fracture surfaces with a dip of 40° parallel the steeply-sloping mountain front. A limestone rock slide developed in 1903 in the joint planes and quickly reached high velocities, falling a vertical distance of almost 1,000 m to flow 4 km across the valley, killing 70 people. This fall came from the South Peak of Turtle Mountain. In a discussion of the rock fall in 1937, it was pointed out that the conditions which existed at the South Peak at the time of the landslide were almost exactly duplicated on the North Shoulder of the same mountain, from which it was concluded that another slide from the North Shoulder could occur at any time. It has not happened yet.

Many of the features of the Sherman Glacier slide are present in the Elm and Turtle Mountain events, but for the latter an alternative mechanism of flow has been proposed, in the suggestion that air is entrained in the falling debris, so that the slide movement assumes the character of a liquefied mass, as in the next case to be described, Aberfan. At Aberfan the fluid phase was water; here it was air.

Many similar landslides or rock falls have occurred in mountainous countries, one of the largest in recent times being the Rossberg or Goldau slide of 1806 in Switzerland. Probably because of the development of high pore water pressures in permeable sloping sandstone and conglomerate layers following a particularly wet summer, a mass of 14,000,000 cubic m of rock was released and slid into the valley of Goldau, burying four villages and killing 457 persons.

In prehistoric times two extremely large landslides took place at Saidmarreh, Iran (20 cubic km) and at Flims, Switzerland, in the Rhine valley (15 cubic km) both in limestone rocks dipping in the downslope directions. It is conjectured that the Saidmarreh landslide was released by earthquake vibrations.

Aberfan, Wales, October 1966

The town of Aberfan (pronounced "Abervan") in Wales lies in a coal-mining area of the Welsh mountains. In the mining process, the debris brought out from the mine is dumped to form an artificial hill called a spoil-heap or *tip*. Since coal is usually found associated with clay and shale layers, and the material is frequently wet, the tip consists of a mixture of various sizes of granular fragments frequently in a muddy or clayey matrix. In the extraction of coal from the material excavated from a mine, a very fine-grained debris called *tailings* is produced, which were included in the material dumped on one of the Aberfan tips that later failed. Seven of these tips occupied the Merthyr mountain slope above Aberfan at elevations of from 40 to 200 m higher than the town up to the middle 1960's. A number of streams run down the slope into the valley. Slides and runs of material from the faces of the tips occurred periodically in the time 1916 to 1966; dumping at two of the tips was terminated following slides.

In October 1966, Tip number 7 suffered a slope failure at 09:15 h and the resulting landslide engulfed a portion of the town of Aberfan. The biggest disaster

Plate 4.6. Failure of Tip No. 7, Aberfan, Wales. (Report of the Tribunal of Inquiry, HMSO)

occurred at the elementary school, standing in the way of the slipping mass, where 116 school children and 5 teachers were killed. In all, 144 people lost their lives (see Plate 4.6).

The tips at Aberfan had been built by dumping mine waste on a 14° slope of sandstone covered with a thin layer of boulder clay. The bedding planes dip downslope toward the valley at a smaller angle than the slope. There are occasional impervious strata in the sandstone, so that the groundwater emerges at lines of springs or seeps along the slope. The tips were formed on top of these springs and also on top of some of the small stream courses leading to the valley. With the high local annual rainfall (about 1,500 mm) and the interruption of the natural drainage, the lower portions of all the tips were saturated with water. What happened to Tip No. 7 on the day of the catastrophe was described by a civil engineer, G.M.J. Williams, to the tribunal appointed to inquire into the disaster:

"On the morning of the 21st October 1966, there were several movements within the tip... and shortly after 9 o'clock (09:00 h) there was apparently another movement, but in this case it took part of the saturated material past the point where liquefaction occurred." (Author's note: The saturated material was in a loose state so that when it was sheared suddenly without drainage, the tendency of the granular structure to contract in volume caused high positive pressures to be developed in the pore water and the soil liquefied.) "This initially liquefied material began to move rapidly, releasing energy (Author's note: Alternatively, the movement of the initially liquefied material caused an increase in the shearing stresses in the remaining material.) which liquefied the rest of the saturated portion of the tip, and almost instantaneously the nature of the saturated lower parts of Tip No. 7 was changed from that of a solid to that of a heavy liquid of a density of approximately twice that of water... The upper part of the tip, not being saturated, did not liquefy but some of it would be carried forward floating on the liquefied material... Being of the nature of a liquid the whole mass then moved very rapidly down the hillside, spreading out sideways into a layer of substantially uniform thickness. As this happened, water was escaping from the mass so that the particles of soil regained their contact and the soil mass returned to its solid nature."

The presence of the fine-grained tailings in Tip No. 7 may have played some part in the failure.

The inquiry by the tribunal revealed that before 1966 no soil mechanics investigations had ever been carried out prior to the construction of tips in Wales, and no inspections of the stability of existing tips had ever been made, in spite of the fact that a number of slides had previously occurred. The fact that no lives had been lost in prior incidents presumably had a bearing on this. It was estimated that each year in Wales three new tips were begun and three additions made to existing tips. It was recommended to the tribunal that a National Tip Safety Committee be established to look into the hazards associated with present methods of disposing of industrial waste in bulk under all circumstances.

Failures of structures involving mine waste or tailings have also occurred elsewhere, including Chile and the United States. In West Virginia, in February 1972, a dam restraining coal mine waste water and tailings collapsed following heavy rainfall causing 118 deaths.

Anchorage, Alaska; Kansu, China

Liquefaction in soils has occurred most frequently during earthquakes. During the Alaskan earthquake of March 1964, the great magnitude of the earthquake and length of fault rupturing caused strong ground shaking in the Anchorage area. Since the City of Anchorage is founded on level ground approximately 20 m above sea level, and the coast is formed of high steep bluffs, the duration of shaking generated many cycles of shearing stress in the coastal slopes. Many disastrous landslides resulted (see Plate 1.1). The soil profile in the area consists of layers of sand and gravel overlying clay of what is called the Bootlegger Cove Formation. The clay also contains saturated sand and silt lenses. A detailed post-earthquake investigation of landslides in the Anchorage area indicated that liquefaction in the sand and silt lenses in the clay or of the clay itself, as a

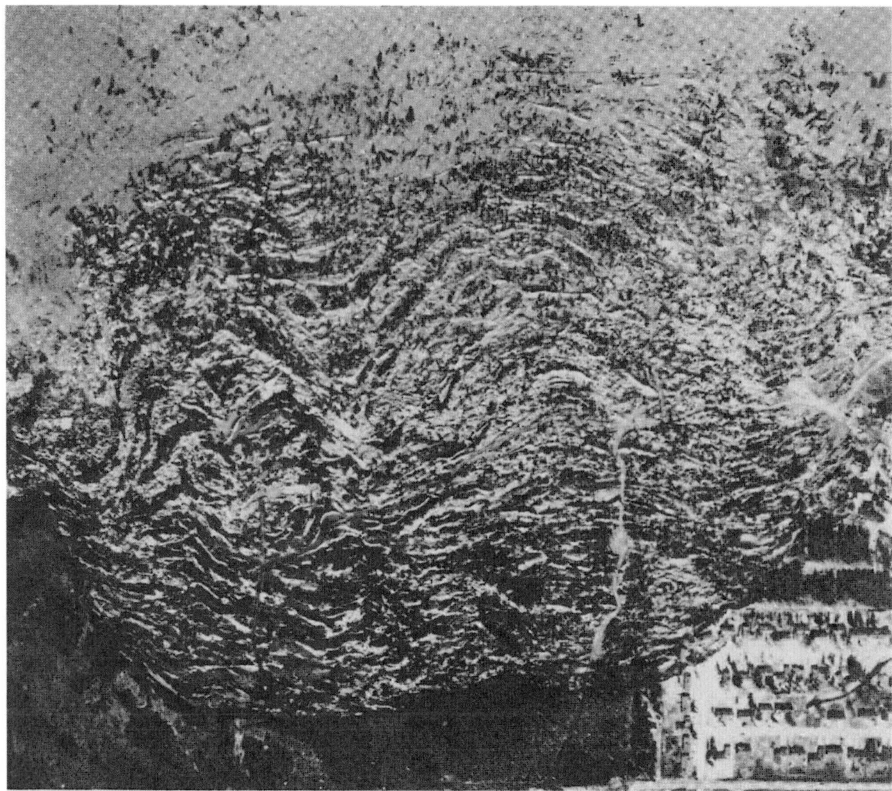

Plate 4.7. Aerial photo of Turnagain Heights landslide, Alaska. For scale, see houses in bottom right corner

consequence of the oscillations, was a cause of the landslides. In a single landslide during this earthquake, an area approximately 2.5 km along the coast broke as much as 0.5 km inland and slid towards the ocean, causing great damage in the residential area of Turnagain Heights, where some of the sliding blocks moved as much as 300 m. Eyewitness accounts, though not entirely consistent, indicate that the sliding movement started some time after the initiation of earthquake movements and continued after the earthquake stopped. Other landslides occurred along the bluffs near Anchorage causing disruption of utilities and commercial structures, although the movements were not as great as at Turnagain Heights (see Plate 4.7).

Stabilization methods were undertaken after the earthquake, including the leveling and flattening of the slopes, and the construction of compacted fill bulkheads in trenches excavated at the toe of the landslides to provide higher shearing strength in this region. A number of attempts to compact the soil in the relatively low density layers, which had been thought to liquefy during the earthquake, were unsuccessful. Another earthquake of comparable magnitude in this area would probably give rise to further slope movements.

At Anchorage the liquefaction involved a suspension of soil grains in water. A case where apparently air was the suspending medium is recorded as having occurred in Kansu, China during the earthquake of December 16, 1920. The hills in the Kansu area are of *loess* formed from silt grains blown and deposited by the winds. Loess deposits are also widespread in the southern and southeastern United States. If the loess is subsequently wetted, it will usually undergo very large settlements since the cementation between the grains is commonly water-soluble.

The earthquake in 1920 broke down the shearing resistance between the soil particles in the loess hills of Kansu, and since the low permeability of the material prevented the air from escaping readily from the mass, the material liquefied. Vast landslides swept over villages and towns; the death toll has been placed as high as 200,000. There is no good technical description of the event in western literature and our present understanding of what occurred during that earthquake is based on a visit two years later by western travelers, whose photographs and stories are not explicit nor technical enough to derive a detailed understanding of the mechanism of the events.

Similar disasters in these loess hills of China have occurred during other earthquakes. In Chapter 1, mention is made of the 1556 earthquake near Hsian in Shensi Province, in which almost a million people died in part through the collapse of loess slopes.

Submarine Slides, Newfoundland, Canada, 1929

Another instance where the mechanism has not been clearly elucidated is that of the submarine landslides off the coast of Newfoundland in November 1929. At this time an earthquake occurred following which, although no landslide was visible, breaks occurred in succession in seven submarine cables laid on the floor of the Atlantic. The order of the cable breaks and the times at which they happened enabled calculations to be made of the extent, length and velocity of the landslide which must have been generated by the earthquake. No mechanism other than a slide or flow seems plausible. From the initial source area to the location of the last cable break was a distance of 470 km. The landslide or landslides, since it is possible that a number occurred, was at least 150 km wide, from which it may be estimated that an enormous volume of material was involved. The maximum speed of the phenomenon was deduced to be 93 km per hour and the flow was still moving at about 22 km per hour at the last cable break.

At the end of the motion, the slopes of the ocean floor were extremely small, of the order of one-half degree or less, and it must therefore be inferred that the sliding mass was most probably a dense liquid rather than solid. In this instance, it seems plausible that the mechanism involved was as follows: the earthquake triggered a landslide on a steeper portion of the sea floor composed of fine-grained, unstable soil. The shearing zone or failure plane must have been at a depth of the order of meters to tens of meters below the sea floor. It seems likely that some liquefaction may have been involved at this depth in order that the developing slide on the relatively flat slope could achieve sufficient velocity to initiate turbulent mixing of the slide material with the sea water. If no mixing occurred, it would seem very unlikely that a block of sea floor material at a density relatively low, but still within the range of those of solid soils, could go the distances observed. At a velocity of a meter or two per second, the upper surface of the landslide mass would become unstable as it flowed under the sea water; waves would be generated on its surface leading ultimately to mixing of the soil and sea water. With the mixing, the landslide mass would expand and become less viscous, its velocity would increase, causing more turbulent mixing, until eventually a stably moving, generally fluid flow resulted. This kind of flow is usually referred to as a *turbidity current*. Although of low density, it was evidently viscous or cohesive enough to cause breaks in the cables it encountered. By the same evidence of cable breaks, a number of undersea landslide-generated turbidity currents of this kind have developed following earthquakes in other parts of the world including the waters off North Africa and Indonesia.

San Fernando Dam, California, February 9, 1971

The February 9, 1971 earthquake (see Section 1.4) in the San Fernando Valley, California caused thousands of rockfalls of all dimensions in the neighboring San Gabriel mountains, but the most noteworthy landslide feature connected with it was a slide which it triggered on the upstream face of a 60 m high earth dam, the San Fernando Dam. Although the landslide was fully developed, the reservoir behind the dam, through a combination of fortunate coincidences, was at an unusually low level so that the portion of the dam remaining after the movement was high enough to prevent the reservoir water from being released downstream. As a consequence of the failure, however, and because of the risk of further slides developing either through aftershocks or because of the rapid drawdown of reservoir water which was begun by the authorities, some 80,000 people were evacuated from downstream areas.

The dam was constructed in the period 1912 to 1920, mostly of hydraulic fill, although at various subsequent dates it was raised and strengthened with the addition of rolled fill. The post-earthquake analysis of the structure indicated that cyclic shearing stresses developed in the dam structure by the shaking caused zones of the fill to liquefy. This led to a failure of the upstream slope of the dam within a short time after the earthquake began. The slumped portion of the dam material moved upstream into the reservoir, a maximum distance of one or two hundred meters (see Plate 1.8).

During this earthquake, a partial landslide also developed in the downstream portion of the upper San Fernando dam, a somewhat smaller and more recent structure, retaining water some distance upstream from the main San Fernando dam. Movement of about a meter occurred as a portion of the dam moved in a partially rotational failure. It would seem likely that, if the earthquake had been of longer duration, this slide might have developed more fully. Around the shores of the reservoir a number of other landslides developed during the same earthquake. The failure of the San Fernando dam caused an extensive re-evaluation to be carried out on all earth dams in California in seismic areas.

Quick Clays: Quebec, Canada; Sweden

The last of the case histories which involve rapid failures to be discussed are those which occur frequently in certain types of clay, particularly common in Canada, Norway and Sweden. These clays are referred to as *quick clays*, and have an extremely unstable stress-strain relation. They are thought to have been deposited in a marine environment and subsequently uplifted above sea level so that the salt water in their pores has been leached out by rain or underground seepage and replaced by fresh water. The clays develop strong bonds in a loosely-packed structure of clay particles, so that the undisturbed strength of the clay is quite high. However, when the clay is disturbed by shearing or remolding, the bonds are broken and the clay becomes essentially a fluid suspension of clay particles in water.

What appears to happen in slopes in these clays is that the shearing stresses exceed the shearing strength of the clay near the toe of the slope. The clay in the sheared zone deforms and becomes weaker with the result that the shearing stresses are increased in the adjacent clay regions which, in turn, shear and weaken. At some critical stage in the growth of such a shearing surface, the overall shearing stress exceeds the diminished shearing resistance and a landslide is initiated. The movement of the sliding mass once exceeding some small value, the material along the shearing plane becomes essentially liquid so that the landslide accelerates.

Although it is likely that observations prior to failure would indicate that an unstable process was going on, most of these flows occur in rural areas, and therefore their movement is unexpected and can be quite damaging. Over 100 persons have been killed in a single slide. They develop in fairly flat-lying topography with slopes of only a few degrees above bluffs or low cliffs along the river bank. The material flowing out into the river may dam the river or be carried off downstream by the flowing water.

Such events have happened periodically along the St. Lawrence River and its tributaries in Canada and along various streams in Norway. A great slide of quick clay destroyed much of the town of Surte on the Gota River in Sweden in 1950. Perhaps triggered by the operation of a pile driver, a volume of 3,000,000 cubic m moved toward the Gota River, almost blocking it. Quick clay slides have been investigated to such a degree that a study prior to construction in a particular region should reveal the potential harzard from the quick clay. Former slides show up clearly in aerial photographs.

Axmouth, Devon, England, December 1839

Among landslides which tend to develop and continue at intermediate velocities are those occurring in coastal areas where a steady process of erosion goes on at the toe of the cliffs or bluffs. Slumps generally follow greater-than-usual storms which remove substantial amounts of material and lead to a sudden decrease in the stability of the slope. At the same time, increased amounts of rainfall raise water tables and increase the pore pressure in the underlying material. One case serves to demonstrate the phenomena in a representative way, the

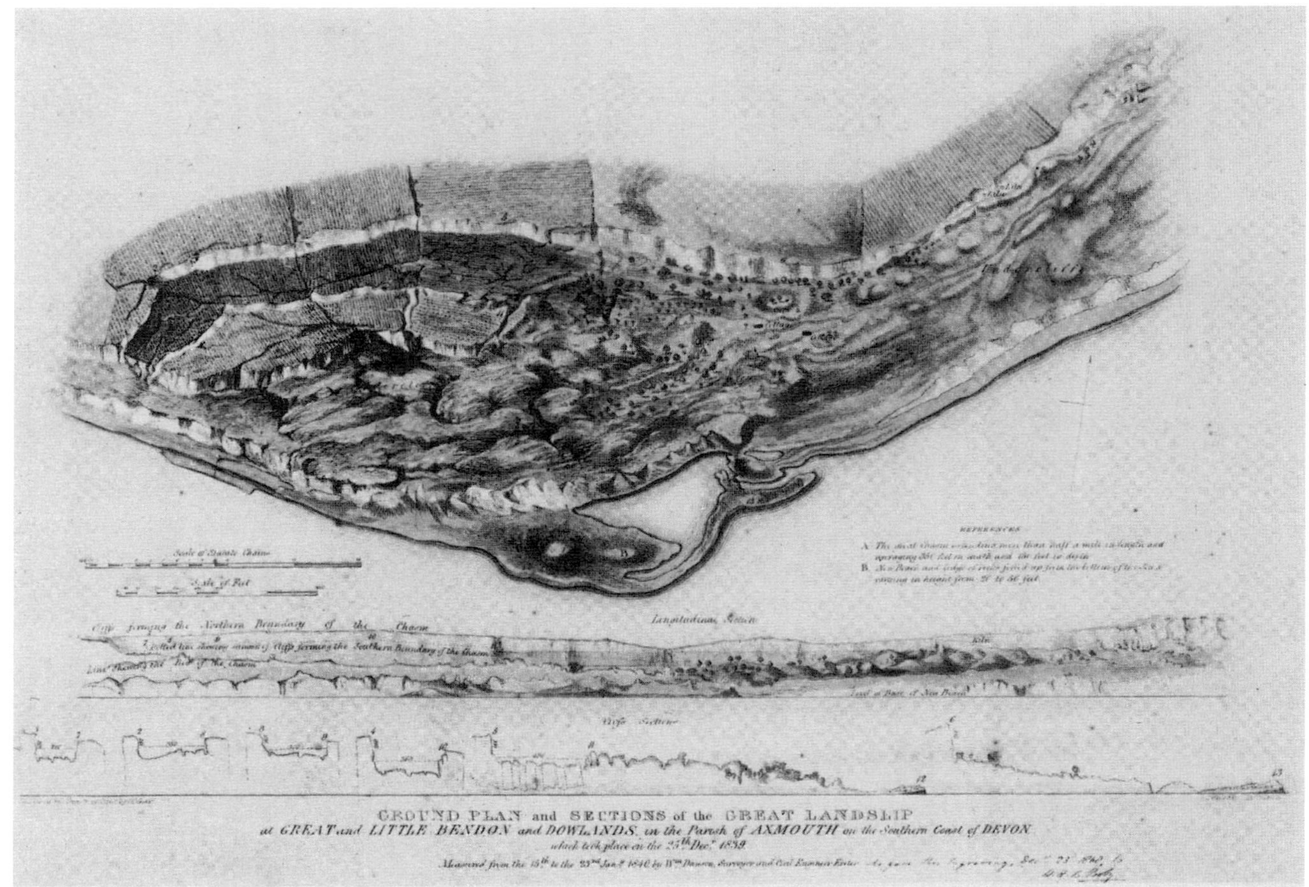

Plate 4.8. Axmouth landslide, Devon, England. (Courtesy of Geology Library, Cambridge University)

landslip on the coast of Devon on Christmas Day, 1839, referred to as the Axmouth or Dowland landslide (see Plate 4.8). The materials involved were a thick sheet of chalk overlying sandstone and clay layers, both chalk and sandstone being relatively permeable so that percolating water tends to flow along the surface of the clay or shale layers to emerge as seeps along the cliff base. The shear strength of the clay is reduced by the saturation and high pore pressures following periods of heavy rainfall, and the clay then acts as the sliding surface for the mass of rock overlying it. A quantity of perhaps 40,000,000 cubic m of material was involved at Axmouth, but because this part of the country was relatively sparsely inhabited, few people were concerned. The landslide was a famous occurrence in its time, and for a period afterwards, it was a popular excursion place for visitors.

The material upthrust at the toe of the landslide has formed an offshore bar which helped to prevent further erosion of the base of the cliffs: the original abrupt discontinuities of fissures and sliding blocks have been worn and eroded into more rounded form. As Archibald Geikie, a Scottish geologist writing of Southern England expressed it in 1885: "Everywhere the rawness of the original fissures has been softened by the rich tapestry of verdure which the genial climate of that southern coast fosters in every sheltered nook."

Such coastal modifications by the sea are, of course, common around the world but have, perhaps, been best studied in Great Britain because of the importance of the loss of land to the crowded island. The cliffs along the south and east coasts of England (see Plate 4.9) have been eroding for many hundreds

Plate 4.9. Landslides and coastal erosion, East Anglian coast, England. (Photo by R.F. Scott)

of years to such an extent that a number of formerly thriving towns have disappeared in the North Sea. Other areas, now on the seacoast, in former days were inland. On the South Kent coast, where the geological profile is similar to that at Axmouth, massive slides have occurred which represent renewals of movement in a mass which must have formerly slid from a 150-m cliff and is referred to locally as the "undercliff." Larger movements occurred in 1877, 1896 and 1915 and lesser events in 1859, 1865, 1937 and 1940. The slides result from marine erosion at the toe but have been inhibited by coastal protection measures. All the movements developed between December and March in the period of locally highest ground water levels.

Pacific Palisades and Highland Park, California

In southern California there are large areas of weakly cemented, relatively unsaturated alluvial deposits of sands, silts and gravels. They are cemented by largely water-soluble cements because of their history of wetting and drying by the seasonal rains. The region is also tectonically and seismically active so that the uplifted alluvial materials are intersected by faults and fractures at all scales of size in addition to the usual sedimentary rock features of bedding planes and joint patterns. During late geologic time in southern California, volcanic eruptions occurred which laid down thin layers of ash on top of the accumulating alluvium. The ash layers were, in turn, buried. Since their deposition, the volcanic ash layers have altered to a montmorillonite clay which therefore exists in thin layers throughout the sandstones and siltstones. This material has a low shearing strength and can almost be considered to consist of a lubricating layer appearing at intervals in the rock formations, forming the base shearing layer in a number of landslides.

Little or no rain falls except in the rainy season, in the months of November through March, when up to 500 mm of rain may occur in three or four large storms. Thus, in the canyons and valleys, erosion, when it occurs, is rapid, and cracks and faults which through much of the year are dry, may, for a short time, become filled with water. Consequently the rainy season is frequently accompanied or followed by landslides in the steep-walled canyons and along the bluffs at the coast.

Along the coast near Santa Monica, in particular, the cliffs are 20 to 50 m high, consist of weak siltstones and sandstones and are plentifully fissured and faulted. A major landslide occurred there in 1959 and was large enough to require the relocation of a major highway which formerly ran along the toe of the slope. Landslides in this area have caused the highway alignment to be changed a number of times (see Plate 4.10). Generally speaking, in these regions the major slides are preceded by creeping movements which are frequently detected because of the density of population and structures in the area.

Highland Park is a community just north of the center of Los Angeles city built in hills composed of the interbedded siltstones and sandstones with the occasional altered volcanic ash layers described above. The hills are intersected with canyons in which increased erosion takes place during the winter. Although

Plate 4.10. Ancient, periodically reactivated landslide, coastal highway 101, Santa Monica, Calif. (Courtesy of Pafford and Associates, Los Angeles, Calif.)

many of the drainage patterns in this region have been altered by man to increase the stability of the slopes, some remain unchanged, even in inhabited areas. During a winter storm in January 1969, a very large amount of rainfall fell in two days and caused substantial amounts of erosion in one small canyon in Highland Park. Towards the end of the rainstorm, a block of material undercut by the stream moved out into the canyon blocking the stream, its movement downhill leaving unsupported an upslope block which followed its movement. This, in turn, was followed by a third block as the movement progressed uphill. Although there were no casualties, a number of homes were badly damaged.

Investigation following the slide showed that the eastern edge of the slide was a fault plane along which the material was in a very disturbed state through repeated movements on the fault. The bedding planes of the rock forming the landslide block were parallel to the surface of the slope at an angle of about 12°. Excavations on the upslope side of one of the sliding blocks revealed that sliding was taking place entirely on a montmorillonite layer, 5 m below the surface. Instrumentation was established at this location a day or two following the slide, and measurements of the details of the creeping motion of the upper block were made as it slowed down after the gross movement, as shown in Fig. 4.11. The character of these movements may be similar to the creeping motions that developed prior to the major motion of the landslide block. The total movement of the lowest sliding mass was about 20 m, and it was brought to rest by the other wall of the V-shaped canyon in which the slide occurred.

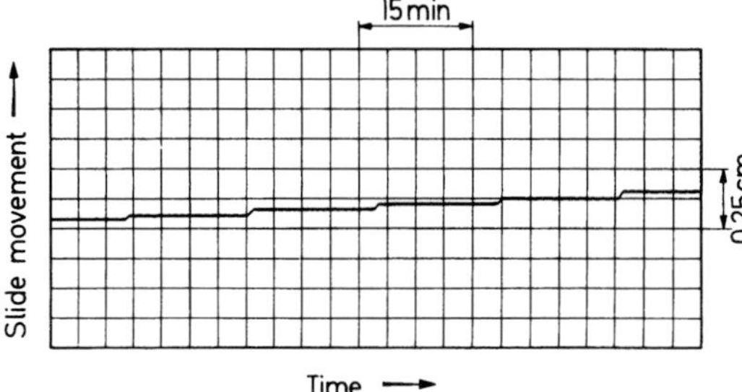

Fig. 4.11. Portion of a record of slide movement, California (R. F. Scott)

Portuguese Bend, California

In aerial photographs of the Palos Verdes Peninsula taken in the 1930's and 1940's before the population increase covered the area with houses, it is possible to see that a large area on the south side of the peninsula, in a region called Portuguese Bend, consists of landslide material. It is probable that the motion of this landslide has gone on at intervals through history; but probably not all of the material takes part in a landslide at any one time. Instead, one portion may slide, causing a reduction in the stability of an adjoining portion, which, a number of years, decades or centuries later, follows the motion. The appearance of the whole area is that of a rumpled carpet thrown on the hills. In spite of the obvious evidences of instability, no measures were taken to prevent the establishment of housing subdivisions and the construction of houses in the area in the 1950's.

In 1956 a large portion of the area, about a kilometer square, began to move on a slope of only 6.5°, and within a few months the movement had reached approximately 20 m. A scarp formed at the top of the sliding area, and intermediate fissures and fractures developed in the sliding mass itself (see Plate 4.11). The highway running across the base of the slide material had to be relocated a number of times in following years as the slide continued to move, houses located on and at the edges of the slide area were racked, and, in the majority of cases, had to be removed. To the present time (April 1974) the slide has moved a maximum distance of approximately 70 m, and is still

Plate 4.11. Portuguese Bend landslide, Los Angeles, Calif. (Courtesy of Metrex Aerial Surveys)

moving at a rate of about 3 m per year. The cause of the landslide was the subject of a legal debate between Los Angeles County and a homeowners' association. It is estimated that $10,000,000 worth of property damage was done by the moving mass.

The majority of houses constructed in the area employed cesspools rather than a sewer system, so that the waste water discharged from these houses went straight into the ground in the potential slip zone. The construction of houses and streets, of course, disrupted the natural drainage of the area, and once slide movements occurred, the breaking of water pipes and drainage pipes gave rise to an increase in the ground water supply. In addition to this, the Los Angeles County Highway Department in 1955 and 1956 was engaged in the construction of a highway over the hills to the north and down the Portuguese Bend slope to join the coastal highway. The steep grades required the placing of substantial quantities of fill, particularly at the top of what became the slide mass. In 1956, in August, construction workers on the highway were the first to observe signs of movement which increased progressively. The property owners in the area formed an association which sued the County of Los Angeles claiming that the damage had been caused by the fill involved in the road construction. Experts were found for both sides in the argument to claim, on the one hand, that the fill was contributory to the failure and, on the other hand, that failure was primarily caused by the change in the ground water pattern in the sloping area. Eventually the judge found for the property owners' association, and they were compensated by the county for the damage. Various estimates have been made of the costs of a variety of measures aimed at stopping the landslide; they are of the order of $10,000,000. There are at present, relatively few houses still left in the area.

Vaiont Dam, Italy

In the late 1950's a thin dome or cupola dam, 267 m high, was built across the Vaiont Valley of the Piave River in Italy, impounding a reservoir between the steep walls of the valley. The investigation prior to construction concentrated on the dam site itself, and only a few bore-holes were made in the rock slopes adjacent to the reservoir. In November 1960 a portion of the side slope of the reservoir moved. The rock movements continued until they reached a maximum of 2.5 m before a large slide occurred on the south bank of the reservoir upstream of the dam in October 1963. A volume of 250,000,000 cubic m of soil and rock fell into the reservoir. It displaced a very large volume of water which overtopped the dam by 100 m and swept both up and down the valley, killing almost 3,000 people (see Section 3.1). In this case, the effects of the landslide were indirect but no less disastrous.

Investigations after the failure revealed that the geological structures of the south bank had a chairlike form, similar to the sliding surface illustrated in Fig. 4.10 (c), and that failure took place in highly fractured oolitic limestone along such a shear surface. The large movements which had occurred prior to the final catastrophe indicate that the material strength had been reduced to its residual value. Analysis indicated that artesian pressures along the failure surface contributed to the instability of the sliding mass. About 60% of the failure surface was found to coincide with bedding surfaces in the limestones. The dam itself survived the disaster but was rendered useless by the presence of the slide debris in the reservoir.

4.7. References

Eckel, E.B. (Ed.): Landslides and Engineering Practice. Highway Research Board Special Report **29**, NAS-NRS Publication 544 (1958).
Hunt, C.B.: Geology of Soils. San Francisco: Freeman, 1972.
Kerr, P.F.: Quick Clay. Scientific American, **209**, 132–142 (1963).
Report of the Tribunal Appointed to Inquire into the Disaster at Aberfan. London: Her Majesty's Stationery Office, 1967.
Scott, R.F.: Principles of Soil Mechanics. Reading, Mass.: Addison-Wesley, 1963.

Sharpe, C.F.S.: Landslides and Related Phenomena. New York: Columbia University Press, 1938. Reprinted by Cooper Square Publishers, New York, 1968.
Terzaghi, K.: Mechanism of Landslides. In: Application of Geology to Engineering Practice. Berkey Volume, Geological Society of America, 83–123 (1950).
The Great Alaska Earthquake of 1964. Engineering Volume, Washington, D.C: National Academy of Sciences (1973).
Zaruba, Q., Mencl, V.: Landslides and Their Control. Amsterdam: Elsevier, 1969.
Zienkiewicz, O.C.: The Finite Element Method in Engineering Science. New York: McGraw-Hill, 1971.

Chapter 5
Hazards from Ground Subsidence

In the twentieth century particularly, uncontrolled pumping of water and oil from underground has led to sharp economic and cultural losses. Too late, the ground surface has been found to have settled or subsided below natural levels. In this chapter *subsidence* will be given a broad definition, as any displacement of a generally level ground surface arising from surface or subsurface causes.

In general, the principal kind of hazardous subsidence is vertical, the observable surface effect usually being a dish-shaped or bowl-shaped region of downward surface displacements. However, if an initially level ground surface is drawn down to this shape for some reason, horizontal surface displacements will also be involved. The extent to which these are dangerous to structures depends on the magnitude of the subsidence and the gradient of subsidence with radial distance from the center of the disturbance. The subsidence discussed in this chapter will be considered to result from surface or internal loading or the extraction or alteration of materials below the surface. Although subsidence associated with landslides or partial landslides does occur, for example, at the head of unstable slopes, that subject has already been treated in the previous chapter.

The effect of subsidence on a structure depends on the relative scale of the structural dimensions and the area subsiding. A subsidence bowl of several kilometers in diameter will have little effect on houses or buildings a few stories high located inside the displaced area, unless the movement increases the flood hazard in the area. However, structures of greater extent, such as canals, or possibly large bridge or dam structures, may well be stressed severely by displacements over such distances. Smaller structures will be affected to some extent by surface ground movements over lateral distances of a few meters to hundreds of meters.

Surface movements also occur on a wide range of time scales. Perhaps the greatest time range is involved in tectonic displacements which may develop in a few seconds during earthquakes (Plate 5.1), in hundreds to thousands of years as a consequence of removal of geological surface loads, or in tens to hundreds of millions of years in the vast mountain- and continent-moving processes. This last scale has the least effect on man's works. However, the movements associated with surface uplift following ice removal after the last Ice Age and, of course, earthquake effects, can both have a significant effect on human structures. The subsidence conditions of most serious consequence to structures are caused by man through his operations at and below the Earth's surface. These occur on time scales comparable to the lifetime of structures. In the assessment of a subsidence hazard, the mechanism, magnitude of displacement and time scale must all be known.

Plate 5.1. Raised beach, Montague Island, Alaska, after the March 27, 1964 earthquake in Prince William Sound (see Chapter 1). Kelp (inset) now well above tide level. (Photo by USGS)

5.1. Classification of Subsidence

A variety of classifications of ground subsidence is possible, including, for example, considerations of dimensional and time scales, mechanisms and possibly effects. Nevertheless, the speed with which subsidence occurs is slow except in a few special instances which will be mentioned, and so velocity or time scale is not such an appropriate unit of classification as, for example, with landslides. The dimensional scale on which subsidence occurs is intimately related to the mechanics causing it, and thus it is convenient to consider classifications based on these two considerations only.

Scale

Following the Alaskan earthquake of March 1964, surveys over a large area of Alaska and the adjacent Aleutian Islands indicated that considerable changes in the ground surface elevation had occurred (see Fig. 1.2). Superimposed on these vast regional movements were local movements due to, among other things, the compacting effects of the vibration on saturated soil deposits. The effect on man-made structures and the indirect effects on man's activities were, of course, widespread and are still continuing. Docks, harbors, and waterfront structures in some areas were elevated far above the new mean tide levels; at other locations these facilities were submerged, while many structures had to be relocated or rebuilt. Long-range effects on the biology of the whole area were also extreme. Beaches uplifted meant the death of certain forms of shell life; the alteration in river régimes, sand-bank formation and coastline topography meant changes in fish movements and therefore affected the fishing industry.

During previous Ice Ages, and the last one in particular, which tapered off about 11,000 years ago, portions of the earth's crust were subjected to ice loads from thicknesses of ice of as much as a few kilometers. The ice was present for tens of thousands of years. Since the areas over which the ice loads were applied were great in comparison with the thickness of the crust, a consequence of the loading was the depression of the crust with time as the less viscous mantle material below moved out laterally under the long-term loads. When the ice melted, the removal of load on the surface caused the crust which had been depressed below an equilibrium level to move upwards to reach equilibrium again as the mantle material moved in below it. Geophysicists refer to the process as *isostatic adjustment,* the time scale involved being tens of thousands of years. Thus, several regions of the earth's crust are still moving upwards as part of the adjustment following the removal of the ice 11,000 years ago. The areas principally affected are Norway, Sweden, Finland and Denmark (Fennoscandian Uplift); portions of the North American continent have also been uplifted. The effect is noticeable as a gradual lowering of sea level with time. Rates of movement at different periods in the past in various areas have been inferred from geological evidence in lakes and indications of former beaches around lake and seacoast margins. The rate of change is as much as 1 m/century on the shores of the Baltic Sea. The movements may contribute to the instability of offshore slopes, and possibly to long-term changes in mass movements of sediment around harbors.

Movement of the Earth's crust can occur on a scale much smaller than the extent of an Ice Age glaciation (see Section 5.4). Movements have been observed around the borders of natural and man-made lakes which reflect the readjustment of the local crust to the changing load imposed by a body of water. The Great Salt Lake in Utah is the remnant of a larger, older body of water to which the name Lake Bonneville has been given. This lake diminished in size through evaporation in the past few thousands of years and thus the local weight on the crust has been decreasing and the area rising. Downward movement of the Earth's crust has been caused by the weight of Lake Mead, U.S.A., the body of water impounded by Hoover Dam on the boundary of Nevada and Arizona, and around Lake Kariba in Africa (see Section 1.3).

Regions of size comparable to those affected by man-made lakes are also subject to subsidence as a consequence of other human activities such as the withdrawal of water, gas and oil and solids from subsurface rock layers. The resulting surface effects are generally in the range of a few meters of subsidence or less over areas ranging from hundreds of kilometers to a few hundred meters in extent. In general, these movements occur over periods of time measured in tens of years.

Somewhat smaller areas are affected by surface loading conditions caused by man. With suitable underlying soil conditions, a building or dam can settle a vertical distance of a few centimeters to a few meters and influence the subsidence of ground over an area two or three times its own dimension. Mining and tunneling activities carried out without sufficient precautions too close to the ground surface have also caused ground surface displacements at about the same scale. Finally, at perhaps the smallest scale of structural interest, displacements of a few centimeters over distances of a few meters can result from the inadequate compaction of backfill in trenches excavated for sewers and other utilities.

Mechanisms

The slow movement of the large plates forming the more stable areas of the Earth's surface causes stresses to develop in the contact regions, so that in appropriate conditions, rupture may not occur suddenly and cause an earthquake; instead, creeping motions develop. As the stresses grow to the point at which an earthquake occurs, surface manifestations of ground displacements can sometimes be observed.

When large loads, such as ice sheets, are applied to the surface of the Earth, the surface crust behaves essentially as a solid elastic material, whereas the underlying mantle at higher temperatures and pressures behaves viscously and flows under the shearing stresses applied to it through the crust by loading or unloading. The slow creeping flow of the mantle material then gives rise to the observed vertical crustal displacements. The shearing stresses caused by the viscous movement of the mantle material underneath the crust also give rise to horizontal displacements in the overlying crust. However, these are substantially smaller than the vertical movements.

The largest surface displacements caused by man-made features result from the removal of material or fluids from below the Earth's surface, the most noticeable effects having been caused by the withdrawal of water, oil and gas at various depths. Sometimes the water withdrawal leads to great cultural tragedy such as the inundation of Venice (see Section 5.4).

Consider a layer of sand situated at a depth of perhaps 100 m below ground surface, overlain by an impervious clay layer, and saturated with water. Under existing natural conditions, an element of sand at a particular depth is at equilibrium. On it acts vertical total stress, which results from the weight of soil and water in the overlaying layers, and a lateral total stress which depends on the vertical stress and the material properties. As described in Section 4.2, a total stress in the soil can be broken down into the sum of an effective stress in the soil structure, and a pore water pressure in the pore fluid. If a well is drilled into the sand layer and permitted to fill with water until an equilibrium height is reached, the water level in the well will represent the pore pressure in the soil at the base of the well.

In Section 4.2 it was explained that soil will respond only to changes in the effective stress condition. For example, a load may be applied everywhere on the ground surface by a layer of fill. In that case, the surface load appears at underlying soil elements as an increase in total stress. If, through drainage in the permeable sand, the pore pressure stays at its original value, there is an increase in effective stress, and thus the soil compresses.

Alternatively, if the total stress acting vertically on the element remains the same, and the pore pressure is decreased, it follows that the effective stress will also be increased and the soil will compress. This illustrates what happens when water is withdrawn from an underground layer by means of a well or wells. To extract water from the well, the water level is lowered by pumping; in other words, the pore pressure in the aquifer from which the water is being withdrawn is decreased. Thus, the soil layer or aquifer compresses vertically if a number of wells are installed in it and all draw the water down equally. If only one well or a cluster of wells occupy a small area compared to their depth, compression again occurs but the effect on the stresses and deformations in the overlying soil is more complicated. In the example above, lowering the water pressure in the sand layer will also cause a reduction in pore pressure in the overlying clay zone. This will compress but more gradually with time as the water drains slowly out of the clay.

The extent to which the ground surface subsides depends on the amount by which the water level is lowered, the compressibility and depth of the layer from which water is extracted, and the stiffness or rigidity of the overlying soil materials. The fluid withdrawn from the layer in question by lowering the pore pressure can consist of water, oil or gas. Commonly, water is withdrawn from shallower layers, consisting of fairly compressible soils, and oil and gas from deeper layers of less compressible material. Consequently, more surface settlement is generally observed as a consequence of water extraction than is associated with oil and gas fields. However, all the processes can give significant ground settlements.

Any process which extracts water from the ground can result in surface movements. Ground displacements have been measured as a consequence of the drying of soil either naturally or artifically as in the vicinity of brick kilns, or adjacent to rows of trees. The trees withdraw the water from the soil and evaporate or transpire it through their leaves. The drainage of marshes, which frequently contain substantial thicknesses of highly compressible organic soils, can cause very large vertical movements at ground surface, for which reason parts of the Fens in eastern England have subsided up to 5 to 10 m since the early 1800's.

At a smaller scale, it can be seen from the above discussion that the application of surface loads can also give rise to surface displacements as a consequence of the increase in effective stress on the compressible layers below; all structures built on the surface cause some surface movements.

A different effect arises if the extraction process involves the actual removal of solid material from the underground, such as in mining or tunneling operations.

Here, the state of stress in the surrounding soil or rock is changed and the material will undergo both volume and shearing deformations as a consequence. In general, the construction of a tunnel close to the surface causes the ground above the tunnel to settle. However, if a tunnel or cavity is driven in a particularly cohesionless material or *incompetent rock* (one which fractures or cracks easily) with inadequate support, a further and more serious consequence can arise, namely the process referred to as *caving* or *stoping* in which the rock in the roof collapses to fill the tunnel and leave a void above it. The material forming the void walls and roof in turn also falls in, so that the void space propagates upwards. When it reaches ground surface there is a final sudden subsidence and a *sink-hole* or cavity is formed. The last stages of the caving process can proceed rather rapidly, and frequently there is little warning of the final event, particularly if the tunnel or underground chamber is an old one no longer monitored by mining authorities.

A similar process can occur naturally where limestone formations exist below the ground surface. Limestone is slightly soluble in water, and particularly in hot and wet climates the solution of underground limestone deposits may proceed actively. The overlying soil deposits then subside or collapse into the solution cavities formed in the limestone to give rise to surface sinkholes or depressions. Where the limestone layers are widespread, this process results in a typical form of surface referred to as *karst topography* after the district in Yugoslavia where the phenomenon is common. Collapse features are also associated with lava tubes in volcanic areas (see Chapter 2).

Another process occurs in certain types of soils or soft rock formations. In Sections 1.2 and 4.6 landslides caused by earthquakes in China were described, where loess was the material involved. Loess has the characteristics of a soft rock, since the fine-sized grains form an extremely loose open structure held together by a water-soluble mineral cement. If a deposit of loess, which is quite firm and hard and forms a good foundation material in the dry state, is subjected to excessive wetting by sprinkling or ponding water on its surface, the dissolution of the water-soluble cement will cause a collapse of the soil structure. For this reason, these materials are referred to frequently as *collapsing soils*. The area of subsidence in this case is markedly confined to the region in which ponding or wetting has occurred. The adjacent area remains largely undisturbed and is usually separated from the subsidence by cracks and fissures.

Loess is not the only soil type in which this process develops. In arid regions, infrequent rains frequently give rise to *flash floods* which advance down previously dry canyons, valleys or gullies (see Chapter 7). The water picks up the dry soil grains and carries them along in suspension, so that the flow becomes a mudslide or mudflow which spreads out as it reaches the lower gradients near the valley floor. The material need not be saturated and can contain a relatively large quantity of air, and is thus deposited as a very loose soil structure under the right conditions of floods, gradients and grain-size of the local soil. Because of the infrequency of these floods, such layers can develop to considerable depths as, for example, along the west side of the San Joaquin Valley in southern California, U.S.A. It has been suggested that freezing may have played a part in the formation of the porous soil structure. When such a soil deposit is wetted in a subsequent rainstorm, or as a consequence of man's activities, the loose structure collapses and large subsidence craters or sink-holes can be formed to depths of a few meters. With these materials the phenomenon is usually referred to as *hydrocompaction.*

Apart from the regional movements that are associated with earthquakes, the vibrations they generate also have local effects depending on the soil type. If the soil is sufficiently loose, and in either a dry or a saturated state, vibratory motion may shake it into a denser configuration; the consequence, as usual, is a surface settling. When the material is saturated, the more serious result of the vibration is the liquefaction of the soil which can cause major damage to buildings and structures located on the liquefying material; once again, the final result after the excess pore pressure has dissipated is a surface settlement of the ground.

One final mechanism that must be discussed involves not the subsidence of the ground, which is the usual consequence of changes in the soil or rock condition, but a swelling or rising of the ground surface, a phenomenon frequently encountered in arid areas of the world in soils containing the mineral montmorillonite. Montmorillonite clay particles are of extremely small size and have a sheet-like structure, with the ability to absorb large quantities of water. Consequently, if a soil containing some proportion of a montmorrillonite clay is present in a dry state and water is supplied to it, the clay will take up the water and in so doing will expand, resulting in an uplift at ground surface.

In general, because soils are not homogeneous and the absorption of the water varies from place to place, the uplift is seldom uniform all over a site and the differential movements can cause considerable distress to structures. Houses may be constructed in a region where the soil has this swelling potential. If the presence of the clay is unsuspected, the phenomenon will only develop after the houses are completed, inhabited, and the residents have begun watering their gardens or using cesspools. The infiltration of water into the ground causes swelling which heaves and cracks the base slabs of the houses and, in so doing, breaks the various connecting utilities, with the consequence that in extreme cases, a house can be rendered uninhabitable.

5.2. Subsidence Analysis

In a number of cases of ground subsidence, for example, where tectonic movements are concerned, it is impractical to consider the analysis of the problem. For instance, the areas of the world where isostatic adjustment is going on are relatively well-known and new cases are not likely to occur within the time scale of importance for human affairs. The rates of uplift have been measured in these regions and can be taken into account for whatever construction demands them, but cannot be altered by human intervention.

The discussion of analysis techniques will therefore be limited to an examination of those classes of ground subsidence or settlement which result from man's activities, so that the important characteristics of a given problem, such as timing, quantities and material properties, are generally known.

Surface Loading; Consolidation

For simplicity, only the case of one-dimensional compression of a layer under an added load will be considered; references are given at the end of this chapter to the more complex two- and three-dimensional problems which may be treated by analysis. In Fig. 5.1 a compressible saturated soil layer of thickness H lying on top of an impervious rock surface is shown. The soil is quite compressible compared with the rock so that it may safely be assumed that no surface settlement will occur as a consequence of the load on the underlying rock material. The problem is, in general, divided into two parts: (i) the determination of the total amount of compression of the soil layer which will ultimately occur under a superimposed load p, caused by the presence of, for example, a fill layer of compacted soil covering the whole site under investigation; (ii) the calculation of how the compression of the soil layer develops with time.

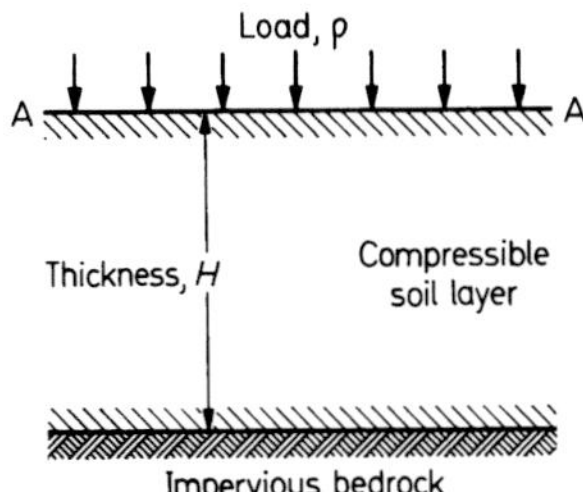

Fig. 5.1. Settlement and consolidation of soil layer

The total compression is arrived at through the product of the compressibility of the soil layer, its thickness and the magnitude of the load p. How the settlement increases to its final value relates to a characteristic of the soil layer other than its compressibility. For the soil to compress under the load p the water has to escape from the soil, that is to say, drainage of the excess water must take place. The rate at which this can occur depends on how fast water can move through the soil and availability of drainage boundaries. The property of a soil to conduct water is called its *permeability*, and this depends primarily on the grain size of the material. Soils such as well-sorted sands and gravels with large grain sizes have a high permeability, and water can escape relatively quickly. Thus, in a sand or gravel comprising the soil of Fig. 5.1, the application of the load p will instantaneously increase the pore pressure in the soil, but the pore pressure can dissipate rapidly through drainage at the surface AA. The time to complete dissipation will vary from a few seconds to a few minutes when the thickness of the soil layer H ranges from centimeters to tens of meters. The surface settlement will therefore take place simultaneously with the load.

On the other hand, in soils such as clays and silts, the permeability can be extremely low and the excess water can only escape very slowly. In these fine-grained materials, the settlement in layers a few meters thick which occurs under the load may take from months to years to reach its final value. The time taken to attain nearly complete settlement depends directly on the compressibility and the square of the thickness of or length of the drainage path in the clay layer and inversely on its permeability. The process of gradual compression of a soil under applied load is called *consolidation.*

If the layer of soil of Fig. 5.1, instead of resting on an impermeable bedrock material, is underlain with a coarser sand or gravel layer through which the water may also drain from the compressible layer, the drainage path length is halved in comparison with the case of Fig. 5.1. The surface settlement or consolidation will take place four times more rapidly because of its dependence on the square of the drainage path.

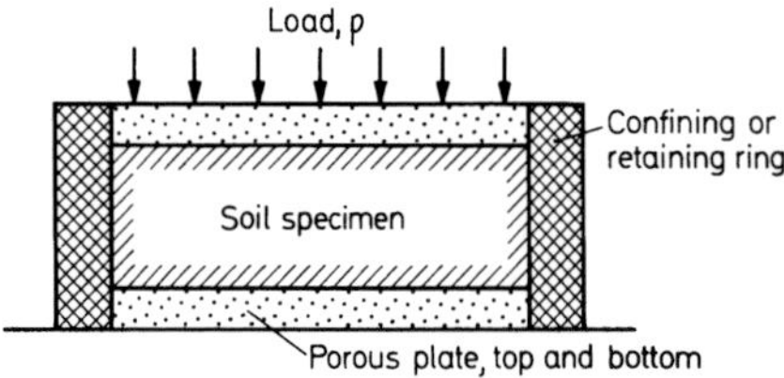

Fig. 5.2. Consolidation test

To calculate the amount and rate of settlement which will develop, it is necessary to obtain samples of the compressible layer and subject them to one-dimensional consolidation tests of the type shown in Fig. 5.2. The disc-shaped samples approximately 6.5 cm in diameter and 2 cm thick in typical dimensions are obtained from borings made in the soil. The discs are placed in the consolidation apparatus, and a porous ceramic plate saturated with water is applied to the top and the bottom of the soil. A small vertical load is applied to the soil sample, and the rate and final amount of compression (usually at the end of 24 hours) of the specimen measured. The vertical load is increased in successive steps, and rate and final compression measured each time, until the load reaches a value somewhat higher than the pressure which will be eventually applied to the ground surface. Because of the variable properties of soil, sufficient samples are usually tested so that average or representative values are indicated for the whole layer. From the tests, the compressive and time-behavior characteristics of the material in the pressure range of interest are obtained.

Fluid Extraction

There are two common simplified soil or rock profile conditions under which fluids are withdrawn from below the surface; these are shown in Fig. 5.3(a) and 5.3(b). In Fig. 5.3(a) a permeable layer overlies an impervious material and contains liquid whose level prior to extraction is shown as AA. This is commonly the situation when the fluid to be extracted is water. In Fig. 5.3(b), the pervious layer containing the fluid, the *aquifer*, is confined above by an impervious region through which wells have to be driven to tap the fluid in the aquifer. Water is frequently obtained from water wells in this configuration and it almost always applies when the fluid is natural gas or oil. In Fig. 5.3(a) the aquifer is said to be *unconfined;* in Fig. 5.3(b) it is *confined.* In Fig. 5.3(b) the fluid exists in the pervious layer under some pressure such that when a well is drilled into the layer, the level of liquid in the well rises to the level shown as AA in the figure. If the level AA lies above ground surface the condition is referred to as *artesian,* and the well will flow. If it is below ground surface, pumping is required to extract the fluid. The consequence of pumping is to lower the height (or pressure) of fluid to level BB as shown in both figures.

The lowering of the fluid pressure increases the effective stresses in the layer from which fluid is withdrawn. In the case shown in Fig. 5.3(a) the effective stresses increase for two reasons: the first is that in the lower zone of material that still remains saturated, the *total* stress increases since the material between AA and BB is changed from a buoyant condition to one in which its total weight is exerted on the underlying zone; secondly, because the pore pressures in the underlying material have been reduced. The increase in effective stress causes the material below the line AA to compress. Since the soil from which the water is extracted is usually relatively permeable, it is found that the displacement of the ground surface follows the lowering of the water table directly and immediately. On occasion, pumping is stopped and natural replenishment of the ground water causes the water levels to rise from BB to the level AA again. Most granular materials exhibit an irreversible behavior upon being compressed and either do not expand or expand very slightly on relief of stress. Thus the increase in water level does not result in the return of the ground surface to its original condition. At best, the settlement of ground surface is halted by the re-establishment of former water table levels.

In the situation represented in Fig. 5.3(b) which might result from water, oil or gas withdrawal, there is no change in total stress in the system as water is withdrawn, since the presence of the impermeable upper layer will cause the permeable layer to remain fully saturated. In this case, the decrease of pore pressures consequent on lowering the effective liquid head from AA to BB causes a corresponding increase in the effective stress, and therefore the fluid-bearing layer compresses. There is also an increase in effective stress in the confining layer and it will compress in addition, although, since it is usually fine-grained, the compression will occur more slowly.

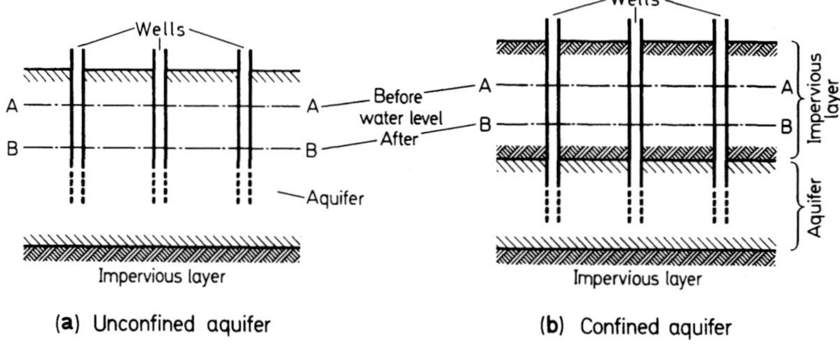

Fig. 5.3. Drawdown by wells in unconfined and confined layers

Water wells are usually located in aquifers in the range of tens to hundreds of meters below ground surface, and water table draw-down has both immediate and time-dependent effects, depending on the presence of fine-grained soils. Oil and gas withdrawals usually take place from zones hundreds to thousands of meters below ground surface. The permeable structures in which the oil or gas is trapped usually have a complicated structure, so that analyses of the ground surface displacements caused by oil and gas withdrawal are complicated. With time, as oil, gas or water is pumped out of the ground in the situation shown as Fig. 5.3(b), the fluid pressures in the fluid-bearing zone decrease so that a diminishing amount of fluid is yielded. Sometimes, to increase the quantity of fluid obtained or to halt surface subsidence, water may be injected around the edges of an oil field in order to "repressurize" the field. Because of the irreversible behavior of granular materials discussed above, this repressurization does not cause the ground surface to move upward; its effect is to retard the subsidence.

Other Mechanisms

In Section 5.1 a number of processes which caused ground surface subsidence were discussed. These included material removal by mining or tunneling, saturation of compressible materials with water, and vibration. In general, calculational methods for the determination of the amount of settlement caused by these mechanisms have not been employed as widely as those previously discussed. In particular, the calculation of surface effects which can arise as a consequence of tunneling operations is difficult, partly because of the material properties involved, the unknown tunneling conditions, and the two- or three-dimensional nature of the problem. The degree of support of the tunnel during excavation has an important effect on the stresses and displacements generated in the surrounding material.

An estimate of the amount of compaction which can take place in a particular soil profile as a result of either vibration in an earthquake or vibrating machinery, or of saturation of the ground, can be determined by means of field and laboratory tests. For vibration compaction, samples of the soil are obtained from different depths in as undisturbed a condition as possible and subjected to vibration tests in the laboratory to demonstrate the possible density or volume change which the material may undergo. The intensity of vibration to which it will be subjected in the field is estimated and calculations of ground-surface subsidence are based on these considerations. It is, of course, impossible to carry out a controlled field test subjecting the soil to realistic earthquake ground motions.

5.3. Detection and Control

The processes which lead to subsidence of the ground surface are subtle and frequently operate for long periods without suspicion arising that surface effects are occurring. Thus, it is quite common to find that the first surface manifestations noticed are those reported by highway or railroad survey crews, and homeowners. Survey crews observe movements of bench marks and other reference points and ultimately that an unusually high amount of maintenance is required on a particular stretch of road or railway; homeowners notice cracks developing in driveways and garage or house walls. The precise detection of subsidence amounts and distribution must be carried out through surveying and leveling networks, usually by surveyors working for a government agency. The measurements obtained in a particular subsiding area have to be referred to bench marks or monuments sufficiently far outside the area to be unaffected by the subsidence. For the ground displacements that accompany a large earthquake, or regions of intensive fluid withdrawal this requires a very substantial effort indeed, usually involving frequent re-surveying of lines extending over many hundreds to thousands of kilometers. Under these latter conditions, it is obviously an immense task to determine if the ground movements are changing or increasing with time.

It is not surprising that many cases of subsidence remain unreported, and that it is frequently impossible to answer questions relating to ground settlement in specific areas or as a consequence of, for example, water, oil or gas extraction in a particular region. By the time subsidence has been detected as a result of visible surface effects, a long period may have elapsed since the process causing the subsidence was initiated. The detailed examination of surface movements therefore requires a study of the results of surveys carried out many years before with generally less precision than is desired. This will be better understood when it is considered that frequently subsidence amounts are of the order of a meter or two taking place over distances of tens of kilometers in periods of years.

A detailed history of the development of a particular subsidence basin requires an exhaustive correlation of surveys carried out at various times by various companies and agencies; consequently a considerable uncertainty frequently attaches to the results.

Where settlement is caused by building loads applied to the ground surface, analyses are usually performed to insure that the ensuing differential settlements do not impose excessive stresses on the structure. If a new structure is to be built among existing buildings, it is also necessary to check that the displacements caused during and after construction of the new building do not have a harmful effect on the adjacent older structures. These displacements can arise most commonly because of the weight of the new structure, but they may develop through the inward movement of the temporary retaining wall of the excavation if it is not carefully constructed. It is also possible for the soil next to an excavation to consolidate and settle because the vertical face of the hole provides a new drainage surface for the soil water. Drainage of the water to the face permits pore pressures to drop with a consequent increase in effective stresses leading to compression of the soil.

Where building settlements are expected to occur, special surveys are usually made periodically to bench marks located in and around the structure in question. Subtle changes in elevation can be measured with a special leveling device employing a water-filled tube. When the settlements are very important or critical for the stability of the building under construction or adjacent structures, fixed bench marks may be established by drilling holes to bedrock and installing pipes in the holes. An inner pipe or rod constitutes the bench mark and is protected from the ground movements by an outer casing placed in the bore-hole. The inner pipe is concreted into the bedrock.

Sometimes it is desired to know how much settlement each layer in the soil profile contributes; in which case, pipe and casing level indicators may be anchored at different levels in the compressing material. The difference in settlement between bench marks embedded at two different depths indicates the compression undergone by the layer between the anchor depths. In the San Joaquin Valley, California, water is withdrawn from various aquifers, and the ground surface settles (see Section 5.4). To determine the amount of compression in each aquifer and in the confining strata, anchor weights were placed at the bottom of casing located in boreholes drilled to various depths down to 700 m. A thin stainless steel stranded cable was stretched from the anchor weights to the top of the hole, where it was kept in tension by a counter weight. A continuously operating recorder measured the subsidence through a wire clamped to the cable. By monitoring the water level in adjacent wells, the proportion of subsidence occurring in each aquifer was directly related to water level or pressure changes. A typical record of water level and associated subsidence is shown in Fig. 5.4. The effect of variation in pumping rates can be seen.

No actual control can, of course, be exercised on the movements of the Earth's crust which are generated by tectonic activity. Precautions, in the passive sense, can be taken in the case of sensitive or important structures which must be located in seismically active zones. The effect on the operation of a nuclear power station located on the coast, for example, of a change of the land surface elevation by a few meters with respect to the sea level can be assessed during the design of the power station. Cooling water from the plant is withdrawn from and returned to the sea; a change in sea level, has implications for the circulation of the cooling water in the power plant system. This consideration is separate from the question of the behavior of the power station during the strong shaking of an earthquake, or its protection from tsunami hazards (see Section 1.5), and bears instead on the post-earthquake operation of the plant.

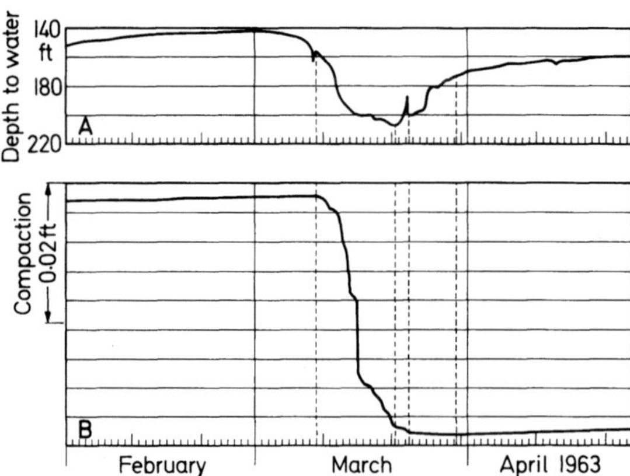

Fig. 5.4 A and B. Land subsidence, Tulare-Wasco area, California. Observed rebound and compaction near Pixley. (A) Water-level fluctuations, (B) Compaction record

When site investigations and analyses indicate that construction of a particular building, embankment, or dam will cause total settlement of more than a centimeter or two, precautionary measures are taken. Sometimes the structure is founded upon piles extending down to a firmer layer. Alternatively, the structure may be designed with a deep basement so that the excavated material weighs the same as the building; because of the similarity of the situation to that of a ship, this solution to the problem is called a *floating foundation.* It may be possible to apply loads to the soil before construction begins in order that settlement will take place of approximately the same magnitude as the building would cause. These loads are usually applied by embankments or mounds of earth fill, and when the desired effect has been achieved, the fill is removed and construction takes place. The consolidating effect of the fill can sometimes be enhanced by the installation of wells to lower the ground water, by the mechanisms discussed earlier.

On occasion it is possible to design the structure in such a way that substantial settlement can occur without impairing the safety or function of the building, either by making the structure extremely rigid, so that it settles as a unit without differential movement, or by making it flexible so that the settlements which do occur do not overstress the members of the structure. A rigid structure might be used for a power station, a very flexible structure for a warehouse for stockpiling material such as newsprint or steel plate. The displacements which occur in the latter case can usually be eliminated by releveling the floor periodically.

Because the compression which occurs in a soil layer as a result of loading or water withdrawal is irreversible, the large scale settlements which develop under certain conditions, for example, from the pumping of water or oil, can only be halted—not turned around. The compression is caused by the fall in fluid pressure; when the fluid pressure is stabilized or raised, the settlement stops. The technique of *repressurizing* consolidated zones has been employed in various locations to halt or slow down surface subsidence. The pressures can be raised or maintained, for example, by pumping water into the peripheral regions of an oil field. In addition to slowing or stopping subsidence, the added water aids in increasing the amount of oil extracted from the field. The cessation of subsidence by raising the water table is clearly seen in Fig. 5.4.

5.4. Ground Subsidence Case Histories

Fennoscandia; Lake Bonneville, Utah, USA

In the last Ice Age the whole of the Scandinavian Peninsula was glaciated with an ice sheet estimated to be of the order of 2 km in thickness and 1,600 km in diameter. The Ice Age continued for about seven to ten thousand years, long enough to depress the crust in the area occupied by the ice and to effect crustal displacements to a distance of many kilometers beyond the ice margin. If the load had been maintained long enough for equilibrium to be reached then, in effect, the ice load and crust would have been floating on the liquid mantle at an equilibrium level much as a ship does at sea.

However, it is likely that the ice load was not present for long enough to achieve equilibrium. It is estimated that the maximum depression of the area under the ice load was as much as 250 m relative to the present at the time when the glaciers began to retreat. The removal of the ice means that this deflection began a slow recovery to an equilibrium level which has still not been reached.

The effect of crustal depression or uplift will develop at any place where load has been or is applied or removed. The magnitude of the elevation changes, and, thus, their perceptibility depends on the size of the load and the extent of the loaded area. Probably crustal movements take place wherever the dimension of the loaded area is greater than the thickness of the crust; this varies from 15 to 50 km in different parts of the world. Thus, any natural or man-made feature 50 km or larger in diameter, which persists for times of 10 years or longer, will give an observable effect with present-day measuring systems. These will detect vertical elevation changes of about a centimeter over distances of the order of 100 km with level surveys run every few years.

There are a number of man-made lakes in the world which are large enough to give observable effects. Among those for which such movements have been noticed is Lake Mead, Arizona (18 cm depression in 15 years), and Lake Kariba, Zambia, Africa, but a number of other reservoirs such as the Bratsk Lake, U.S.S.R., and Volta Lake, Ghana, are probably large enough to show the effect. The dewatering of the polders (see Section 7.4) of the old Zuider Zee in the Netherlands has probably caused some uplift.

The best-known natural feature for which surface elevation changes have been measured is the area around Salt Lake, Utah, which was formerly a much larger natural lake called Lake Bonneville. For many thousands of years the rate of evaporation of the former Lake Bonneville has exceeded the inflow from its catchment basin, so that the lake has been steadily shrinking in size. In this case, the removal of the water load has meant that an upward motion of the crust developed. The rate of movement is currently about 20 cm a century as measured by careful surveying techniques. Previous rates of movement have been somewhat higher than this value estimated from the position and dates of former raised beaches. Movements of this order of magnitude at inland reservoirs and areas such as Lake Bonneville are not likely to give rise to geological hazards.

Crustal subsidence is also undoubtedly related to large basins of active sediment accumulation. However, the deposition rates are on the order of 0.3 m per 1,000 years or less so that subsidence would be extremely slow, and would also be masked by the continuing compression of the sediment itself.

Surface Loading; the Leaning Tower, Pisa, Italy

When the underlying soil is a coarse or granular sand and gravel in a relatively dense state, the amount of vertical movement that occurs for any normal structure is usually of the order of a centimeter or less, and takes place as the loads are applied because of the high permeability of these materials. For most construction work, these amounts can easily be taken into account in the design of the structure. However, erection of a building on top of a deep layer of compressible clay can cause settlement of 1 m or more which may take tens of years to centuries to develop.

Since the mechanics of consolidation and settlement were only elucidated in the 1920's and 1930's, there exist a number of famous buildings built before the principles involved were recognized, of which the most notable is the Leaning Tower of Pisa, Italy. Construction of the tower was begun in 1173 at a site about 50 m distant from the cathedral of Pisa for which it was intended to be the bell tower. The separation distance probably indicates a recognition on the part of the builders of the problems involved with the subsoil at that area, as the cathedral itself, built 100 years earlier, shows unmistakable evidence of substantial differential settlement and subsequent repairs.

Some time after construction of the tower commenced, it was observed to be leaning and construction was stopped just above the third gallery or level in 1185. The tower remained in this condition for almost 100 years, until 1274 when work recommenced. Another 10 years of effort saw the structure up to the seventh level and almost finished. Only the highest level, the chamber for the bells, still remained to be added, but this period of construction had caused the tilt of the tower to accelerate greatly, so work was stopped again. By this time, it is estimated that the cornice at the seventh story was out of plumb with respect to the base by almost a meter. This had increased to over 1.5 m by 1350 at which time a final 5 years' effort saw the construction of the bell chamber or gallery and completion of the entire structure.

The tilting has continued, and now amounts to slightly over 5 m at the seventh floor cornice, while still increasing at a rate of about 0.02 mm a year. The angle of inclination of the 59 m high tower (measured from the foundation base) is nearly 5°. Each time that construction was recommenced an attempt was made to straighten the tower. Thus, in its final form, the tower is slightly banana shaped, with the form of a tree situated on a hillside which is creeping downslope. A fact which is less often observed about the Tower of Pisa is that it has also settled on the average more than 2 m since it was built, and the entrance to the structure is now this distance below ground surface.

The tilting of the tower is attributed to the presence in the substrate of a layer 2 m thick of compressible clay whose upper surface is only half a meter below the massive foundation. The clay overlies a 4 m sand layer, which, in turn, rests on a thicker clay stratum. For such a massive structure, the original foundation was placed at a very shallow depth of only 2 m below ground surface. The consequence of the tilting is that the pressures on the soil on the downhill side are increased and on the other side decreased. The maximum and minimum pressures are now 9.8 and 2.7 kg/cm^2 respectively. Greater settlement accompanies the higher stress and the tower movement is therefore inherently unstable.

It seems likely that the foundation suffered a partial bearing capacity failure in the early stages of construction, perhaps because of shearing in the thin clay layer or in the clay and the underlying sand, which would have caused a sudden tilting immediately apparent to the builders and may have been the cause of the first halt in the work. With the passage of time, the excess pore pressures dissipated, the clay consolidated and became stronger under the load, so that it was able to take the increased stresses of the second period of construction. This must have been the most exciting time, with the increase in the rate of tilting and settlement. At about the time of construction of the 5th story in 1278, the structure must have been very close to complete collapse due to foundation failure. It was saved only by the extreme slowness of the progress, which permitted the clay to pick up strength as it drained and consolidated under the increasing load. The continued movements now are probably the consequence of viscous flow or creep of the underlying soil material at a relatively constant volume under the very high stresses to which it is subjected.

A number of other towers of medieval age in Italy and elsewhere also lean, but to a lesser extent than that of Pisa.

Settlement problems exemplified by the case of the Tower of Pisa are common in many other parts of the world where deep beds of relatively soft clays underlie cities, for example, Mexico City, Boston, Massachusetts, and Chicago, Illinois. Mexico City is founded on the bed of an ancient lake whose sediments include volcanic ash deposits emitted by the surrounding volcanoes during prehistoric periods of eruption. The ash, as in the southern California area, has weathered to montmorillonitic clay, which has extremely high water contents (the ratio of the weight of water to the weight of solids in a unit volume of clay), and will readily compress when subjected to external load or when the water is removed by pumping or drainage. Large structures founded on the Mexico City clay have undergone settlements of a few meters, frequently accompanied by substantial differential movements. The Basilica of Guadelupe, for example, is tilted about 5°. For this reason, many of the newer structures in Mexico City are founded either upon piles or with the assistance of modern foundation techniques.

One of these techniques consists of the floating foundation mentioned earlier, but in Mexico the clay is so compressible that the presence of the excavation and its drainage surfaces causes settlements in the soil adjacent to the hole. To prevent this from occurring, wells have been sunk around the excavation in some cases, and water has been pumped into the wells to replace the water being drained from the excavation face, which has been successful in inhibiting the peripheral settlements. In Boston, Massachusetts, larger building have settled up to 1 m over periods of twenty to thirty years and most modern large structures in that city are now founded upon piles reaching approximately 35 m down to bedrock.

Water Extraction; Mexico City and Venice, Italy

One of the most common, yet hidden, causes of subsidence of the surface has been the pumping of water from wells located in permeable, compressible strata below ground or in permeable strata adjacent to impermeable but compressible clay layers. Thus, pumping from water wells may result in both an instantaneous settlement of the ground surface and a settlement delayed in time as a result of consolidation.

This process has occurred in many parts of the world, most notoriously in Mexico City, Shanghai, People's Republic of China, and Venice, Italy. In Mexico City, with foundation soils as described previously, within the volcanic clay layers there are other beds of water-bearing sands and gravels. These have been used by the city and private industry as a water supply which has been tapped by wells. The extraction of water from these layers has caused the whole area of Mexico City to settle a total of 7 m in the period 1880 to 1970, and by 1952 settlement was occurring over this large area at a rate of about 30 cm a year. Under these conditions, it becomes apparent that the foundation expedient of locating a structure on piles was not wholly successful, as the subsidence of the surrounding ground causes such structures to appear to rise above the surface at the subsidence rate observed. The rate of subsidence did not slow until measures were taken to supply Mexico City with water imported from outside the city, so that now (1974), the rate of subsidence has slowed to a few centimeters a year.

In Shanghai also, the ground settled as a consequence of the withdrawal of water from wells to supply industrial needs. The settlement, first noticed about 1921, reached a maximum of about 2.4 m by 1973, when severe inundation along the dockyard areas occurred in heavy rains and high tides. Quaternary sediments are 300 m deep, with the first aquifer at 75 m depth. Much effort has been put into halting the settlement in recent years by dikes around the waterfront and by withdrawing water (largely used for cooling in factories) only in the summer, and replenishing the aquifers by pumping rain water into boreholes in the winter.

Venice in Italy has also been settling for the same reason, but in this case, water withdrawal is primarily controlled by the surrounding industrial areas. Whereas subsidence in Mexico City causes a great deal of trouble in terms of construction and engineering difficulties, the potential of subsidence in Venice is disastrous, since the entire city is built at sea level. The rate of subsidence in Venice has been proceeding at about 1 cm per year, and still continues, since a solution to the water supply problem has not yet been achieved. Portions of Venice are now submerged at high tide (see Section 7.4), and a combination of tide and weather conditions is causing immense damage. The change in conditions between 1908 and 1961 can be seen from the difference in areas inundated at high tide as shown in Plate 5.2.

Outside of cities, of course, ground-water withdrawals are frequently followed by subsidence but, in general, the effects are less noticeable and cause less trouble. One area where substantial effects have been observed, mostly because it is traversed by a canal system, is the San Joaquin Valley of California, where water has been withdrawn from the ground for agricultural purposes since the 1880's. In some areas of the San Joaquin Valley, surface settlements due to this withdrawal have reached 4 m; 1 or 2 m of subsidence are measured over areas of thousands of square kilometers.

Careful measurements of ground subsidence have been made in the San Joaquin Valley by the United States Geological Survey through the use of the compaction meters described previously, indicating that approximately three-quarters of the total settlement observed takes place between the surface and the depth of 250 m. In the region of the San Joaquin Valley where the largest subsidence has been recorded, a number of cracks have appeared at the ground surface, probably caused by tension developed by the curvature of the surface resulting

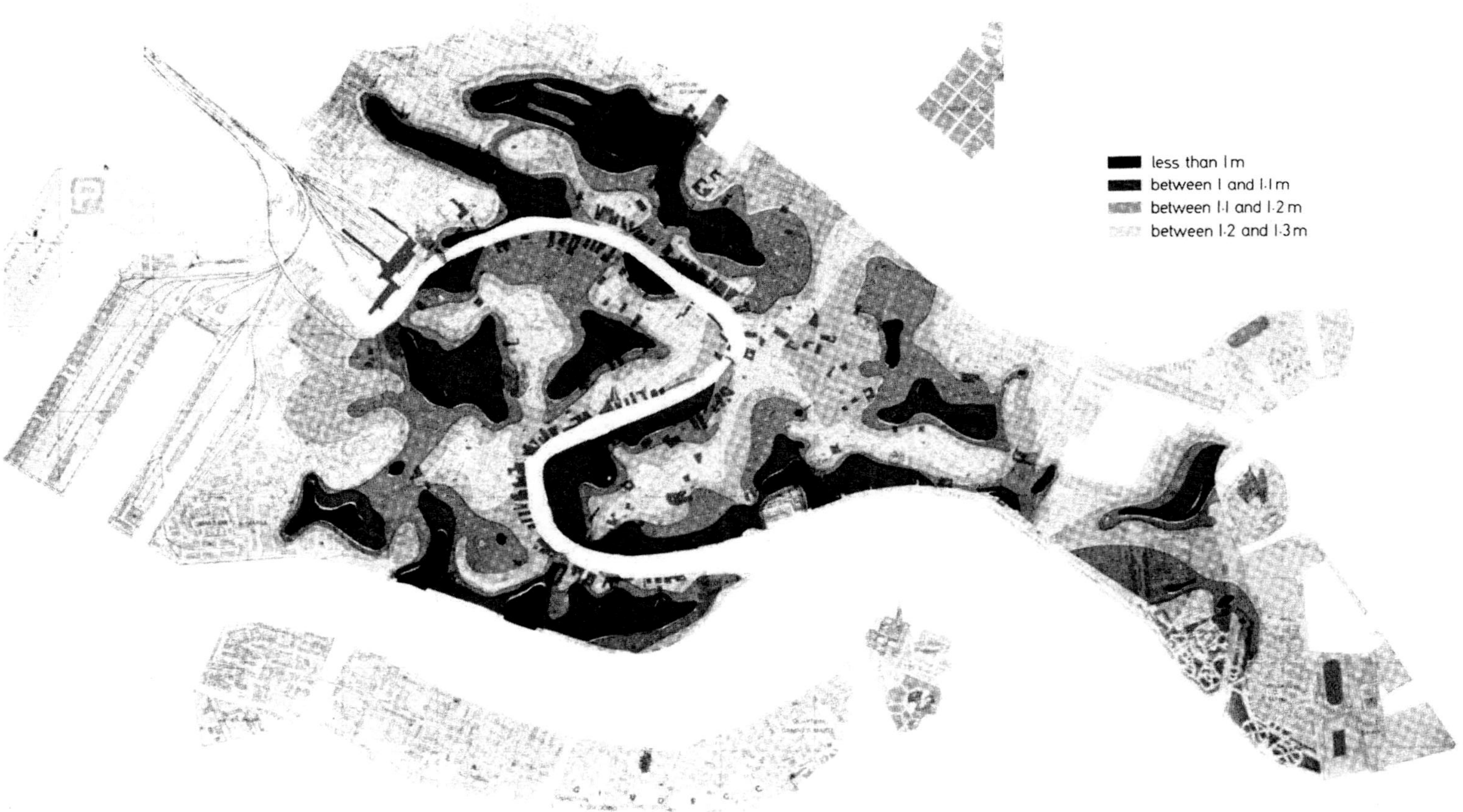

Plate 5.2 a and b over. Areas of Venice inundated by different sea levels: (a) 1908, (b) 1961 (from P. Colombo, Rivista Italiana di Geotechnica **6**, 7–30, March 1972)

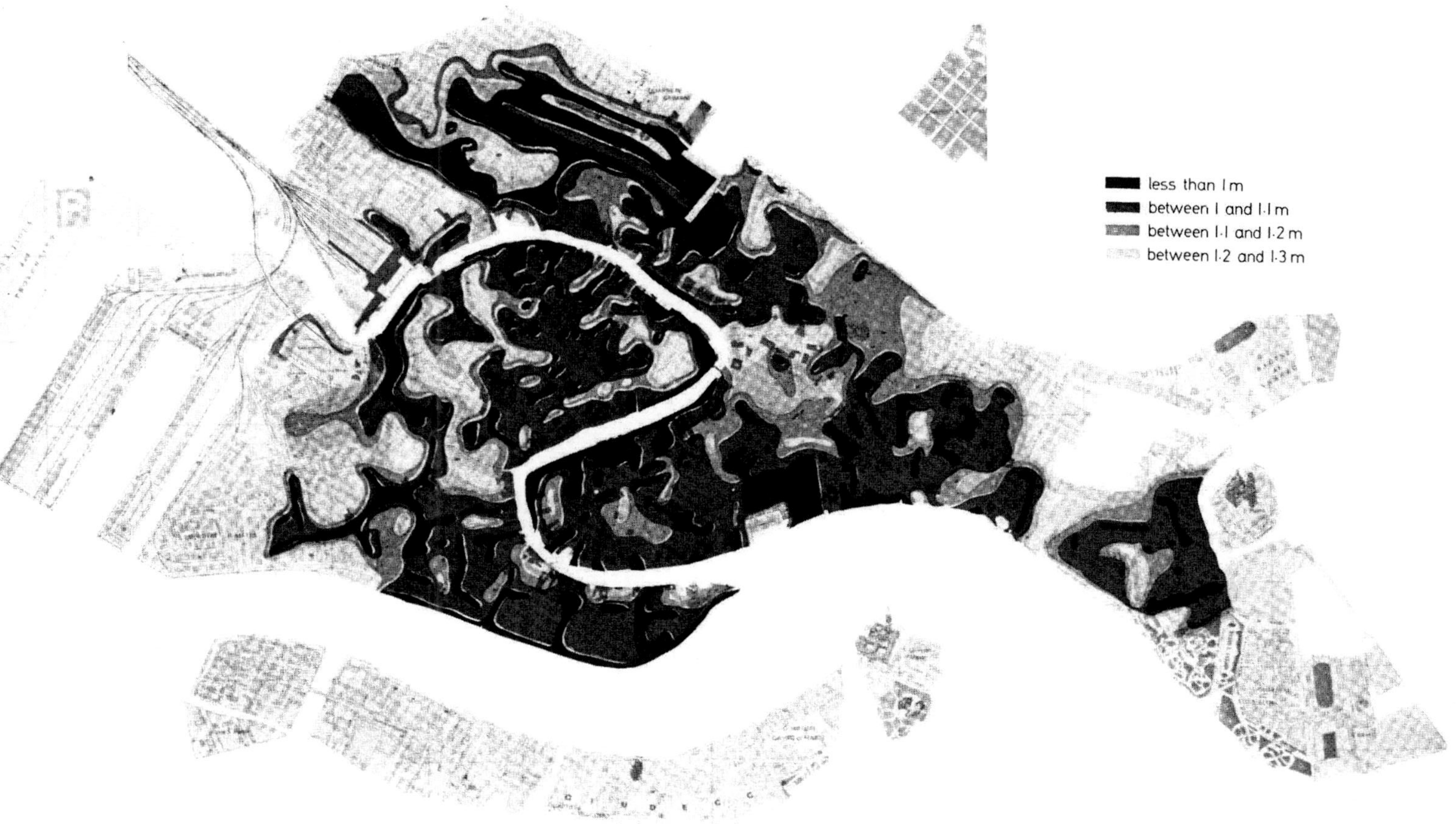
less than 1 m
between 1 and 1·1 m
between 1·1 and 1·2 m
between 1·2 and 1·3 m

Plate 5.2b

from settlement. Such extreme changes in the local stress distribution in the soil offer potential hazards for building construction.

Oil and Gas Withdrawal; Long Beach, California

As pointed out earlier, the situation with oil and gas extraction is similar to water withdrawal, but the observed effects are usually smaller, as greater depths and less compressible materials are involved. Most of the observed cases of surface subsidence are in areas where oil and gas have been withdrawn fairly close to the ground surface and, in general, from depths shallower than 1,000 to 2,000 m. A famous case is that of Long Beach, California, where a large portion of the city is underlain by oil fields which were first developed in the 1920's and 1930's. Surface subsidence had been observed by the 1940's and a growing concern about its magnitude was manifest by about 1950, by which

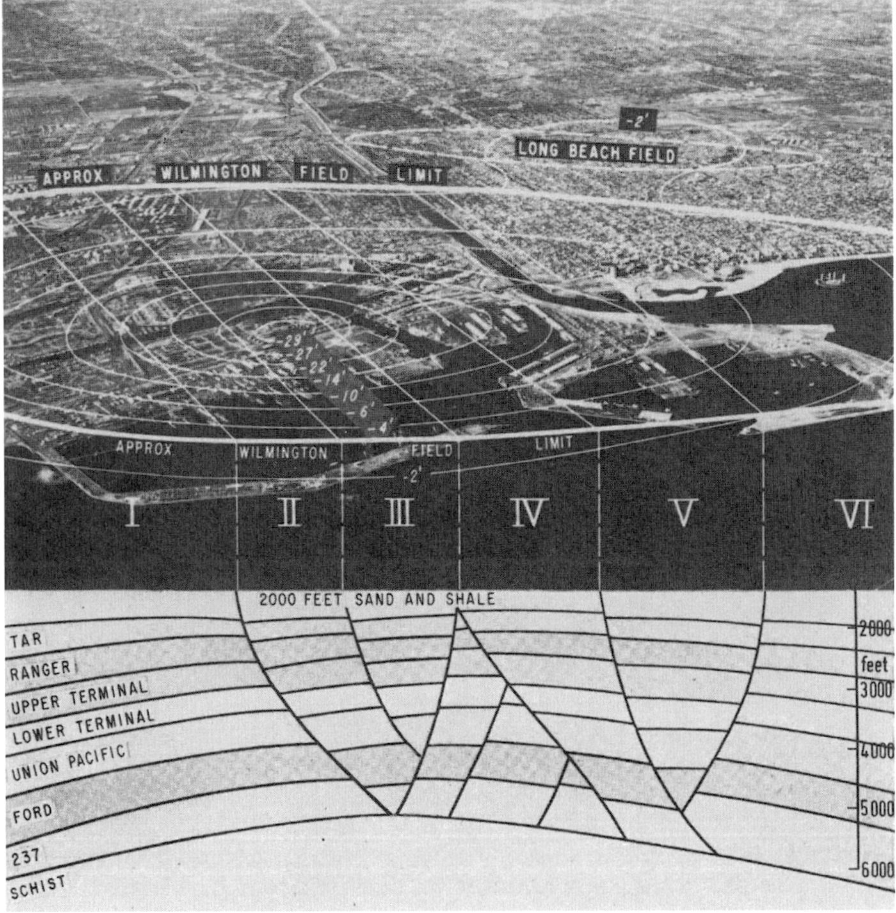

Fig. 5.5. Subsidence contours and the underlying geological profile at Long Beach, California

time the vertical surface movements had reached the magnitude of meters. Since much of the area involved is at or close to sea level, subsidence had reached such proportions in some areas that dikes and sea walls had to be built to protect structures in danger of inundation. The vertical movements were so great that horizontal shearing occurred between soil layers below the surface; many oil wells were damaged by these shearing deformations. On occasion, sudden episodes of shearing at shallow depths caused earthquake vibrations. Subsidence contours and the underlying geological profile at Long Beach are shown in Fig. 5.5.

By 1960 the maximum surface settlement had reached almost 10 m and remedial measures were undertaken. These consisted of injecting water into the peripheral areas of the underlying oil fields in order to raise the pressures in the field and stop the pressure decline that was giving rise to subsidence. These were largely successful in stopping further subsidence although, as explained above, they did not restore the ground level to its original position. At present, little or no vertical movement is taking place in the Long Beach area.

Another region of relatively shallow oil and gas withdrawal lies in the Baldwin Hills, Los Angeles, California, where surface settlements of approximately 3 m maximum took place between 1925 and 1960. Ground surface cracking also occurs locally in this region, but it is not known to that extent it is associated with the ground subsidence. In the region where maximum tensile stresses may be expected to have developed as a consequence of the subsided area, the Baldwin Hills Reservoir was built in 1950. In common with many other seismically active regions of southern California, the Baldwin Hills are seamed with geological faults, some of which ran through the reservoir. In December, 1963, the reservoir failed by internal leakage which developed through its dam (see Section 7.4).

Ground surface settlements have also been observed in many other areas of the world where oil or gas fields have been tapped, such as in the vicinity of Niigata, Japan, and Lake Maracaibo, Venezuela.

Mining Activities

The extraction of solids such as ores, but most particularly coal, from subsurface mines and the construction of tunnels also gives rise to surface movements. The area affected lies immediately above the mine or tunnel and subsides because of stoping or caving of the soil or rocks above the mine cavity. The resulting characteristically pot-holed irregular ground surface, wavy roads and water-filled depressions are common in certain parts of Scotland and England as well as in regions in the eastern United States. The principal dangers arise as a consequence of the continuing decay of old and largely unmapped mining structures. Modern mining or tunneling methods rarely give rise to surface movements large enough to cause distress to surface structures.

A special instance of subsidence due to extraction of material from below is the collapse tube associated with lava flows, which develops when the surface of a lava flow congeals as it cools to form a lid of solid rock under which liquid lava continues to flow. Under appropriate conditions, the source of the lava flow may be cut off and the lava runs out from under the solid skin (see

Chapter 2). When this happens and the roof is thin enough, it will collapse, forming a sinuous depression of characteristic shape. The collapse may occur some time after the lava flow as a result of weathering or chemical changes in the rock materials.

Where limestone layers exist below the surface and annual precipitation is heavy, the limestone may be dissolved and carried away in the ground water to form underground cavities which also, on occasion, collapse or cave, forming surface depressions, the sink-hole-studded topography formed by this process being common in parts of southern Europe and in the south and central United States. Hazards to engineering structures develop if they are to be located above the cavities in the limestone. Thus, where this geological structure is suspected to exist, careful subsurface borings must be carried out to determine the integrity of the limestone underlying the soil profile. Special precautions may be taken, for example, in the driving of piles to insure that the tip of the pile is not located above or near such a cavity.

A special case of this kind of mechanism is represented by some features observed in Arctic areas, where glaciation in past times has occasionally resulted in the deposition of ice masses or ice boulders on the ground, which have been covered by granular material as the glaciers retreated. In effect, the ice blocks are rocks in arctic climatic conditions and are constituents of the soil, like any other rock, so long as the temperature remains low enough. However, if the temperature rises subsequently, the blocks of ice melt; when they do, the ground surface above them subsides, forming a sink-hole. Sinks formed by this particular mechanism are known as *glacial kettles*. In Arctic areas underground ice must be looked for in foundation investigations, since ice would give rise to difficulties should a heated structure be erected above it.

5.5. References

Legget, R.F.: Geology and Engineering. 2nd ed. New York: McGraw-Hill, 1962.

Legget, R.F.: Cities and Geology. New York: McGraw-Hill, 1973.

Poland, J.F.: Land Subsidence Due to Fluid Withdrawal. Geological Society of America, Eng. Geol. Review, Vol. **2**, 1970.

State of California, The Resources Agency, Landslides and Subsidence, Geologic Hazards Conference, Los Angeles, Calif., 1965.

Chapter 6
Hazards from Snow Avalanches

6.1. Introduction

The sagas of mountaineering contain many accounts of sudden deaths from snow avalanches. For instance, the treacherous Nanga Parbat (9,000 m) in the Himalayas claimed 16 lives in 1937 when a German mountaineering team was trapped by an avalanche. Climbers of the Himalayas, Andes, Alaskan Ranges and even the Alps must always be on the look-out for dangerous snow slope conditions. For this reason the Alpine countries particularly have, since last century, established a tradition in research on avalanche dangers. Specially notable is the work at the Swiss Federal Institute for Snow and Avalanche Research.

In a number of ways, avalanches of snow resemble certain kinds of landslides, but important differences are involved. As a material, snow has some mechanical properties in common with soil, such as cohesion, angle of internal friction, and density, all related to one another in a complex way. When snow falls on a sloping hillside and accumulates, the conditions which govern whether or not an avalanche occurs are similar to those in the case of soil landslides of shallow depth on a long slope. That is to say, the snow will remain in place if the shearing strength at all depths exceeds the shearing stress developed by the weight of the snow and the angle of the slope. If the shearing strength is exceeded by the downslope component of the weight, an avalanche will develop.

The depth of snow accumulation before a failure takes place depends on the properties of density, cohesion, and friction in the snow and the slope angle in the same way as these conditions apply to soil on an infinite slope as discussed in connection with Fig. 4.8(a) and (b). The snow depth at which the stresses are exceeded is usually referred to also as the *critical depth.*

However, here the similarity between snow and soil slides ends. There are no deep-seated snow avalanches taking place on a shearing surface which forms a portion of an arc of circular or other shape, as is the case with certain soil and rock movements. In addition, the physical processes affecting the mechanical properties of snow are entirely different from those on which the shearing strength of soil depends. Although soil is a complex material, some aspects of the development of its strength are fairly well understood, and the strength of a particular layer or zone changes only over many years as a consequence of chemical effects. Faster changes in the strength are generally due to changes in the pore water pressure or in the degree of saturation of the material. By contrast, the condition, structural arrangement and inter-relationship between the particles of ice that constitute a snow mass are changing continuously.

Pressure, temperature, and the migration of water vapor constantly modify the material properties at all depths in the snow cover. For example, snow which is newly fallen under very low temperature conditions may have essentially no cohesion at all. Its shear-strength behavior is equivalent to that of a dry sand possessing friction alone. It will therefore not remain at rest on a slope steeper than its angle of internal friction, so that surface sloughing and surface slides will occur on steep slopes during a snowstorm. If the slope is less steep than the angle of internal friction of the material, the snow will initially stay in place. Insolation, changing temperatures and the migration of water vapor through the snow pack will alter its properties, giving it a density and cohesion which change constantly with time and also vary with depth in the snow pack. At some subsequent time, a combination of these properties can cause a snow layer to become unstable and it slides.

The weather and topography determine to a high degree the occurrence, frequency and range of avalanches in a given region. The various combinations of meteorological conditions give rise to a variety of avalanche types, which may develop during, shortly after, or a long time after a particular snowstorm, or may take place as a consequence of the conditions following a number of winter storms. As in the case of landslides, the variety of possible avalanche forms has spawned a number of classification systems.

6.2. Classification and Mechanisms of Avalanches

Since avalanches do not possess the wide divergence of morphologies demonstrated by landslides, the classification schemes which have been proposed all resemble one another in their basic structure, only small differences appearing in the details and descriptions or mechanisms. In comparison with landslides, avalanches are evanescent features which develop each winter in hilly or mountainous regions subject to heavy snowfalls. The final product eventually vanishes (see Plate 6.1).

In some regions, avalanche zones endanger extensive forests while logging of mountain pine stands may increase the risk from avalanches. Structures that are built on hillsides which may be traversed by avalanches are few and are confined to ski-lift support structures, transmission towers and occasionally highways, railroads and mining structures. Because the paths of avalanches, with rare exceptions, are the same from year to year, the structures which are built avoid them, as far as practical. The principal hazard of an avalanche is to skiers or climbers traversing its slopes, or slopes immediately below it, and the danger to dwellings and other structures at the lower end of its traverse, should it run farther than usual. It will be recalled (see Chapter 4) that landslides develop over periods ranging up to hundreds or thousands of years and can occur in regions where structures have been located, so that effort is concentrated on post-failure analysis or the calculation of the stability of slopes in regions where construction is proposed.

A different emphasis is necessary for avalanches. It is necessary to design protective structures for highways, railroads or utilities in regions where it is known that avalanches will occur, and it is, in addition, common practice to estimate the stability of snow-covered slopes in ski areas daily in order to initiate

Plate 6.1. A snow avalanche cascading down a mountain side (From "Avalanches" Bratislava: Slovakian Academy 1967)

protective measures. These usually involve causing the avalanche to occur at a time when its hazards can be minimized. Only on rare occasions are landslides triggered deliberately, and then only in construction works to check the actual field performance of the material for comparison with the laboratory and analytical investigations. The classification schemes reflect these differences.

Table 6.1. Classification of avalanches

Movement	Snow type	Conditions for occurrence	Scale	Characteristics
Primary (During snowfall)	Dry	Slope steeper than friction angle of snow	Small	Surface runs of snow
Secondary (After snowfalls)	New snow on crust	External trigger	Small to intermediate	Snow block debris
	Crusted surface	Settling of underlying snow	Intermediate	Large slabs
	Wet surface layer	Solar radiation, warm wind, rain	Small	Wet mass, slow
	Depth hoar	Constructive metamorphism	Intermediate to large	Large blocks entire layer
	Wet snow	Warm weather, weak base	Intermediate to very large	Large blocks slow to medium velocity

An avalanche classification system, shown in Table 6.1, will be described briefly here. The type of sliding that develops depends heavily on the process of snow metamorphism taking place in the fallen snow layer. If the snowfall is dry and very cold, surface movement occurs by the periodic sloughing or flowing of surface layers, whereas under warmer conditions, the snow possesses some cohesion as the individual flakes and granules adhere to one another on deposition. In this case it is possible that the accumulation of snow on the slope during a snowfall will reach a critical depth at which the surface layer will slide off to cause a slab avalanche. If the slide does not occur during a snowstorm, then a number of processes can develop in the snow during the following period; they may tend either to increase or decrease the stability of the layer. Under appropriate weather conditions, *destructive metamorphism* takes place in which the snow particles break down into smaller grains. As they do so, the snow becomes more dense and compact and consequently more cohesive. In general, the effect of this destructive metamorphism is to stabilize the layer so that avalanching is less likely to occur. Metamorphism does not occur in the same way at all depths in the snow.

The contrary process is called *constructive metamorphism* when initially small particles of snow grow larger by sublimation. The snow pack becomes less tightly packed with larger voids between the bigger particles. The worst case of this develops, under conditions that are still not well understood, at some depth below the surface of the snow cover so that a layer forms, consisting of coarse rounded pieces of ice with little cohesion, referred to as *depth hoar*. The shearing resistance of such a depth hoar layer is extremely small. In extreme circumstances, such large voids form in the snow at depth that the snow layer above is eventually supported for the most part around its periphery by contact with the adjoining snow pack. Under these conditions, cracking and shear planes will develop around the edge of a large slab of snow on the steepest slope on a hillside, and when these cracks reach a critical length, the slab will break away from the adjoining material and form an avalanche. Initially, motion will take place as a unit, but as the hillside is traversed, the slab breaks up into smaller chunks of snow which pour down the hillside. If the snow is weakly cohesive, an extreme development of this type is the formation of a flowing current of snow particles suspended in air. This form is equivalent to the liquefaction phenomenon described earlier in connection with landslides. The avalanche becomes a density current flowing down the hillside. Depending upon associated meteorological conditions and local topography, the heat generated as a snow mass slides can also change its character from possibly a dry, granular powder at the beginning of sliding to a much wetter and denser condition near the terminus.

Different categories of slab-snow avalanches arise depending on when the slide occurs. A slab may break loose during or shortly after a snowfall, and consist of one deposit of snow; or, alternatively, following a number of separate snowfalls with intervening periods of *föhn* (warm, dry wind), an unstable slab may consist of a number of snow strata. In the latter case, the snow mass is considerably denser, and thus the avalanche tends to be more destructive. The melting and refreezing of the surface snow layers in sunny periods can also provide a low friction surface on which sliding of subsequent snowfall is facilitated.

The remaining subdivisions of the classification system depend, as can be seen, on the meteorological conditions accompanying and following the snowfall, so that the snow itself can be relatively dry, damp, or wet. The topography in which the avalanche takes place also dictates some of its characteristics. An open flat slope yields avalanches which advance on a broad front, whereas the presence of natural gullies or channels concentrates the flow into a narrow and thick zone of flowing material.

From the above discussion, it can be seen that the size of avalanches ranges all the way from small harmless events of a few meters in size on open slopes, through medium-size flows with sufficient energy to kill a man who may be traversing the slope, to large avalanches which can be immensely destructive to life or property in their paths. At the upper end of the scale lie those relatively rare avalanches that are classified as climax events; they occur under extremely rare and unusual meteorological conditions on certain slopes, such that through successive snowfalls on the slope during an entire winter season a very considerable depth of snow accumulates without exceeding the strength of the base layers. Then, as a result of processes involving the increasing solar radiation at the end of the season, an avalanche develops which includes all the snow on the slope down to the ground or rock surface. An enormous mass of material of relatively high density can consequently move downslope.

In Japan, a classification of the size of an avalanche has been based on the weight of material involved in metric tons. The classification is somewhat similar to the Richter Magnitude Scale for earthquakes (see Section 1.2) in that a *mass magnitude* (MM) is defined which is equal to the logarithm (to base 10) of the mass of snow in tons. On this scale, small avalanches have an MM number less than 1; medium avalanches have numbers in the range of 1 to 3, and large events are described by MM 3 to 5. One or two events have occurred with mass magnitudes possibly as high as 7.

Although, on a slope, snow can creep at a rate of meters in weeks to months, the movement is only of significance in indicating where a future avalanche may develop. In contrast to landslides, potentially destructive avalanches all travel at relatively high velocities typically ranging from a few kilometers per hour up to as much as 200 to 300 km per hour. At the Snow Research Station in Japan artifically released avalanches were observed by movie cameras; average velocities ranged from 40 to 60 km per hour. The velocity depends, of course, on the angle of slope, the snow density, the shearing strength of the material, and the length of path traveled.

6.3. Analysis, Investigation and Control of Avalanches

Analysis

Because many avalanches carry a significant amount of rock debris, they are important geological agents. Repeated avalanches transport large amounts of unsorted rock fragments to the base of the slopes and aid erosion. Avalanches occur on slopes at all angles, although the more important ones are on slopes from 25 to 40°. Avalanching is more likely on the steeper slopes, but a compensating effect comes into play as the slope angle steepens. Less snow can accumulate on slopes steeper than 45°, since the snow slides off continually during the storm, and, thus, extremely steep slopes are not prone to massive avalanches.

The analysis of an avalanche or avalanche mechanisms depends, as in the case of landslides, on a determination of material properties in place at the avalanche site. However, avalanche investigations present some peculiar complications. The studies are rendered more difficult because, first of all, everything — the density and shearing strength of the snow and the depth of accumulation — is changing very rapidly in time. The softness of the material and its sensitivity to temperature make the determination of its in-place properties extremely difficult. If samples of the material can be taken, it is difficult to transport them back to the laboratory and maintain them so that no property change takes place, as to do this would require maintaining the sample at the same temperature and vapor-pressure conditions at which it existed on the slope. However, those conditions would have to be measured first, and it is not easy to measure the correct temperature at various depths in a snow cover. The thermal balance of the snow depends on the air and ground temperature, wind velocity, and the thermal radiation condition at the snow surface. Except for special research purposes, snow temperatures are not normally measured, nor are laboratory tests performed.

In field investigations emphasis is placed on the determination of as many properties of the material in place on the slope as is practical. A variety of tools is used for this purpose, including sampling tubes and penetration tests which give information on the density and shearing strength of the snow. It is, of course, difficult to do shearing-strength tests on snow similar to those carried out on soils, so that it is not possible to ascertain clearly the quantitative effects of the different variables on the strength of the snow, and analysis of slide potential of an avalanche area tends to depend less on a calculation of the forces involved than in a landslide. Instead, more emphasis is placed on the qualitative relationships known to exist between the meteorological variables and the depth of snow accumulating, together with the association of special conditions with particular forms of avalanche. The possibility of an avalanche is assessed by comparing the conditions in one snowfall with those associated with past avalanches on the same slope.

For example, snow falling on a steep slope at low temperatures and little or no wind will slough readily, while if the conditions are the same except for a higher wind velocity, the snow will develop a greater degree of packing and some cohesion so that a *dry snow slab avalanche* may ultimately be expected. When snow accumulates rapidly at relatively warm temperatures on a slope whose surface is covered with an ice layer formed from melted and refrozen snow, the situation is ripe for a *wet slab avalanche*. Because of local peculiarities, some experience with the meteorological and slope conditions in any one region are required before a prediction of the avalanche hazard can be made.

Besides all these difficulties, one further point must be emphasized: the study of the avalanche potential of a particular slope is hazardous to the investigator, since at present the relevant properties must be measured in place. The snow may let go while the slope conditions are being examined.

Control

There are both passive and active methods of avalanche defense. In the rupture zone, wind baffles and snow fences may prevent avalanches by influencing the snow settlement on leeward slopes, while retaining steel and concrete walls may provide balancing forces that avert ruptures. On the slopes, terraces and breaking barriers may slow the snow movement or redirect the avalanche. Where structures must be located in regions subject to avalanching, the structures can be protected by fences or walls arranged to deflect the snow mass (see Plate 6.2). Measurements of the pressures generated by flowing snow have been made so that the information necessary for the engineering design of protective structures is available. In regions of frequent large avalanches, however, the best protection is to bury the structure below the ground surface, for example by putting a highway or railroad in a tunnel through the mountain. In the long run, the economics of tunnel construction may be more satisfactory than the expense of continual maintenance, repair and interruptions caused by avalanches on an exposed alignment.

The active method of avalanche control consists of triggering the avalanche artificially at a chosen time under safe conditions rather than permitting it to

Plate 6.2a

Plate 6.2a and b. Protective barriers for avalanche defense. In (a) concrete wedges are shown as permanent braking and deflection devices on the slopes above a town in the Carpathians. In (b) modern steel braking barriers are shown. (From "Avalanches", Bratislava: Slovakian Academy 1967)

occur naturally and unpredictably. Thus, at the ski centers of the world, continuous avalanche patrols are maintained to monitor the slopes and the weather conditions. Instead of trying to keep skiers and climbers off the slopes when the conditions are hazardous, the practice has developed of causing avalanches to occur at times when the slopes are clear, thereby removing the danger.

The methods of triggering avalanches usually involve vibrations, such as from loud noises or by impacts, and the usual method at present is to fire explosive projectiles to impact at the head of potential slides. When the topography permits, it may be possible to set charges in the snow at the top of a slope or on an adjacent ridge in order to blow off overhanging cornices or release the potential avalanche. A variety of launching devices has been used, including

Plate 6.2 b. Legend see opposite page

Legend see opposite page

conventional artillery, recoilless launchers, and air guns. The practice has become so well developed in various areas of the United States that avalanche casualties are relatively rare, in contrast with Europe where on the average about 100 people are killed by avalanches of all types every year.

6.4. Avalanche Case Histories

Avalanches lack the diversity of landslides in some respects; all forms of avalanche occur in mountainous regions with high snowfall at all scales of size. In this case it is most instructive to select large and rather spectacular events to give some idea of the magnitude of the problem which certain avalanches can pose. In the Alps, for example, it unfortunately can happen that a village can be destroyed by avalanches following heavy snow, with many fatalities.

Mount Huascaran, Peru, 1962

Near the north end of the Andes in South America there rises the 6,700 m peak of Mt. Huascaran, situated 9° south of the equator and named after an Inca prince, Huascar. On the peak of Mt. Huascaran lie the remnants of the last glacial epoch in the form of a summit glacier and snow field some hundreds of meters thick. Seasonally, masses of ice and snow break off from the edge of the glacier and form avalanches down the high slopes.

On January 10, 1962 a very large chunk, later estimated to have contained 3,000,000 cubic m of snow and ice, separated from the main glacier and fell downwards a vertical distance of almost 1 km. The impact broke up the ice and snow to form a dense cloud of suspended particles which traveled down the valley below the mountain with enormous speed and energy; on the way, the moving mass entrained boulders, earth and soil to form a combined avalanche and mud slide. At an elevation of 2,700 m lay the small town of Ranrahirca and six villages directly in the path of the moving mass. By the time the avalanche reached the town, it had fallen a vertical height of about 3,500 m and had expanded to an estimated volume of about 13,000,000 cubic m. The town, the villages, bridges and highways were largely destroyed and a local river was dammed when the slide finally came to rest.

Eventually the debris dam was broken by the rising river, and a further disastrous flood took place which wrecked all the bridges downstream. Approximately 4,000 people were killed in this disaster. The distance of approximately 20 km was traversed by the avalanche in approximately 7 minutes, so that its average velocity was about 170 km per hour.

This catastrophe was almost exactly duplicated eight years later in the earthquake of 1970, as described in Sections 1.4 and 4.6. In 1970, rock from summit strata on the mountain seemed to have participated in the slide or flow to a greater extent than in 1962. These are only the latest events in a long sequence of ice and rock falls from Mt. Huascaran.

Camp Leduc, British Columbia, Canada, 1965

Camp Leduc was the name given to a mining camp in the Canadian Rocky Mountains, where, in 1965, work began on the development of a very large copper ore body. In February, 1965 there were 154 men in the camp engaged in the preliminary work of driving tunnels into the mountain. The ore body had been revealed by the recession of a glacier, and was located high in the mountains in a region where the annual snowfall frequently exceeds 15 m; the buildings of the mining camp were located on a ridge in the belief that avalanches which might occur would be split by the ridge and sweep around on each side of it.

On February 18, 1965, however, a very large avalanche moved down the mountain slope above the camp and destroyed the southern half of the camp area before wrecking the buildings at the portal to the mine tunnel. The actual amount of snow that moved was only a portion of the potential avalanche volume and consequently did less damage than could have developed. Even so, seventy men were buried in the mining camp. Of these 43 were rescued in the feverish search and excavation which followed; 27 were killed.

To an expert in avalanche detection and technology, this disaster could have been prevented, since it was possible to predict the avalanche occurrence and path at the area, and the camp could have been located in a more protected position. The addition of some artificial protective structures would have increased the safety.

Carpathians, Czechoslovakian Region

Special studies of the causes of avalanches have been made in the Carpathians. Here most originate in the mountain chains of the Low Tatra, Liptovske Tatra and the Velka Fatra. Typically, snow rupture zones are situated in these mountains at about 1,700–2,000 m elevation. The lengths of avalanche tracks range from less than 1 km to 3 or more km.

Two winter periods 1955/56, 1961/62 were chosen for special analysis. In both cases there was a large quantity of new snow. In March 1956 there was inconstant cloudy weather with rain or snow in the lowlands and substantial snow in the mountains. The weather conditions then changed to stormy winds which formed large snowdrifts and cornices on the leeward slopes providing fertile conditions for avalanches. Many large snowslides followed in the Tatra and Fatra with, in some cases, considerable damage. Among them was the avalanche on March 8, 1956 from the slopes of Zdiar alpine meadow in Low Tatra which killed 16 forest workers.

In 1961/62, snow conditions for avalanches did not occur until February 1962. For that month, cold moist air produced such a rapid increase in snow cover that a balance of traction and tension forces on the critical slopes between the old and new snow was difficult to reach. These unstable conditions produced many great avalanches over nearly the whole Carpathian area.

6.5. References

Atwater, M.: The Avalanche Hunters. Philadelphia, PA.: Macrae Smith Co., 1968.

Knapp, G.L. (Ed.): Avalanches, A Bibliography. Wash., D.C.: U.S. Dept. of the Interior, Water Resources Scient. Inform. Center, WRSIC 72–216 (1972).

Quervain, M.: Avalanche Classification. Congress IUGG 1957, Toronto. International Association of Scientific Hydrology, Publ. **46**, Vol. IV (1958).

Seligman, G.: Snow Structure and Ski Fields. London: Macmillan 1936.

Swiss Federal Institute for Snow and Avalanche Research: International Symposium on Scientific Aspects of Snow and Ice Avalanches. Davos, Switzerland, 1965.

U.S. Forest Service: Avalanche Handbook. Washington, D.C.: U.S. Government Printing Office 1952.

U.S. Forest Service: Snow Avalanches. Handbook No. **194**, Washington, D.C.: U.S. Government Printing Office 1961.

Chapter 7
Hazards from Floods

7.1. Introduction

The Great Australian Floods, 1974

Stark contrast between "droughts and flooding rains" has long been known in Australia. In January 1974, monsoons rolling in from the Timor Sea moved across the northern part of the continent bringing a torrential deluge to the northwest and the Gulf of Carpentaria region (see front map). In Western Australia, 48 cm of rain were recorded in a 17-hour period in mid-January, with Broome and Darwin being partially flooded and evacuated. Flood waters spread from horizon to horizon in the outback regions, where in normal times dusty, dry river systems are exposed to the burning sun.

By January 20, flood waters had risen over telephone poles in northwestern Queensland. People isolated by rising waters waited desperately for help in what was the worst flood the region had suffered in this century and Australia's greatest national disaster. Six major towns in western Queensland became isolated by the floods. On January 31, heavy rainfall in the far west of Queensland amounted to 14.3 cm. Copper production was cut in half at the great Mt. Isa mines to conserve coal supplies. As flood waters moved toward the Gulf of Carpentaria, rivers merged and spread over 150 km across the Gulf country, while farther south, in New South Wales, the rain continued week after week, flooding large areas of the northwest, where inundated grazing land was littered with the carcasses of hundreds of thousands of sheep. Aerial food drops were made to Alice Springs and other isolated communities in central Australia and Queensland.

Disaster grew as, toward the end of January, cyclones moved down the Queensland coast. The Brisbane River, that runs through Brisbane (population 800,000), the capital of Queensland, burst its banks. By January 30, this normally placid river presented an incredible sight from the air (see Plate 7.1); it was now more than 3 km wide, spreading from the university area of St. Lucia, through the industrial suburbs. Upstream from St. Lucia, water stretched for kilometers along the river's flood plain as far as the eye could see toward the city of Ipswich. Debris from houses, farms, and industry bobbed along the top of the water, headed for the ocean.

Flood damage in Ipswich and Brisbane was great. In Ipswich, 1,200 houses were ruined and the heart of Brisbane was paralyzed by the floods, with 20,000 persons made homeless. The casualty toll was at least 15 persons dead.

Strikingly, there had been repeated warnings that another flood disaster in Brisbane could occur. In the event, the 1974 flood did not quite match the torrential fury of an earlier one in 1893, when in three weeks, 10,000 of Brisbane's

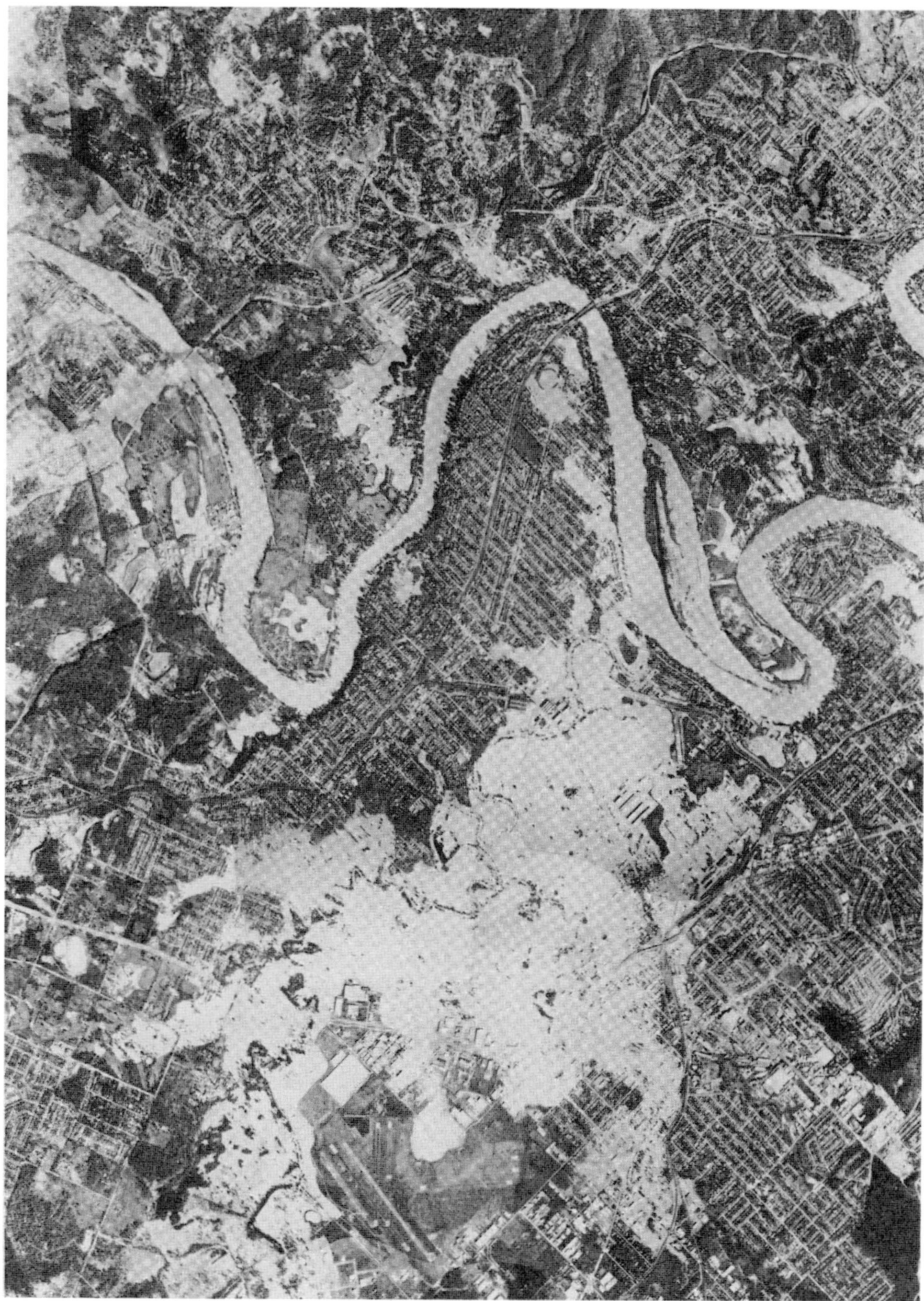

Plate 7.1. An aerial photograph showing the inundation of Brisbane, Queensland, Australia in the great flood of January 1974. (Photo mosaic by Mapmakers Pty. Ltd., Brisbane, Australia)

90,000 people were left homeless and a legacy of destruction was left which took years to correct. In the 1893 havoc, there had been a long period of heavy rain, and as many as five cyclonic centers moved down the Queensland coast and deluged the area extending over Brisbane and southeast. The heaviest rains started on February 1, 1893 and by February 4 there were 2.5 m of water in Brisbane.

Swirling flood waters in 1893 swept many wrecked houses down from Ipswich, along with household debris of all kinds and dead animals. On February 5, the water washed away the iron Indooroopilly Railway Bridge which had had debris piled against it; ships and dredges were dragged from their moorings and carried downstream. On February 6, Victoria Bridge crumbled and its northern end sank into the river. By February 11, heavy rains started falling again. On February 17, a waterspout tornado shaved a path through forests, houses and boats, inflicting severe damage. A third floodcrest hit the Brisbane River and people were once again forced out of inundated homes. By February 21, the floodwaters subsided, leaving behind 35 dead.

Geological evidence and aboriginal legend indicate that bigger floods than 1893 and 1974 have occurred before European settlement in the Brisbane area. Today, the hazard seems more acute, as forests and grass lands that help absorb the heavy rains have been removed and the rain now hits pavements and roofs and runs off at increased velocity into creeks, ponds and gulleys. Many creeks have been filled in, thus increasing the burden on the remainder.

Flood control structures, such as the Somerset flood control dam on the upper Brisbane River, were not large enough to save the city in the major 1974 flood. Public demand has now led to investigation on how disasters in the proportions of 1893 and 1974 can be prevented in the future. Laws on land-use of the city's flood plains are being reexamined. The 1974 operation of the Somerset dam as a flood barrier has been called to question and laws to prevent hills from being stripped are being considered.

One reaction toward flood hazards was given by the Minister for Environment and Conservation of the Australian government, Dr. Moss Cass. He stated, "Flood plains are for floods, although this is not to say that there is nothing we can do to minimize the damage which flooding can cause. Australians should remember that floods and droughts are in many areas the norm; it is good years which are the exception. Serious floods have occurred in these areas not only in 1974, but also in 1971, 1961, 1956, 1955 and 1950—six serious floods in half a working lifetime. Except in relation to small areas, there is simply nothing that man can do to prevent widespread flooding and drought. It is simply impracticable to build dams to prevent inundation of all floodplain areas in periods of exceptionally high rainfall. There are often perfectly sound reasons for using land subject to flooding for agricultural and other activities. However, it is essential in planning such enterprises that full provision is made for foreseeable losses. This may require private investment to mitigate the effects of flood by providing refuge for stock and moveable objects."

Mitigation plans are, however, also going forward nationally. The cost to the Australian Government for rebuilding and repairs after the 1974 floods is at least \$75,000,000. Queensland Government housing grants to home owners were made up to $4,500 each, since little flood insurance was available. One plan is that the Federal and State governments join in the establishment of a national disaster organization backed by a national fund. Insurance details remain to be worked out. (See Chapter 8 for the details on United States flood insurance.)

Destruction of Works of Art, Florence

Like fires, floods are great destroyers of cultural treasures. A notable example is the repeated flooding of Florence, Italy, one of the foremost art centers of the world. *Firenze bella*, beautiful Florence, lies on the Arno River at a site where the river is still narrow after leaving the Apennines and before crossing

Plate 7.2. Mud and wreckage inside church S. Remigio in Florence after the disastrous flood of November 1966. (Courtesy of Jürgen Schulz)

the rich flood plains toward Pisa (see front map). The river has inundated the city many times, the floods of 1333, 1557, 1844 and 1966 being particularly disastrous.

On November 4, 1966, the river's torrent engulfed the Renaissance city, as unwarned and unaware, the city slept as the furious stream rose over the high watermarks of past calamities. At 07:26 h, electric clocks stopped throughout the city as the full fury of the waters washed away the San Niccolò bridge and began to turn the narrow streets into cataracts laden with rubble and automobiles.

On November 3 and 4, in the watershed of the Arno, one-third of its average annual rainfall fell in two days. In northern Italy, 750 villages and 5,000 km of highway were flooded; a hundred people and 50,000 cattle were drowned southward from the Po valley into Tuscany. On November 3, water was released from the large Penna and Lévane flood control reservoirs on the Arno, releasing an enormous volume of water downstream. After the disaster, a government commission was appointed to investigate whether blunders had been committed in controlling the dam release gates.

At its peak, the flood in the city reached a height of 6 m. It deposited large amounts of debris and soil, severely damaged many churches and buildings of architectural value (see Plate 7.2). The Archivo di Stato and the Biblioteca Nazionale Centrale were deeply immersed and affected by the modern hazard of floating oil—*nafta*—from central heating units.

In the Biblioteca, Italy's largest library with more than three million books, more than a million and a half volumes were damaged, many dating from Renaissance times. After the flood subsided, volunteers wearing gas masks against the stench of sewage and rotting leather bindings worked to recover thousands of these priceless books from basements filled with black mud.

The most famous works of art ruined were the Etruscan collections in the Museo Archeologico and Cimabue's "Crucifix" in the Museo di Santa Croce. As a result of the international rescue and restoration operations, the total loss, although severe, was probably less than in some earlier floods.

The Danger Grows

Of all nature's rampages, floods are particularly destructive, because people and their works are usually in the way. We pay a high price, both in damage sustained and in cost of flood control, for building on alluvial plains and living in the troughs of valleys. Many civilizations have struggled to win a living from the fertile land adjacent to mighty rivers. Our early ancestors lived along the Nile, Tigris, Euphrates, Indus and Yellow rivers in an environment that nurtured man and his crops, but at the same time was a source of disaster.

In the growth of industry in Europe, North America and Australia it was convenient and often necessary to reside on the flood plains along rivers and streams, for these provided drinking water, food and transportation. As a result, towns sprang up along river banks, on fertile deltas and low coastal plains. While the communities were small, damage was correspondingly small and little effort was made to guard against it, but as the towns grew the flood hazard became more acute.

Untold time, effort and money have been spent in many countries in building flood control facilities to reduce the hazard, yet each year many thousands suffer loss of life and possessions. Each year the totals (see Fig. 7.1 and Table 7.1) grow, for few areas are immune to flood hazard. Without doubt, flood losses will continue, but, hopefully, with increased knowledge and technology, the losses will decrease.

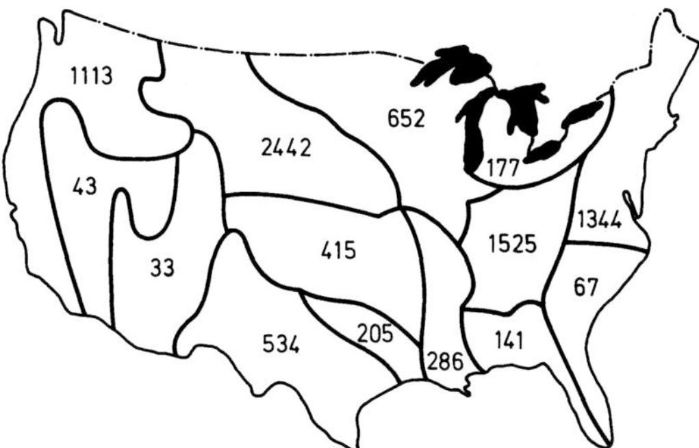

Fig. 7.1. Estimated flood losses in millions of dollars in the United States for the period 1925 to 1968 (Source: Department of Commerce, 1969)

Two general solutions are evident, but their accomplishment is difficult: (i) recognize a hazardous area and avoid it as a location for habitation, and (ii) if industrial and agricultural investments are so vital to the economy that they must be protected from floods, engineering must be provided to control and reduce the flood. Even then, we must recognize the likelihood of a super flood that, despite the flood-control works, will take its toll.

Types of Flood

Floods are river flows or ocean elevations that cause or threaten damage, and are produced in a number of ways. The most common, and the major subject of this Chapter, is the *rainstorm-river* flood; a second major type is a *coastal* flood, resulting from an unexpected relative increase in sea level that may be caused by a storm, by a tsunami (see Chapter 3) or by land settlement.

Other but less commonly experienced floods are: snowmelt floods; ice jams and ice thaws; floods from structural failure, such as that of a dam or a levee; and floods that result from rupture of glacial ice, landslides, or release of a volcanic lake (see Fig. 2.8). Inadequate storm drainage in a community or *sheet flooding* from high intensity rainfall can also inundate homes and agricultural land. The major floods of the United States since 1844 are listed in Table 7.2.

Excessive rains on the drainage basins of a stream from thunderstorms, migratory storm systems, or tropical air masses increase flow of rivers so that they

Table 7.1. Annual flood damages in the United States. (Source: U.S. Water Resources Council, 1968). (Values in millions of dollars; Upstream refers to those streams above a point where the total area drained is 100,000 hectares or less; downstream refers to the stream pattern below that point)

Region	1957		1966		1980[a]		2000[a]		2020[a]	
	Down-stream	Up-stream	Down-stream	Up-stream	Down-stream	Up-stream	Down-stream	Up-stream	Down-stream	Up-stream
North Atlantic	64.3	62.6	63.1	70.7	75.6	91.2	89.8	120.9	116.2	163.3
South Atlantic-Gulf	46.7	109.6	44.1	123.8	55.8	183.2	74.8	267.3	90.4	383.7
Great Lakes	12.3	29.8	13.0	33.7	15.8	43.8	21.0	57.2	27.7	76.1
Ohio	78.7	49.2	73.9	55.6	99.5	68.3	151.0	90.9	237.0	116.6
Tennessee	3.5	27.3	4.9	30.9	7.6	42.6	8.3	58.3	[b]	80.2
Upper Mississippi	60.0	52.0	64.5	68.5	96.0	101.9	151.0	143.2	218.0	197.2
Lower Mississippi	66.0	38.5	86.8	43.5	117.2	55.3	164.2	73.5	224.5	100.1
Souris-Red-Rainy	5.8	13.5	5.6	15.3	6.4	18.4	7.5	24.1	8.6	32.2
Missouri	101.4	148.1	44.0	167.3	69.0	222.3	118.0	302.7	221.0	430.1
Arkansas-White-Red	48.6	129.2	50.0	146.0	61.6	184.0	90.6	245.3	127.0	330.0
Texas-Gulf	32.5	49.5	28.2	55.9	39.5	86.1	59.3	125.3	86.4	178.4
Rio Grande	12.2	10.4	14.7	11.8	14.8	19.5	15.8	30.9	18.8	44.9
Upper Colorado	0.9	16.9	13.0	19.1	19.0	27.4	30.3	42.1	57.0	62.1
Lower Colorado	5.4	25.8	10.0	29.1	20.2	59.3	42.2	93.3	96.7	141.3
Great Basin	3.0	8.4	4.1	9.5	6.7	17.5	10.2	27.5	14.1	42.0
Columbia-North Pacific	52.2	106.7	52.1	120.6	73.6	170.1	120.6	235.3	197.7	325.8
California	36.9	67.2	61.6	75.9	102.1	134.3	185.9	211.0	262.6	311.2
Alaska	3.2	[c]	4.3	[c]	5.6	[c]	8.4	[c]	12.4	[c]
Hawaii	1.2	10.6	1.8	12.2	2.2	16.8	2.8	23.7	3.6	34.0
Puerto Rico-Virgin Island	2.6	3.8	2.9	4.3	3.2	6.0	3.5	8.5	4.0	12.0
Total[d]	637	959	643	1,094	891	1,548	1,355	2,181	2,024	3,061

[a] Projected damages based on existing flood control works.—[b] Not reported.—[c] Not available.—[d] Rounded.

may overflow into *floodplains*, the wide flat areas adjacent to the channel. Typically, a river such as the Po in Italy, the Nile in Egypt, and the Mississippi in the United States, uses some portion of its floodplain each year. Once every twenty years or so it may inundate the major portion of the floodplain and perhaps once in a century reach a significant flood depth.

The flow in the stream and the elevation it reaches depend upon natural factors, such as a change in temperature resulting in release of snow pack, or upon engineering factors, such as a storage dam or confining levee. Its recognition as a major flood depends upon the extent of damage to property and disruption to life along its banks.

Perhaps the most frequent severe rainstorm floods occur in Asia and the Far East. At Fukiko, Taiwan and Baguio, Philippines world-record storms in 1913 and 1911 produced 200 cm of rainfall in 72 hours and 63 hours, respectively. Parts of India are flooded every year and often suffer considerable damage from southwest monsoon rains. These storms, moving from the Indian Ocean onto the land mass during June to September, deposit large volumes of water. The major streams affected by the monsoon rains are the Brahmaputra and the Ganges Rivers, but all of southern Asia's streams share in periodic rainstorm flooding. In 1953, the Godavari floods were unprecedented in 50 years and in 1955 many of India's rivers reached their highest stages on record.

Tropical air masses and cyclones (typhoons) of tropical origin accompanied by heavy precipitation influence southern China. In July 1931, torrential rain produced the greatest flood since the beginning of hydrological observations in the Yangtze valley. One of the most densely populated regions of the world annually faces the flood waters of the Huang Ho (Yellow River) which, because of devasting floods to settlements along its banks for 4,000 years, has been called "China's Sorrow" (see Section 7.4).

The orderly and controlled floods of the Nile in Egypt have been used for thousands of years for irrigation and soil fertility replenishment, but the magnitude of the annual flood cannot be foretold until the moisture-laden air masses blowing from the south Atlantic across Africa hit the headwaters of the Blue Nile in the highlands of Ethiopia.

Rivers of the United States are particularly influenced by rainstorms. The southern states are exposed to tropical disturbances moving from the Gulf of Mexico or mid-Atlantic Ocean to the land mass. The western states are struck by migratory frontal storms from the Pacific Ocean that result in long periods of rainfall. Mid-continent states are affected by the convergence of the moist tropical Gulf air with the continental polar air. The Kansas River flood of July 1951 was one of the most severe of the latter types experienced in this century.

Coastal floods from hurricanes are common occurrences in both the West and East Indies and, although not occurring as frequently, high tides and onshore winds often strike northern Europe. In 1099, a storm-tide flooded the coasts of England, the Netherlands and Belgium resulting in the loss of nearly 2,000 lives. Tropical storms, such as Hurricane Carla that struck the Gulf coast of Texas, USA in September 1961, did damage estimated at $850,000,000, mostly from high-velocity winds and high tides.

Table 7.2. Major flood disasters in the United States (1973 damage in excess of 100 million dollars) Source: after W.L. Horn.

Year	Stream or location	Estimated damage in million dollars (1973 value)	Type of flood
1844	Upper Mississippi	2,030	Rainstorm
1861	California	NA[a]	Rainstorm
1889	Johnstown, Pa.	150	Dam failure
1900	Galveston, Texas	175	Hurricane
1903	Passaic-Delaware	480	Rainfall, Dam failure
1903	Missouri River	NA	Rainstorm
1913	Ohio River	900	Rainstorm
1913	Brazos-Colorado, Texas	600	Hurricane
1921	Arkansas River	100	Rainstorm
1926	Miami, Florida	225	Hurricane, tidal
1927	New England	300	Rainstorm
1927	Lower Mississippi	2,550	Rainstorm
1928	St. Francis Dam, California	100	Dam failure
1936	Northeastern U.S.	650	Rainstorm
1936	Ohio River Basin	650	Rainfall, snowmelt
1937	Ohio River Basin	1,700	Rainstorm
1938	California	500	Rainstorm
1942	Mid-Atlantic	180	Rainstorm
1943	Central States	1,250	Rainstorm
1944	South Florida	200	Hurricane, tidal
1944	Missouri Basin	720	Rainstorm
1945	Hudson River Basin	130	Rainstorm
1945	South Florida	170	Hurricane, tidal
1945	Ohio River Basin	100	Rainstorm
1947	Missouri Basin	640	Rainstorm
1948	Columbia River Basin	400	Rainstorm
1950	San Joaquin River, Calif.	100	Rainstorm
1951	Kansas River Basin	2,500	Rainstorm
1952	Missouri River Basin	550	Rain-snowmelt
1952	Upper Mississippi Basin	100	Rainstorm
1954	New England	380	Hurricane
1955	Northeastern U.S. from New England to North Carolina	1,550	Hurricane
1955	Calif.-Oregon-Nevada-Idaho	700	Rainstorm
1957	Southeastern Kentucky	150	Rainstorm
1957	Texas	330	Rainstorm
1959	Ohio and adjacent states	220	Rainstorm
1961	Texas Coast	850	Hurricane Carla
1963	Alabama to West Virginia and Ohio	200	Rainstorm
1964	Florida	200	Hurricane
1964	Ohio River Basin	200	Rainstorm
1964	Oregon-North Calif.-So. Wash.	820	Rainstorm
1965	Southern Florida	250	Hurricane
1965	Upper Mississippi Basin	280	Rain-snowmelt
1965	Platte River	340	Rainstorm
1965	Arkansas River	100	Rainstorm

[a] Not available

Table 7.2 (continued)

Year	Stream or Location	Estimated damage in million dollars (1973 value)	Type of flood
1965	New Orleans	590	Hurricane, tidal
1969	Southern California	300	Rainstorm
1972	Rapid City, South Dakota	120	Flash flood
1972	Entire East Coast of U.S.	3,000	Hurricane Agnes
1973	Mississippi River Basin	420	Rainstorm

Unusual sets of astronomical coincidences involving the position of the Earth, moon and sun have produced extremely high tides about 20 times in the last 300 years. Low-lying coastal areas, such as the city of Venice, could be seriously flooded if unfavorable winds occur simultaneously with such high tides. In March 1962, onshore gale-force winds struck the coast of New York, Delaware and North Carolina, USA during a similar high-tide period, resulting in extensive damage and the loss of 40 lives.

Coastal floods from tsunamis have inflicted great damage to Japan and on numerous occasions to Hawaii, as in April 1946; to Lisbon, Portugal, as in 1755; and to Crescent City, California, USA in 1964. Tsunamis are discussed more fully in Chapter 3.

Snowmelt is most frequently a contributing factor to a rainflood but on occasion it can even be the major source of the flood. A sudden warm spell accelerates the melt and the water accumulates as runoff. Snowmelt floods, however, unlike rainfloods, develop peak flows slowly, providing time for both flood control measures and warning to those in jeopardy.

Snowmelt floods of March and April 1961 in the upper Mississippi River basin are typical of a northern latitude flood. A warm period in mid-February melted a portion of accumulated snow, and the water was absorbed in the upper crust of the soil and refrozen; heavy snows then occured in March, to be followed by rising temperatures. A rainstorm began the melt, which was accelerated by further higher temperatures, and the combined rain and snowmelt caused runoff over a wide area, with the accumulated flow from tributary channels and streams resulting in extensive damage along the main streams. A tragic snowmelt flood and mud flow from the crater of Ruapehu Volcano, New Zealand occurred in 1953 (see Section 2.2).

When a thaw develops suddenly after a long cold winter, river ice may not soften for easy breakup in the normal way. Instead, the rapidly swelling river ruptures the ice deck into large hard cakes and sheets that become jammed on riffles, river narrows, or on bridges, so that the river is dammed by the ice. The dam creates flooding upstream and when the force becomes great enough to break the ice-jam, a flood wave will move downstream, often causing damage.

Rupture of glacial ice can create a flood with an extraordinarily high rate of flow. Jokulhaup is an Icelandic term for the sudden release of the water impounded behind a glacier. Tulsequah lake in British Columbia, Canada, is formed in a valley tributary to a glacier, and when the lake level rises sufficiently to lift the glacial ice and start a flow of water through small crevices and tunnels in the ice, the lake suddenly spills. The jokulhaup of July 1958 resulted in a lake-level drop of 60 m in three days.

In Western Europe and the United States about 34 dam failures have occurred this century; the sudden collapse of such an engineering structure can cause a major flood and a devastating wake. St. Francis dam, 62.5 m in height, on a tributary of the Santa Clara River about 70 km north of Los Angeles, California, failed at about midnight on March 12, 1928. Its failure released 4.65×10^7 cubic m of water in one hour and created a flood wave of between 11,000 and 14,000 cubic m per second. The failure of Baldwin Hills dam, a much smaller structure in California, is described in Section 7.4. A dam at Malpasset, France, in 1959 also failed from causes related to geology.

Massive earth slides into a valley may produce a temporary dam, which, when filled with water, can cause a flood by over-topping or failure. Thus an avalanche from Mt. Huascaran in 1962 produced a temporary debris dam with subsequent serious flooding (see Section 6.4). In the case of the Vaiont dam in 1963 when an avalanche of earth and rock slid into the reservoir (see Chapters 3 and 4), the sudden surge of water plunged over the dam and swept down the narrow stream channel.

In Section 7.2 some important factors involved in disastrous rainstorm-river floods are described. Why do only some storms produce damaging floods? Some areas rarely see a major storm, yet overnight a flood disaster occurs; while in other areas, a flood threat can be forecast within weeks and days.

The Influence of Geology on Floods

Floods cause damage by inundation, erosion, and by the impact of detritus against man's structures or by its deposition on valuable property. As each of these kinds of damage is influenced by the geology of the watershed contributing to the flooded region, the areal geology map is therefore a useful supplement to meteorological and hydrological data and aerial photos in assessing flood-related hazards.

For any given storm, the quantity of runoff is modified by the vegetation and the characteristics of the soil and rock surface upon which the precipitation falls. The erodibility of that surface is determined by its physical properties and the vegetation. Furthermore, the nature and quantity of the materials transported down the slopes to the channels and thence carried away by the streams is influenced by the properties of the surficial rock and soil. The nature of the debris deposited by floodwaters is also affected by geologic factors; for example, a watershed consisting of poorly consolidated, finely-grained sedimentary formations will produce debris consisting of mud or silt; whereas, one composed largely of harder rock may be expected to yield coarser erosion products. The ratio of coarse to fine material is governed by the intensity of precipitation and the steepness of the slopes as well as by the physical characteristics of the rock composing the drainage area.

A portion of the precipitation falling on an area may be intercepted by vegetation which entraps raindrops on leaf surfaces; light showers may be entirely intercepted in heavily wooded or grassland areas, and subsequently returned to the atmosphere through evaporation. Interception losses decrease as the foliage becomes saturated, and become insignificant with prolonged storms. Such losses by interception are not related to geology except to the extent that the soil types comprising the drainage basin influence the kinds and density of foliage.

Rain may fall directly on impermeable surfaces such as rock outcrops, roofs, or pavement in which case it flows longitudinally. Parts of this flow are (1) retained in crevices or surface depressions, (2) evaporated or (3) absorbed by permeable soil or fractured rock surfaces which lie in its path. Another fraction of the downpour may fall on permeable areas and infiltrate directly into underlying formations, while some of the water that seeps into the ground may be returned to the surface by capillary action or by root systems, to be lost through evaporation or transpiration. Some seepage may reach the water table or perched water bodies and thence move laterally, such underflow usually contributing to surface streams at lower elevations. Its movement through the soil interstices or rock fractures is generally much slower than that of the surface runoff so that it arrives later at the streams and consequently does not augment flood flow.

In some regions highly permeable deposits are underlain at shallow depths by impermeable surfaces. In such instances, the water that infiltrates may move laterally at a velocity comparable to that of the surface runoff, this type of underflow being known as *subsurface storm flow*. When such flow arrives at the streams concurrently with the surface runoff, these two sources combine to augment flooding. Subsurface storm flow occurs through coarse deposits such as some rock slides or talus slopes. Regions underlain by carbonate rocks including limestone and dolomite may also transmit such rapid flow. These formations dissolve in ground waters that are slightly acidic with the consequent formation of solution channels and caves with openings to the surface. These *karstic* features, as they are known, frequently accommodate underground streams; in some regions entire rivers disappear underground where they traverse such rock types.

The infiltration capacity of a watershed or any portion thereof is the maximum rate at which water can be absorbed under a given condition. Some significant factors that affect infiltration of the surface and underlying shallow formations include: (1) their moisture content, (2) their permeabilities, (3) their thickness, and (4) the extent to which they may be perforated by the burrows of animals, worms or insects or by decayed roots. Of these factors the soil moisture content is usually the most influential. Therefore, in most regions infiltration capacity is greatest at the start of a storm when the soil pores are empty and it decreases as rainfall continues.

Overland flow can occur only where and when the rate of precipitation exceeds the infiltration capacity. When this condition is satisfied the surface depressions fill to overflowing and runoff commences. The volume of water required to fill the surface irregularities and cavities is known as *depression storage*, and is ultimately returned to the atmosphere or is subject to slow infiltration depending upon the characteristics of the surface.

Of the total storm precipitation, part is lost through evaporation or transpiration and part moves through the soil interstices or rock fractures. The fraction that exceeds the quantities lost through interception, evaporation and transpiration and detained by slow infiltration constitutes the rapid runoff which contributes to flooding.

Streams may be classified as *erosional* or *aggradational* depending on whether the flowing water is removing or depositing sediments. The capacity of a stream for carrying suspended matter depends upon its discharge volume, depth, and velocity. Scour will occur where streams transporting less than their carrying capacities pass over erodible formations. Intact igneous and metamorphic rock and well-consolidated sedimentary formations generally erode slowly and consequently the works of man constructed on these formations are seldom undermined by floods. Conversely, loosely-compacted sediments, weathered, and decomposed rock are less resistant and provide questionable foundations for structures located in areas which may be overrun by rapidly moving water.

Overland flows may dislodge rock and soil from the watershed slopes and transport them to the stream channels. Where slopes are steep and consist of loose, broken rock or poorly consolidated soils, avalanches, slides or mudflows may occur following downpours (see Chapter 4). Structures located in the paths of these moving masses are subject to damage from impact. The solid materials derived from the watershed slopes lodge in the channels to be swept away by the currents, usually during times of flood.

Streams occupying steep canyons cut in weak formations may be blocked in this manner. The rivers traversing the northern coast ranges of California occasionally experience such blockage. These mountains consist largely of Franciscan formation, a complex of weathered and broken shales, sandstones, cherts and igneous rocks, whose tendency to sliding has given rise to the term *restless topography*. The potential for a large slide and impoundment of a sizeable reservoir remains a distinct possibility in this region. The overtopping and consequent collapse of a high slide dam could release a destructive surge on to downstream developments.

Earthquake lake, a tourist attraction on the Madison River, Montana, is the result of the blockage of that river by a great slide which was triggered by the Hebgen lake earthquake of 1959. This slide contains both coarse and fine material with the greater portion of finer material located near its upstream face. Thus nature fortuitously provided a filter for her dam which has prevented its destruction by piping and erosion when the dam has been overtopped. Other landslide dams have not shared such good fortune. Some, notably in the Andes mountains of South America, have failed, releasing catastrophic floods consisting of rocks, mud and water.

7.2. Some Facts about Floods

Storms

The Earth's rotation deflects heated air as it rises near the equator and drifts toward the poles. As shown in Fig. 7.2, as the cooler air approaches about

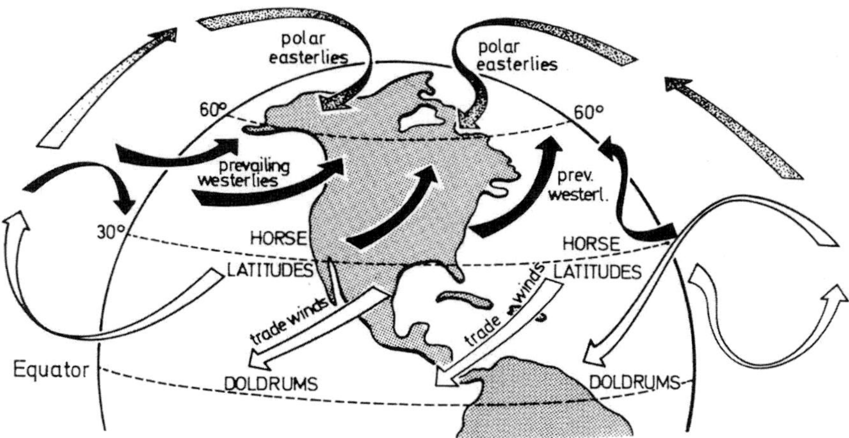

Fig. 7.2. General air movements in the Northern Hemisphere

latitude 30° some of it is carried east in the Northern Hemisphere to form the subtropical jet stream. Much of it slows and descends in the horse latitudes and replaces the rising equatorial air. In the Northern Hemisphere, the southward air flow turns west and forms the trade winds, while some of the descending air moves northward and is deflected east, forming the prevailing westerlies which blow over the middle latitudes.

Not all the high air sinks to the surface at mid-latitudes; some continues north, cooling by radiation as it goes, finally sinking near the North Pole. As this polar air moves southward, it runs into the prevailing westerlies and takes the form of huge surface eddies, the collision producing high- and low-pressure cells. High-pressure areas of cool, heavy air slowly form, by the forces of the Earth's rotation, into a clockwise-spiraling cell. When such cells form in the north they sweep southward and, for North America, may extend to Mexico in winter. They are usually several hundred kilometers in diameter, but the largest ones may cover the entire United States east of the Rockies. Between the high-pressure cells are low-pressure cells which rotate counterclockwise in the Northern Hemisphere. Associated with these low-pressure centers are the weather fronts, which are boundaries between warm and cold air masses and along which clouds and precipitation are formed.

Such principles of the Earth's weather machine are well understood by meteorologists, but the ability to use the available data to forecast a moisture-producing storm is still perhaps more of an art than a science. Each year brings new data and better physical models, and satellite photographs of vast areas of the Earth's surface on a continuous basis have made a major contribution to aid the weather forecaster (see Plate 7.3). From a series of these photographs he is able to track the high- and low-pressure cells, plot frontal activity, and make evaluations of precipitation amounts, duration, and likelihood of a storm.

In California, for example, rainfall occurs mostly in the winter half of the year, from about October to April, during which period, the Pacific high-pressure

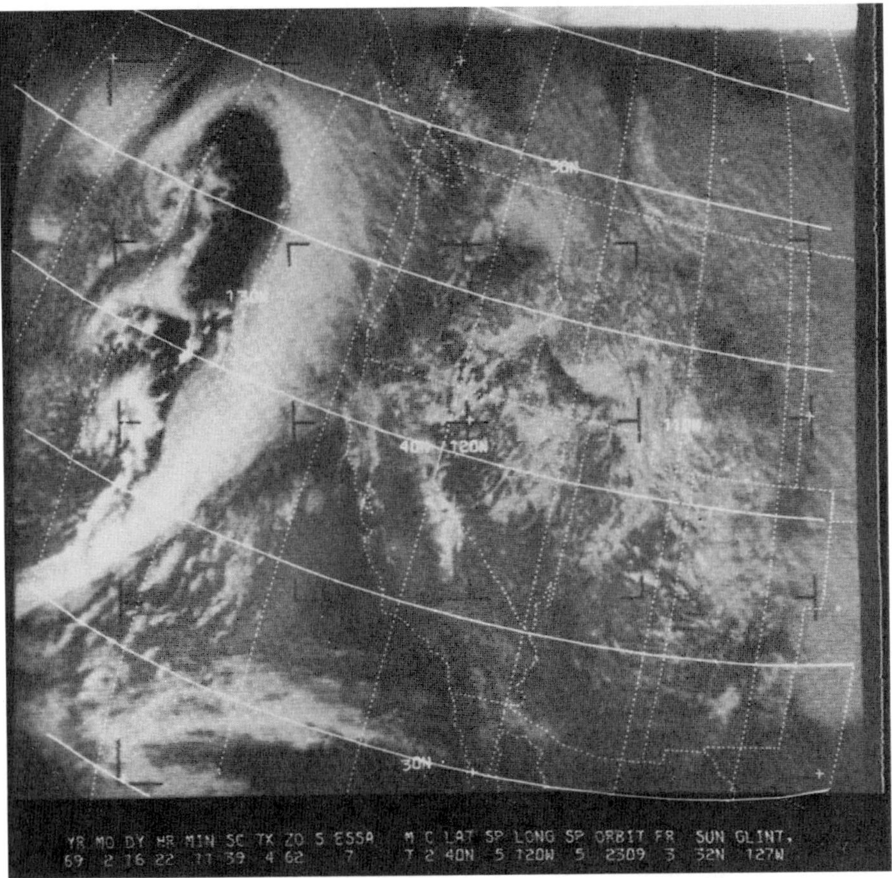

Plate 7.3. Satellite photograph showing a well-defined front approaching the west coast of the United States. (Courtesy of National Weather Service, NOAA)

cell weakens and moves south, while the Aleutian low-pressure center strengthens. Migratory storm systems, forming in the central and western Pacific, feed into the Aleutian low-pressure center, and the track taken by these storms moves farther south as the mid-winter and spring months are reached. The passage of low-pressure centers and the associated weather fronts through the State bring the winter rainfall into California.

Because of the topography of California, with the mountain ranges oriented north-south, rainfall results principally from the rising of the moist air masses as they pass over the mountain barriers. As the warm air mass rises, it expands, cools, and the water vapor condenses as rainfall. The lifting mechanisms associated with the fronts and low pressure centers also add to this *orographic* (mountain-induced) rain.

Heavy rains in California are associated with weather patterns that displace the Aleutian low-pressure center southeastwards, causing migratory storms to move in a more southerly track in the strong flow toward California. The passage of several closely-spaced storms brings prolonged periods of rainfall. Since lifting over the mountains is so important, the presence of strong winds (especially in the lowest 3,000 m of the atmosphere) and high moisture content result in heavy rains.

The level in the atmosphere where rain changes to snow varies with the storm and geographical location. The freezing level may be as high as 3,000 m or as low as 600 m. Warmer storms are critical in causing floods, because the precipitation falls as rain at high elevations and becomes available as immediate runoff instead of being stored as snowpack. The existence of a previous snowpack in the mountains when a warm rainstorm arrives leads to melting of the snowpack and augments the rain runoff with snowmelt water.

Runoff from the mountains of California is rapid because of the steep slopes of the river canyons. The peak flow at many of the foothill stream-gaging stations occurs in a period of a few hours after the ending of the heaviest rainfall in the mountains. The high runoff reaches the valley floor, and if the stream channels are incapable of carrying the high flows, or if a levee failure occurs, flooding of adjoining lands results.

Heavy runoff is also dependent on the condition of the soil in the mountains. If previous recent rains have saturated the soil mantle, then less of the rain from the new storm soaks into the soil and more is available for runoff. The percentage of rain that appears as direct runoff varies from about 15 per cent in storms following an extended dry period to 60 per cent in storms where the rain falls on a wet soil.

Flooding over limited areas may result from intense localized storms, some of which have convective clouds imbedded in the general storm cloud mass which lead to thundershowers. Torrential thundershowers over 25 to 130 square km produce heavy local runoff in the streams. Thundershowers occur over the mountains and desert regions of southern California commonly in late summer or in the fall. The path of the air mass in these storms is from the Gulf of Mexico across Texas, New Mexico and Arizona into southern California. In the southern end of the San Joaquin valley (see Plate 7.6) significant mud flows result from these thundershowers.

For those working in flood control anywhere in the world, a forewarning of storms would offer more efficient preparation. If the predictions were in terms of amounts of precipitation, steps could be taken to provide more available empty volume in upstream reservoirs, evacuate certain danger areas, and prepare flood-fighting equipment.

Stream Channels

In our inquiry about the storms that result in flood hazard the *stream channel* is of primary significance. It not only must convey the flood waters to a non-damaging location, but along its banks lives a great portion of the world's population, because the streams, over geologic time, have built the alluvium deposits that make the best agricultural lands.

Plate 7.4. In Los Angeles County, California, February 1969, flood flows came roaring out of Big Tujunga Canyon, destroying Foothill Boulevard Bridge (upper bridge) and Wentworth Peace Bridge. (Courtesy of Los Angeles County Flood Control District)

Stream channels are constantly moving laterally, as shown by maps of stream courses made years ago and compared with present channel locations (see Plate 7.4). Erosion picks up some material and deposits it in a different location. The rise and fall of river level with an increase or decrease in velocity encourages the cutting action.

In addition to the movement of material back and forth along the channel, streams carry eroded material from the watershed area into the channels at lower elevations. Thus, streams flowing through a nearly flat valley are often situated on alluvial ridges. The land adjacent to the river is higher in elevation than the land farther away (Fig. 7.3). River-flows greater than the carrying capacity of the channel spread into the adjacent lateral basins. Under natural conditions, the river channels are of moderate section and convey but a small fraction of the flood discharges. The lateral basins carry the greater portion of the flood waters and act as *flood storage reservoirs,* also receiving and retaining a large portion of the sediments delivered to the valley floor. This has been drastically altered by the activities of man.

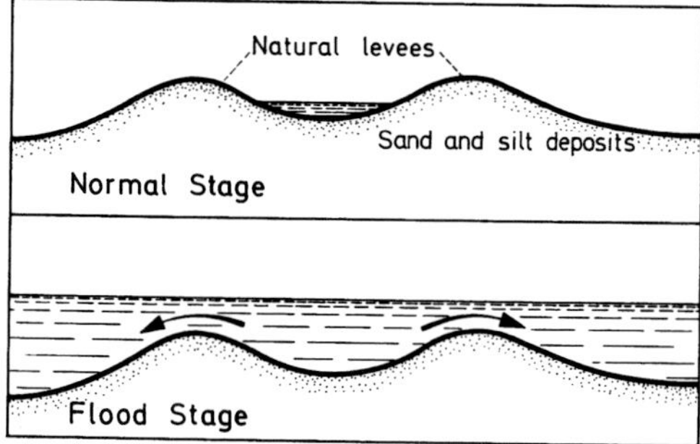

Fig. 7.3. Stream channel on alluvial ridge

An example of the increase in elevation that occurs in a stream channel is shown by the low water readings in Table 7.3 for the mouth of the Yuba River at Marysville, California. The movement of hydraulic mining debris from the watershed area along the flank of the Sierra Nevada was in this case the main cause of the elevation change of 6 m during a period of 55 years. (Hydraulic mining was outlawed after the turn of the century in California.) Progressive deposition such as this slowly reduces the carrying capacity of the channel for flood flows, so that in order to compensate for the rising bottom elevation, levees have to be raised to provide the same carrying capacity.

Table 7.3. Progressive record of stream channel elevation, Yuba River, California

Year	Elevation (m)	Year	Elevation (m)
1850	12.1	1905	17.9
1860	13.4	1910	17.0
1870	14.7	1920	16.1
1880	16.8	1930	14.6
1890	16.8	1940	14.4

Stream channels in arid areas are likely to be unstable. The process of erosion moves material onto the flood plain in the form of an alluvial fan. Buildings located on these fans are highly susceptible to damage by floods of frequent occurrence. Occasional floods erode the unconsolidated material, but the next flood may or may not follow this channel. The Yellow River in China in 1852 broke from its enormous system of inner and outer dikes and changed its course 600 km to the north, to empty into the Gulf of Chihli, instead of the Yellow Sea.

A continuous record of flows in natural stream channels is of the utmost importance, not only to flood hydrologists, but to all who utilize water resources. A *stream gaging station* is a system for continuously recording the relative elevation of the water surface in the stream channel. It should be located at a cross section of the channel where a minimum of changes will occur, because a stable section is necessary for a continuous related record. Traditionally, the gage consists of a float mechanism in a stilling well adjacent to and interconnected to the stream. By periodic measurements of the velocity, a relation between flow and stage is established. Unfortunately, the location of such a gage makes it extremely vulnerable to damage during flood periods, though recent innovations in methods of detecting and recording the water surface have reduced somewhat this hazard.

A network of stream channels provides natural drainage for a land area or basin, developing over the years the capacity to carry most of the water that flows from the basin. In the headwaters, the channels are smaller, in the steeper areas more incised, and on the valley floor the channels and overflow areas receive and regulate the accumulated flows of the network. In a flood, the water height at any place is the accumulated result of inflow from the various tributaries. Knowledge of location and movement of the maximum flood wave downstream is essential for flood control operation on the river system or in warning or preparing for flood.

Flood routing is the term used when the peak water elevation is followed downstream as it accumulates flow from tributaries and picks up reservoir releases. Various techniques have been developed for this purpose, but in all cases the problem is much the same: what are the water volumes and speeds of travel of the flood peaks as they flow in and out of storage along specific reaches of the stream? The data and time available for solution limit the number of places along the stream channels that can be specified, but the use of high-speed and large-memory computer facilities has greatly enhanced this ability.

Channels of a major stream possess a significant volume of storage to hold and carry flood flows. If this storage is combined with storage in by-pass channels by engineering schemes, a significant capacity for flood control is gained. Channel storage may be very great in large river systems. In the Ohio River basin, it is about 7×10^{10} cubic m, and in this river system the total flood runoff may be in storage at one time.

Rainfall

One phase of the water cycle—rainfall—is so commonplace that we are apt to think much is known about its occurrence, intensity, and variation. But this is often not the case.

Records of precipitation are obtained from the amount of water collected in a can open to the atmosphere. The rain gage only samples the rainfall of a drainage basin at one place and, because it is affected by wind, surrounding vegetation and terrain, it may provide a poorly representative value. Use of such data over a greater area involves risky extrapolations, and thunderstorm activity is often not recorded at all.

Thunderstorms in California, for example, augment the spring runoff of Sierra Nevada streams, but data were not available until recently to show the extent. In the last few years, recording rain gages have been located at high elevations in the Sierra (3,400 m). To give an indication of thunderstorm activity over 230 square km, these rain gages are placed by helicopter in April or May as the snow melts and removed late in the fall before the snow season begins.

Standard installation practices are to locate the precipitation gage to meet specific exposure criteria. Many of the records available for study and for operational use are unfortunately not from representative sites, but are from such places as tall metropolitan buildings or airports; priority for location is often convenience, not exposure.

In the United States, the National Weather Service is the agency responsible for the formation and maintenance of precipitation networks and for the collection, compilation and publishing of the basic data, the work being accomplished by a cooperative effort with local agencies and individuals. The data collected are available to those agencies directly responsible for flood operations and flood control. The most common collecting period for rainfall is 24 hours. The collecting period for operational data is hourly or in six-hour periods. The six-hour quantities are *synoptic data*, so called because the data are collected by observers at the same time throughout the world without consideration for time differences. Only at a very limited number of stations are data collected in small enough

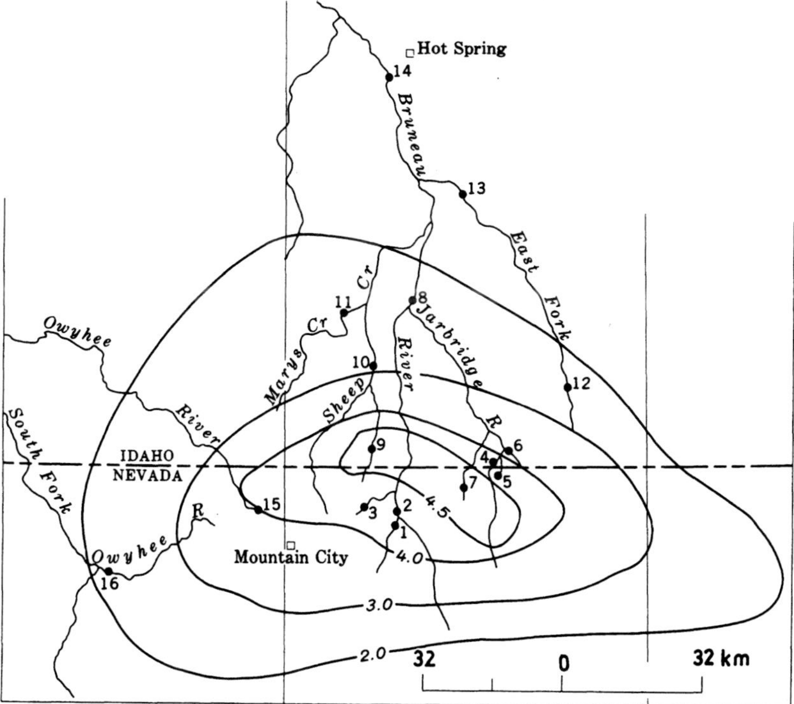

Fig. 7.4. Isohyetal lines (lines of equal rainfall) on a small watershed for the total period of a storm. (Courtesy of U.S.G.S.) Numbered dots refer to gage locations

increments so that the record actually represents a true pattern of the rain as it falls.

Storm *magnitude*, *duration*, and *intensity* are basic factors in flood flow calculations and are the subject of detailed study and evaluation where engineering works are planned. Of equal significance are the characteristics of the watershed or catchment area and the capacity of the stream channel to carry predicted flows. It is evident from the reports of major flood disasters that interaction of these factors is rather complicated. Precipitation during the December 1964 flood on the north coast of California was about 75 cm. There basins are heavily incised because of frequent frontal storm activity and the erosibility of the terrain. A lesser amount of precipitation over a great basin such as the Mississippi could easily be flood-producing.

Numerous methods are used for organizing storm precipitation data for analysis. Maps and charts are common ways of presenting specific rainfall measurements over an area. The *isohyetal* map (a rainfall-depth contour map) is frequently used to depict the variation of total storm rainfall over a basin (see Fig. 7.4), and may be used also for a particular time interval of the storm that is significant. If the storm area contains a number of stations, the isohyetal map will give a reasonably accurate indication of the rainfall distribution, although generally it is only a rough approximation of the rainfall pattern.

Cumulative rainfall curves are an effective method of presenting the variation in rainfall as well as the total storm amount for a precipitation station. By careful selection of stations, the storm variation over an area is quickly and vividly depicted. Fig. 7.5 shows the progress of Hurricane Agnes in 1972 in the United States (see Section 7.4). The curve for Tallahassee, Florida, is representative of the first day's rainfall on June 18 and 19. For the 24-hour period ending on the morning of June 21, rains were widespread over much of the eastern seaboard.

The curve for Wellsville, New York, shows heavy rains in an early burst of up to 15 cm. For the next 24 hours ending the morning of June 22, the

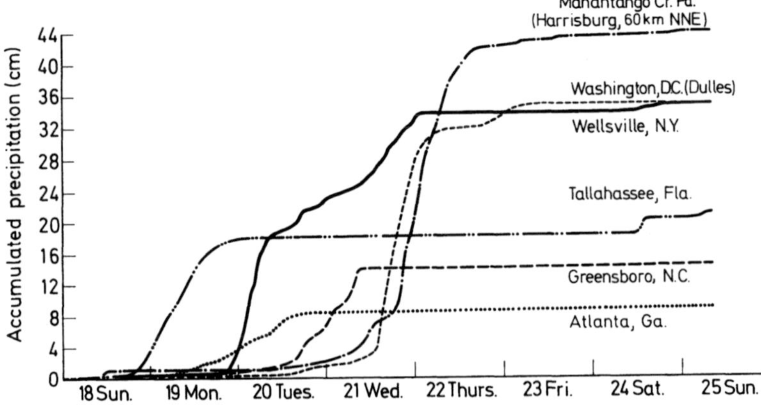

Fig. 7.5. Cumulative rainfall curves for selected locations during Hurricane Agnes (Eastern U.S.A.), June 18–25, 1972. (Courtesy of National Weather Service, NOAA)

heaviest amounts occured along a north-south swath near central Pennsylvania, while on the last day of the storm, June 24, few rainfalls were greater than 1.3 cm.

Snow and Snowmelt

Melting snow may be the sole source of floods and flooding or a contributing factor, as rain falls on the snowpack. By contrast the snowpack may serve, as a significant delaying influence on a likely rain flood. Snowmelt, however, is an acute flood hazard wherever heavy snowfalls occur or when extensive winter snows accumulate. In Europe, for example, significant snowmelt floods occur in countries around the Alps, particularly in the Po basin of northern Italy and along the lower Rhine.

A particular type of snowmelt flood experienced in California occurs mostly in the southern Sierra Nevada where the mountain ranges reach more than 5,000 m. After winters with a heavy snow accumulation, as in 1938, 1952, 1967 and 1969, the onset of warm temperatures in April to June leads to melting of the snow and large volumes of runoff in the mountain streams. Snowmelt floods are not characterized by sharp peaks, as experienced in rainfloods, but rather by sustained large volumes of water spanning periods of weeks to several months. Upstream and foothill reservoirs in big snow years are generally unable to absorb the volume of melt, so that flood control engineers must release large volumes of water into the valley rivers.

Vast amounts of water are held in snow storage. In temperate regions such as Germany, the normal annual total precipitation ranges from 500 to 1,000 mm. By contrast, in the California Sierra the snow accumulation is frequently 5 to 6.5 m in depth. As an illustration, in Table 7.4 the water content equivalent of the snow is given for four California river basins. The water volume reaches the equivalent of tropical rainfall, but the rate and timing of the melt is the key to the flood hazard in such cases.

Table 7.4. California snow data—April 1969

River basin	Elevation (m)	Depth of snow (cm)	Water content (cm)
Stanislaus	2,300	401	198
Merced	2,650	368	224
Kaweah	2,620	498	257
Kings	3,140	361	175

Numerous mathematical models have been developed to describe and forecast the resultant snowmelt from drainage basins. The melt rates in Table 7.5 were used for a study of snowmelt in the Columbia River basin. Rain is frequently identified as intensifying the melting process, and warm rain will, by erosion and by quickly carrying heat to a high-density pack, contribute to the melt. Frequently, however, the rain is absorbed by the pack and is not a significant contributor to its melt. More important as a major source of heat in melting snow are warm winds.

Table 7.5. Melt rate factors for Columbia River basin, USA

Period		Melt rate
from	to	(cm/°day[a])
Dec. 1, 49	Jan. 21, 50	0.127
Jan. 21, 50	Jan. 26, 50	0.508
Feb. 15, 50	Feb. 28, 50	0.508
Feb. 28, 50	Mar. 1, 50	0.254
Mar. 1, 50	Mar. 22, 50	0.114
Mar. 22, 50	Apr. 18, 50	0.178
Apr. 18, 50	May 8, 50	0.305
May 8, 50	July 1, 50	0.254

[a] Degrees in Fahrenheit (mean daily temperature minus 32).

A practical and simple instrument for measuring snow depth is the aerial snow depth marker. It is a vertical post to which horizontal crossarms are welded at intervals. If painted black and located in a mountain clearing, it can be photographed from a flying airplane to provide intermittent data from a snow course or to extend the observation to areas not sampled on the ground.

Recently-developed instruments, as well as satellite surveillance of snow fields, have added a great deal to this form of hydrologic knowledge. A rapidly-expanding program of *snow pillows*, an instrument that senses the pressure of snow, has been designed and used to record and transmit by radio the snow accumulations or melt from a mountain location. This instrument is a valuable asset for forecasting the effect of a rainstorm or of melt on downstream flows.

Drainage Basins

The area drained by a system of rivers and tributaries makes up a *drainage basin*. The pattern of streams in the basin depends upon its geologic history, rock structure and resistance to erosion, and the steepness of the slopes.

The Amazon in South America is the greatest river in the world in volume and has the greatest drainage basin; its valley is a great subsidence trough filled with Tertiary sediments. The river rises gradually from November to June and then falls until spring. Flood levels reach 15 m above low river and silt is deposited into the settling basins on a gigantic scale during each flood season as the river refills the valley with its own alluvium.

The nature of the land surface on which rainfall accumulates or from which snow is melted provides character to flood flows. Quite clearly, water runs off more quickly from a steep slope than from a flatter one, a basin with dense vegetation and deep soils delaying the flow into the creeks, while a rocky surface quickly yields churning rivulets.

The change in water level in the streams (the *hydrograph*) is a clue to the general run-off characteristics of the basin, but, like all single measurements, does not fully explain them. The other significant factors are the precipitation itself, that is, the intensity and duration of the rain, the stream channel and its ability to carry the water entering it, and the wetness of the basin.

The antecedent condition of the basin is a most significant factor in runoff produced. On the U.S. Pacific coast, storms of October, November, or in some years December, occurring as they do after a six-month period of no rainfall and summer heat, barely wet the stream channels, while equal amounts of precipitation in January, February or March produce peak flows often on already flooded channels.

Flood forecasters use a factor, called *Antecedent Precipitation Index (API)*, that recognizes antecedent conditions of the basin. A simple method of computing the *API* for a basin is to apply a decay factor to the accumulated precipitation, involving a daily accounting procedure that adds the daily rainfall and subtracts a factor estimated to equal evaporation and other losses.

Flood Flow Frequency

The subject of frequency of recurrence of flows or rainfall amounts is always raised after each major flood. People ask, for example, "Have we just experienced a river stage that occurs only once in 100 years?". The economist must reduce in turn flood losses to an average annual basis to justify new projects. Therefore, with the growing need for improved flood plain management, and with the public interest in flood insurance, it is exceedingly desirable that the frequency of flood flows have a common meaning and that the computational techniques produce reliable results.

The basic objective of a frequency analysis is to derive a flood magnitude-frequency relation from historic data, and then to project the relation beyond the period of record, particularly in the study of unusually great floods. In most statistical work, the data available for analysis are very numerous, as, for instance, in life insurance (see Section 8.7). But in the study of flood records, where the maximum flood occurring each year is considered, the number of cases available is limited. Seldom in newly settled countries do records extend over more than 50 years (although in Europe and Asia partial records extending over centuries may be found as, for example, in the sea floods of Holland; see

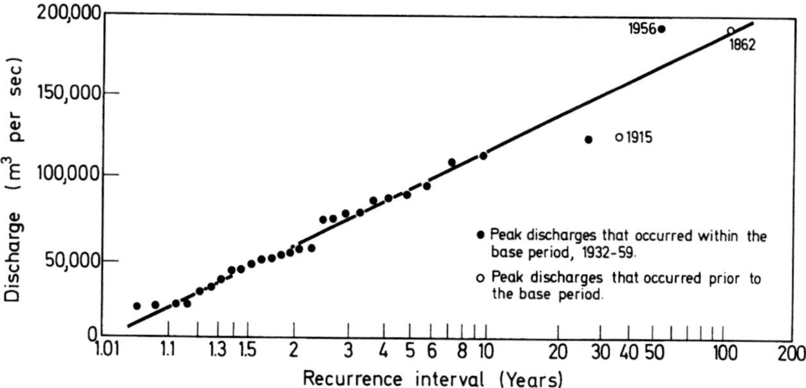

Fig. 7.6. Flood frequency curve for Eel River at Scotia, California, U.S.A. (Courtesy of U.S.G.S.)

Section 7.4). Thus, generally there is a difference between statistical methods used in flood risk analysis and the more commonly known methods.

There is not, at present, universal acceptance of any one single statistical method. Agencies charged with managing water resources or planning new projects have developed favorite procedures. In view of the importance of flood-flow frequency estimates in the expanding field of water resources development and related programs for managing flood losses, studies of desirable flood-flow frequency methods should be extended in the coming years. Fig. 7.6 is an example of a flood frequency curve for the Eel River at Scotia.

7.3. Fighting Back

The Watershed

The biblical story of Noah and the Ark recounts an attempt to do something constructive about surviving in the face of impending disaster. More modern approaches to the problem of living with floods are to undertake either corrective or preventive measures. In the former approach, engineering works endeavor to make the stream flow fit the channel or to make the channel fit the flow of the stream. The second approach recognizes that if we must live with damaging high water, but wish to avoid major damage, we must manage the use of flood plains. In this category, also, is the concept of accepting damage, but protecting a portion of the investment through insurance. An effective flood control program, of course, combines all approaches. The elements of a comprehensive flood damage prevention program are outlined in Table 7.6.

Because a significant portion of a flood is produced in the watershed, some flood control can be developed there. Efforts have been made to build dams, and to undertake contour plowing and terracing in order to retard surface runoff and increase infiltration. The degree of effectiveness usually depends on the steepness of the terrain, ground cover, storage possibilities and geologic factors such as the permeability and porosity of the soil. However, it is generally recognized that although flows of moderate intensity or lesser floods on small streams may be reduced, this work is relatively ineffective in controlling major streams.

Table 7.6. Flood damage prevention program

Corrective measures		Preventive measures	
Flood control	Other	Flood plain regulation	Other
Dams and reservoirs	Evacuation	Zoning ordinances	Development policies
Levees	Flood forecasting	Subdivision regulations	Open space
Channel improvement	Flood proofing	Building codes	Tax incentives
Watershed treatment	Urban development	Health regulations	Warning signs Flood insurance

Not all corrective measures in the watershed are directed toward reducing runoff, but may be directed toward erosion control. Thus, in addition to saving watershed soils, the measures minimize sand and silt movement and deposition in the downstream channels or other structures. The downstream channels thus are able to retain their capacity to carry flood flows.

An example of an upstream-downstream relation is the flood control program under construction for Los Angeles County, California. Here, 106 debris stabilization dams in canyons feeding the Los Angeles River were constructed as part of an overall plan that consists of downstream flood-regulating dams, concrete lined flood-channels, spreading grounds, storm drains and pumping plants. Stabilization of mountain channels to reduce debris flow into storage and other structures is achieved through this series of check dams.

The Soil Conservation Service, a U.S. federal agency, has, since the dust-bowl days of the 1930s in the mid-west, been charged with a program of watershed management for the conservation of soil and water and flood flow reduction. In 1954, the Watershed Protection and Flood Prevention Act extended and consolidated their activities. The methods used to carry out the program consist of one or a combination of items such as land treatment, cropping practices, channel structures, channel lining, and small reservoirs that serve the multi-purposes of livestock watering, recreation and flood control.

Reservoirs

Some dams are built solely to create a flood control reservoir; that is, a reservoir that will delay the peak flows in the channel by storing water during a critical high stage. Most reservoirs, called *multi-purpose reservoirs* (see Plate 7.5), serve many objectives. They may provide a power head, store water for the irrigation season, or simply create a lake for recreation. When flood control is also required, some of the storage space is dedicated for this use. Some of the storage space of the reservoir reserved for other purposes may also serve flood control, but not on a planned basis.

Indeed, requirements of reservoir operation are often fundamentally in conflict and this has led to many public debates and sometimes tragedy (see Section 7.1 on the Florence flood):

(i) Regulation for flood control is best accomplished when reservoirs are kept empty in anticipation of floods. They should be emptied again as rapidly thereafter as possible.

(ii) Conservation for irrigation or domestic use requires that flood waters be held in storage, sometimes over a period of years in semiarid areas, and that release of water be in conformity with seasonal demands.

(iii) Regulation of stream flows for the production of hydroelectric power requires a reservoir be kept as nearly full as practicable, that it never be emptied, and that water release be made in accordance with demands for electricity.

(iv) Maintenance of a stable reservoir level is most favorable to fish and wildlife protection and propagation and a reservoir must never be emptied.

(v) Recreational purposes are best served when a reservoir is kept full, simulating a natural lake.

Plate 7.5. The Shasta dam and multi-purpose reservoir on the Sacramento River in northern California. The snow-capped volcanic peak of Mt. Shasta is in the background to the north. (Courtesy of U.S. Bureau of Reclamation)

These conflicting reservoir uses must be resolved by the engineers and hydrologists in charge of the control system. The usual overriding objectives are to derive maximum income from the water, and to minimize the damage downstream.

Operation of a reservoir during flood period is clearly a balance between available flood control space, required release commitments—such as power production—and effect of the storm on tributary streams. All this in the face of the unknown quantities: how long will the storm last? what will be the rates of inflow?

Shasta dam in Northern California (see Plate 7.5) in January 1970 helped to regulate the largest flood in the history of the area since records began in 1903. During the storm period of January 8 through January 27 more than 86 cm of rain fell at Shasta dam and 3.1 billion cubic m of water ran into Shasta lake. The peak inflow to Shasta lake was 6,000 cubic m per sec.

The four charts in Fig. 7.7 indicate the flood control accomplished at Shasta reservoir during the 1970 flood. Chart *a* shows the flow at Redding (15 km downstream of Shasta dam) as it actually occurred and as it would have been if Shasta had not been available to regulate the runoff. This chart indicates

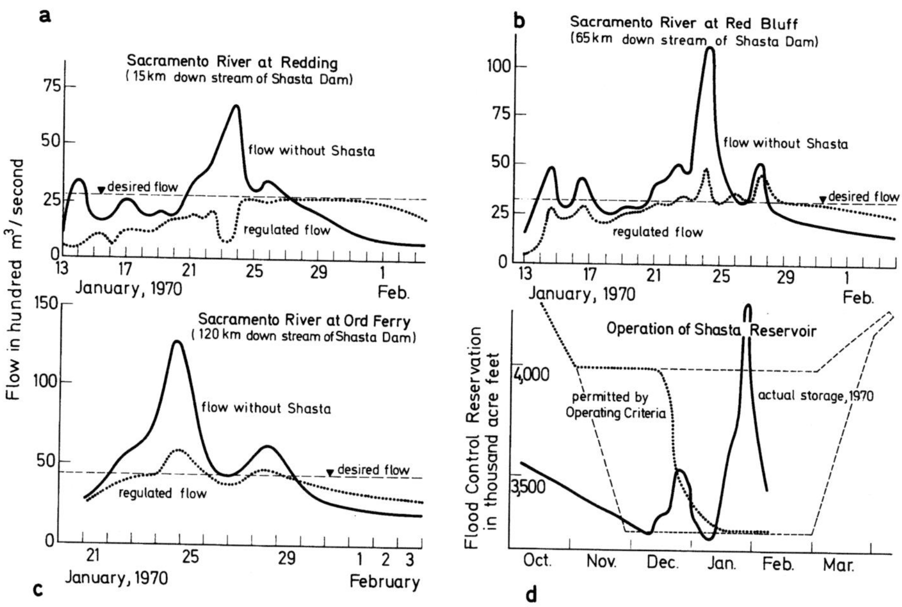

Fig. 7.7. Graphs showing the operation of Shasta reservoir to control flooding of the Sacramento River, California, January 1970

that through the operation of Shasta it was possible to regulate the flow of the Sacramento River at Redding to 2,250 cubic m per second even though this was the largest flood during the 70-year period of record.

Charts *a* and *c* show the flows at Red Bluff (65 km downstream of Shasta dam) and Ord Ferry (120 km downstream of Shasta dam) with and without Shasta. In these two cases it is clear that although Shasta contributed materially toward reducing flood flows and damage, it was not sufficient to regulate the Sacramento River to the desired flow. The reason is that the *tributary inflow* below Shasta, which is unregulated, exceeded the desired flow at the two points and even if all the water flowing past Shasta dam had been stored, the channel capacity at Red Bluff and Ord Ferry would have been exceeded.

Chart *d* shows the use that was made of Shasta reservoir during the flood period. It indicates that the lake was at prescribed flood control level at the beginning of the flood period and that it was necessary to use most of the flood storage space in regulating this particular flood. It will also be noted from the chart that most of the flood water that was stored during January was evacuated during early February so that empty space would be available in the event another flood should come along before the 1970 flood season ended.

The Federal government maintains operating control over Federally-authorized flood control projects in the United States, administered through the U.S. Army Corps of Engineers. An example of this program is the utilization of Oroville reservoir on the Feather River in California for flood control, even

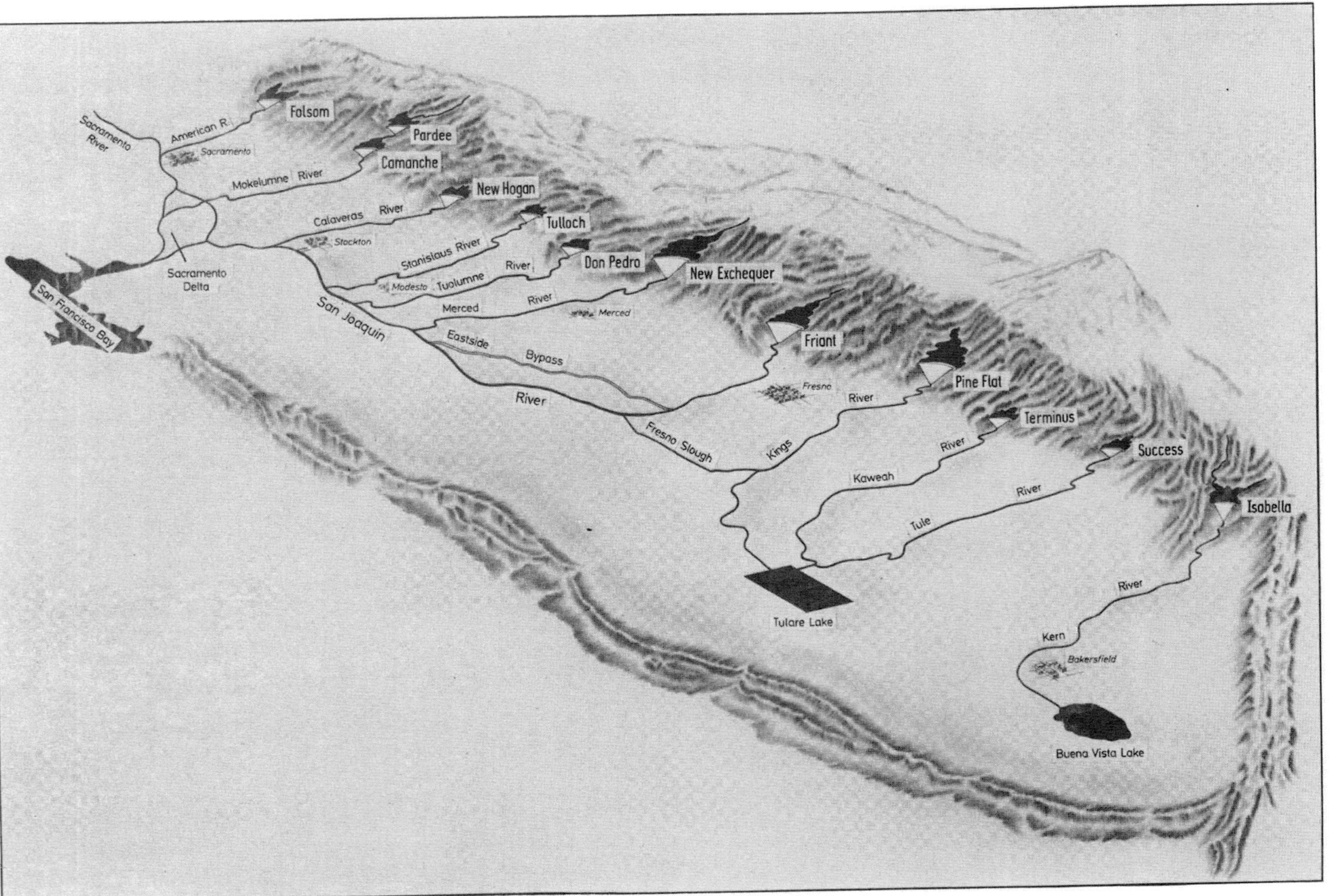

Plate 7.6. Flood control system for the San Joaquin valley, California, using dams in the foothills of the Sierra Nevada and holding lakes in the valley. (Courtesy of California Department of Water Resources)

though this dam was built by the State of California for water conservation, power generation, recreation, and fish and wildlife enhancement. As authorized by the Flood Control Act of 1958, a contract was executed between the State and Federal governments in which the Federal government agreed to contribute 22 per cent of the construction costs of Oroville dam and lake, exclusive of the power features, for protection of downstream areas from flood flows originating in the Feather River. The total Federal cost was about $70,000,000.

Flood operations in more complex river systems than the example given for Shasta dam are interrelated with other water uses and with the operation of reservoirs on tributary streams. The solutions, for example, in the operation of systems in the Columbia basin, Mississippi River, Tennessee Valley Authority and San Joaquin valley, California (see Plate 7.6) are far from obvious in a given flood condition. Recently, computer control and hydrologic data telemetered back to the flood control center from the field gages has made the work more feasible and even allowed the optimization of benefits for the conflicting uses.

River Levees and Training

The most common flood control works throughout the world are structures that attempt to contain the stream flow in its natural channel. A man with

Plate 7.7. The breach of a levee on the Feather River, California near Yuba City, during the floods of December 1955. (Courtesy of California Department of Water Resources)

property along the channel, as shown in Plate 7.7, would be much aware that more material on a natural levee would increase the carrying capacity of the channel and prevent it from overflowing on his land. Dikes were originally constructed by groups of farmers seeking to defend their plots against floods. Along the Yellow River in China are dikes that have been in existence since 603 B.C. In 69 A.D., during the Han Dynasty, the individual works were standardized and unified into a flood control system. In most regions subject to river floods the same pattern is followed today, and as land development expands, the need for a unified flood control system becomes more evident.

River levees are never high enough and their channels never seem to have enough space for all flood flows. On the Yellow River in China dikes were placed 25 km apart. The wide foreshores were cultivated. So, in order to protect this reclaimed land from flooding, auxiliary dikes were constructed nearer the river. In time, these became the main dikes which proved difficult to maintain.

Plate 7.8a. In the foreground can be seen the Colusa by-pass weir in California. In the background to the east are the Sutter volcanic buttes. In the dry summer, the land used for storage is cultivated. (Courtesy of California Department of Water Resources)

A solution to this common problem in many countries nowadays is to provide by-passes that can be farmed during summer months, but carry the flood flows during the winter rainy season (see Plates 7.8a and 7.8b). The main levees, built closely adjacent to the river contain fixed elevation overflow weirs over which higher floods flow into the by-passes. The weirs act as safety valves, limiting the level in the river.

Besides defense against floods by dikes, another method is termed *river-training*. As an illustration, on the Pyuntaza plain of the Sittang River in Burma, river-training without embankments has been developed successfully. Fences made of bamboo stakes parallel the stream, with cross fencing whenever required, the object of the side fences being to direct the flow of the river along a selected line in the early stages of the flood, and control the overflow or bank spill in the latter stages. The modern engineering counterpart of this method is lined and straightened channels, a method frequently seen in densely populated areas where land values are high.

The characteristics of the flood plain have influenced markedly the historical development of protective works. The Ganges plain of India, for example, with a very dense population, has only a few short lengths of levees. This is explained by the curvature of its basin, which is quite unlike those of some other major

Plate 7.8b. A similar view of the Colusa by-pass weir to that above showing water passing over it into storage during high water on the Sacramento River. (Courtesy of California Department of Water Resources)

rivers such as the Yellow, Yangtze and Pearl Rivers in China (see Section 7.4), the Ganges river basin lying between high land on either side. On the north is the high Himalayan mountain wall, while to the south is a system of ranges running east and west. The general cross-section of the plain is a depressed concave valley, with the river flowing along its deepest track. The flow of a river in such a valley is stable, so that flood flows simply attain a higher water level and spread over the banks in a restricted belt along the river course.

In South America, the flood plain of the Amazon is essentially a demographic vacuum. In the Brazilian Amazon, the population is less than 3,000,000, in part, because the soils outside the flood plains are so poor and, in part, because agriculture on the flood plains would demand massive and costly flood control engineering. The great alluvial plains of Argentina are extensively flooded in the rainy season (October to April). The Rio de la Plata is the mouth of an enormous drainage basin which contains the Paraguay, Parana and Uruguay Rivers with many tributaries. So large is the role of erosion of the Andes that the mass of silt carried by the Parana River through Argentina gives the drainage basin a number of peculiarities. When in flood, waters cover a width of 50 km across the plain, while in dry periods the river is dissected and shifting with many sandbars.

Flood Plain Management

At some stage in the economic development of a society it may be acceptable to restrict certain types of development, to limit further growth, or to regulate the use of lands that are subject to flooding, or *flood plains*, as they are called. Such proposals are now making headway in Australia (see Section 7.1), the United States, and elsewhere because even though billions have been spent for flood control works, annual damage continues to mount. On the other hand, in Asia, South America and other parts of the world the demands of the population for land, food and habitation are so great that this form of flood prevention does not now seem practicable.

Flood plain regulation seeks the optimum use of flood-prone areas while minimizing losses in lives and property in situations where this cannot be more economically accomplished by engineering works for flood control.

Flood plain management has a number of merits. First, land within the flood plain of a stream will, with little doubt, be flooded eventually by an extreme deluge, regardless of physical structures built to prevent it. Secondly, flood control does provide a sense of security which encourages development, so that a levee built for flood protection of agricultural land against a flood with a recurrence interval of 20 to 25 years may in a few years be protecting an industrial plant. Then the levee is usually strengthened and enlarged at significant public expense.

Flood control dams and, for that matter, any dams above communities may be a public hazard, for throughout history dams have failed (perhaps from earthquake shaking such as the San Fernando dam, Los Angeles, in 1971) thus creating a flood or flood hazard by their very construction. Modern regulations emphasizing rigid controls of design and construction coupled with regular inspections have done much to reduce the incidence of failure.

Essentially, there are two basic types of flood plain management; the maintenance of an adequate channel by preventing development from constricting flood flow, and restricting the use of the flood plain.

Some of the preventive measures that may be taken are: (i) restricting land uses to those relatively undamaged by water, such as parks, golf courses, and gravel storage; (ii) restricting the areas to uses that can be quickly evacuated, such as trailer parks and agricultural development; and (iii) flood-proofing structures.

Flood Forecasting and Warning

In many parts of the world where damaging floods occur, forecasts are made of the magnitude of impending storms. Forecasts may simply be a warning of danger, but in some regions, as the storm develops, more quantitative statements may be made on the storm runoff. On regulated streams and on those which contain flood control facilities, the magnitude and time of inflow to regulatory reservoirs, as well as flood stages throughout the length of the channel may be given (see Fig. 7.8).

Flash flood warnings associated with heavy thundershowers are often very uncertain. Radar, spotter networks, and flash flood alarm gages all aid somewhat in providing very limited warnings. Hurricanes or typhoons that move inland can only be generally evaluated in terms of likely damage. Ample warning time, however, can be given, with the help of satellite photographs, of the place where a hurricane will move onto the land and its time of arrival, with accompanying information on wind intensities and the expected rainfall amounts.

Probably the greatest success in flood forecasting occurs on a large river system such as the Mississippi (see Fig. 7.10), where the effect of rainfall on

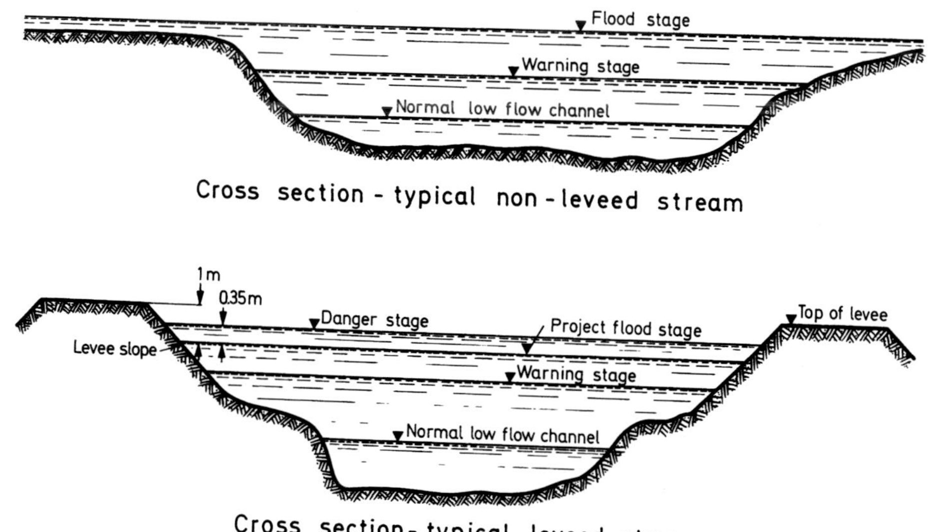

Fig. 7.8. Flood warning stages for non-leveed (above) and leveed (below) streams

reservoirs and channels is known weeks in advance. Snow-melt floods also allow a measure of success for the forecaster, because the quantity of water held in snow storage can be measured. There is thus time for reservoir storage to become available and channel capacity to be improved.

A flood forecasting service that is accepted by the public and operating agencies is obviously one that provides both reliable and timely forecasts. To accomplish this most difficult task, a "Flood Forecasting Center" must have access to meteorologic and hydrologic data, must develop forecast procedures with a well-trained staff, and possess a communication system that permits rapid dissemination of the prediction. In addition, there must be an effective local organization and plan of evacuation to make use of the forecasts.

River forecasting methods are developed from past flood histories and consist of evaluating the characteristics of past storms, compiling and correlating rainfall data and hydrographs of stream flow, and studying the geologic and physiographic features of the basin. Such a study enables the forecaster to develop a mathematical relation between the various items into which he can insert operational data. The forecaster's skill contributes to the result through his experience and judgment in evaluating the average basin precipitation for each time period and in incorporating the expected rainfall during a forthcoming period.

Plate 7.9. Repair work to levee along Sherman Island, California. (Courtesy of California Department of Water Resources)

River forecasting methods vary for each part of a river system. For the headwaters, early forecasts and warnings are based on satellite photographs, radar observations, meteorological judgment, measured rainfall and precipitation forecasts. Recent developments have led to a computerized hydrologic model or a river basin from which continuous flow forecasts can be made. Such dynamic methods have the advantage of the continuous updating of effects of moisture changes in the basin and permit the forecaster to observe quickly the results of varying some of the parameters.

For downstream forecasts, the upstream estimates must be extrapolated downstream. Success requires knowledge of anticipated releases from upstream reservoirs, or the effect of tributary inflow being held in storage.

The benefits from flood forecasting and warning are not easy to evaluate. Outstanding examples of benefits from a flood forecasting service are discussed earlier relating to the January 1970 operation of Shasta reservoir in California and in Section 7.4 concerning the Mississippi River flood of 1973. In the first instance, advance knowledge of the magnitude and timing of inflow into the reservoir permitted the release of water from the dam to be withheld until tributary floods below the dam passed downstream. Forecasts of Mississippi River flows were a valuable tool in planning the flood fight activities. Work crews were able to raise critical low level levees, strengthen weak sections, prepare sand bag reinforcement, and make floodways and by-passes ready (see Plate 7.9).

7.4. Flood Disasters

The River Nile

Since 622 A.D., the rise and fall of the Nile has been continuously recorded at the Nilometer on Roda Island, opposite Cairo, Egypt. Markings in other locations dating back to 1750 B.C. have left additional fragments of evidence of this annual event—the Nile Flood. Usually, the rise and fall is so regular that Egypt became a storehouse of food from which neighboring countries were supplied in time of drought and famine. Sometimes, however, because control was negligible, there were years of very high flood and great disaster resulted. Also, there were years of low flow, but even then the Nile carried sufficient water to prevent starvation. It was the custom of the Pharaohs to announce each year the time when the river had reached the proper level for the dikes to be opened for irrigation of the agricultural land. The low lands were watered when the Nile reached 16 cubits on the gage (a cubit is 54 cm). The middle lands were flooded at 18 cubits and the high ones at 20 cubits. The level today, because of silt deposition, is $20^1/_2$ cubits. As the river reached certain fixed levels there were special festivals, the 16-cubit level—the *wafa*—being the stage that ensured plenty and so was an occasion for great rejoicing because the prosperity of the country and the amount of taxes which could be collected depended upon the river reaching this elevation.

The Nile is an unusual stream for in its 6,000 km length it flows out of a tropical region to a very arid one; included in its watershed are great areas of humid, semiarid and desert lands. The main Nile is formed by the junction of the White Nile and the Blue Nile at Khartoum, 3,080 km above the mouth. Through the remaining reaches, the river flows through very arid country and, except for the Atabara River, has only negligible tributary inflow.

The water is derived from two main sources: (i) the lakes and swamp areas of central Africa, drained by the White Nile, with tropical rainfall through much of the year; and (ii) the highlands of Ethiopia, with high rainfall from July to September, drained by a tributary to the White Nile from the east, the Blue Nile, and the Atabara River.

The flow in Egypt ordinarily is at a minimum in May, then it begins to rise from rains in Ethiopia. First the Atabara, then the Blue Nile, rises in flood, which increases, reaching a maximum in September or October, and falls again as the Ethiopian summer rains diminish; the White Nile maintains the flow during the winter. The collecting system integrates the variable flows in a very effective way, so that a Nile flood is characterized by uniformity from year to year, the long channel providing great volume of channel storage to further smooth out the flow.

The entire Nile valley, from the great Aswan dam to the Mediterranean Sea, is formed of river sediments. The slope of the water surface, 10 cm per km, is remarkably uniform as far as Cairo, both at low flow and in flood. As the valley level has risen by deposition through the centuries by about 13 cm per century, the channel has maintained its depth, width and slope.

At the present time, at Cairo, the river has a mean range between high stage and low stage of about 7 m. The two highest floods of the last century occurred in 1874 and 1878. The former, the larger of the two, reached a stage of 9.15 m at Cairo.

The Huang Ho and Yangtze

Two great river systems drain China. The Yellow River (Huang Ho) taps a 1,250,000 square km area (see Fig. 1.6 and Fig. 7.9), taking the water across the great northern China plain to the sea, while further south, the Yangtze drains the 1,940,000 square km central part of the country. Both rivers have brought natural disasters that stagger the imagination.

The unruly Yellow River is one of the world's mightiest rivers, the artery into the heart of north China, by whose waters 60 million human beings live. For the student of soil erosion and hydraulic engineering, the Yellow river presents by far more challenging and complex problems than the Mississippi valley. The lower Yellow River has no important branches and for 800 km flows on a raised bed built up by its loads of silt. The river has periodically cut new outlets to the sea, moving its mouth 1,100 km up and down the coast during the last 150 years.

Just prior to 1927, a new effort was being made in China to solve some of the flood problems. A few foreign specialists were involved and accounts were published by Walter C. Loudermilk, a soil conservationist, John Freeman, a consulting hydraulic engineer, and O.J. Todd, formerly Consulting Engineer

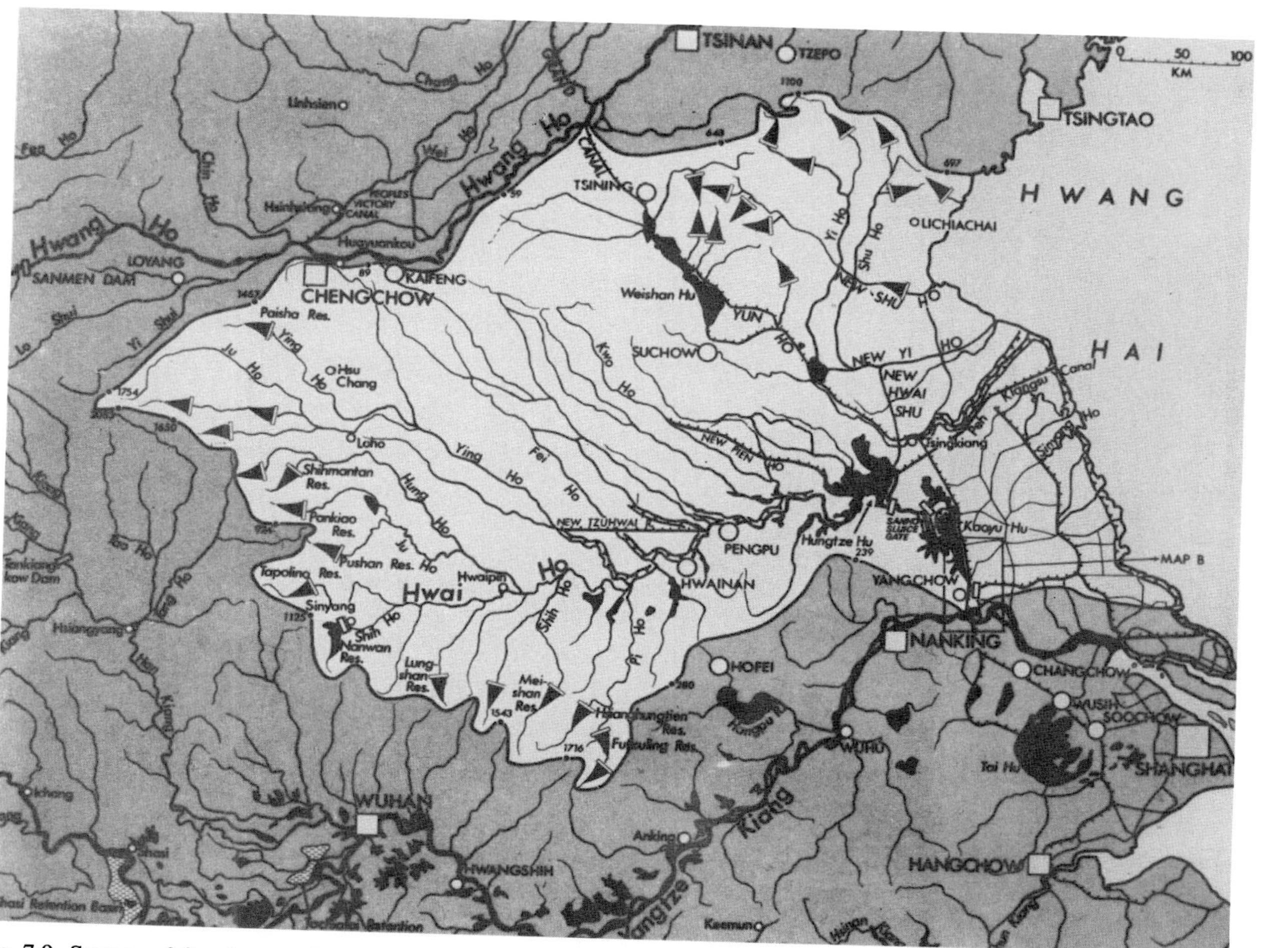

Fig. 7.9. System of flood control dams and canals built or under construction in central China (Courtesy of Professor J. Humlum, "In China, 1972", Gyldendal, Denmark, 1974)

to the Yellow River Commission. Following are some excerpts from their writings, selected to give a historical view of the causes of the Yellow river floods and the impact floods have on the people.

O.J. Todd wrote, "There is probably no river in the world which is of so little use to mankind as the Yellow River, considering the populous district through which it flows. Thus far, people have been only partly successful in protecting themselves from its ravages. Even as a communication artery it is unimportant. The river is an enemy instead of a helpful agent."

As he traveled across the flat plains of Honan province, Loudermilk saw "a great flat-topped ridge, reaching from horizon to horizon. This was the outer dike. We climbed this 12 to 15 m ridge and looked out on another vast plain some 3 m lower than the crests of the dike. About 10 km farther on rose another flat-topped ridge which we climbed. This was the inner dike. Before us lay the Yellow river, flowing quietly, with the low gradient of 20 cm per km in the delta plain, and silently dropping its burden of silt.

"Here the channel was fully 12 or 15 m above the surface of the plain. It has been held in this uplifted position by the hand labor of millions of men—without machines, or engines, without steel or timber, and without stone. These millions of farmers, with bare hands, carrying poles with little baskets at each end, have built, through thousands of years, a stupendous monument to human cooperation and the will to survive. I meditated on what these Chinese farmers had endured, toiling on by the millions in a situation that was hopeless. For there was no end to the demand of the river for higher and higher dikes. As it annually dropped its burden of silt, it lessened the capacity of the channel between dikes to carry flood waters."

John Freeman commented specially on the Yellow river delta. "Through millions of years this delta cone has been built up by deposits of silt. The flatness of the surface of this vast delta is the cause of the great width to which a flood may spread when it escapes from the river's dikes. The particles of silt are extremely fine and derived from erosion of vast beds of loess, supposed to be accumulations of wind-blown dust that originally came from the Gobi Desert. The Yellow River during freshets can carry a silt load that may reach 40 per cent of its weight where it enters the alluvial plain."

In June 1938, the Japanese army captured Kaifeng, the capital of Honan Province. In order to block their move westward, the Chinese government cut a dike and sent the waters of the great river directly southeast. The countryside was deluged by the tawny tide and the river found an old abandoned course to the sea. The diverted river formed an effective barrier against the Japanese armies, bogging down their transport for a considerable period but, reportedly, at the cost of a great many thousands of drowned local persons.

In recent decades, a unified plan of flood control in the Yellow River basin has been announced by the Chinese government. In 1956, there was a call for the construction of 46 dams on the main stream within several decades and the building of numerous reservoirs on the tributaries for flood control, silt detention, flow regulation, irrigation, navigation, power generation and industrial water supply (see Plate 7.10). It will be decades before all these projects are completed and although, at the present, seasonal flood problems have been alle-

Plate 7.10. Adjustment of water flow through a weir at Tai Ping Kou flood detention basin on the Yangtze River, China (from R. Alley, 1956)

viated and irrigation water supplies increased, engineering is still not advanced enough to remove jeopardy from major floods.

To the south flows the second great river—the Yangtze. The river rises amongst the Kunlun mountains in the high portion of far western China, in the same region as the other great rivers of Asia, the Yellow river, the Mekong, the Irrawaddy and the Salween. It is some 4,800 km in length and is navigable up to Chungking. Hankow, one of the triple cities of Wuhan, has modern wharf facilities for ocean-going ships, 1,030 km from the coast. On a tributary of the Yangtze, near its mouth, stands the city of Shanghai, with 7,000,000 people.

The great Yangtze flood of 1931 affected the lives of some 60,000,000 persons, with a loss of life over the whole period of the flood and in the ensuing winter that ran into millions. The Yangtze is a humid-region river with an intricate system of tributaries and a large flow at all times. High rainfalls make it subject to seasonal flooding in some portions of its basin. Active erosion is severe at the western margin.

Torrential rains fell over the whole basin in the summer of 1954, causing one of the highest floods in history. The worst affected area was the Province of Hupei, where hectic engineering efforts were made to maintain the dike systems against the flood. In the Province, there were about 1,800 km of main dikes and there was grave risk to millions of peasants from the encroaching waters. At Hankow, the flood level was reported as high as 30 m. Dikes had been raised during the previous summer and continuous repair and work on them during the floods by an enormous army of workers protected the city throughout the emergency.

Mississippi Flood, 1973

In 1973, the Mississippi River, the major stream of the United States and third in length in the world, experienced the greatest flood in the recorded history of the river.

Flood-hazard control of the drainage basin of the Mississippi (see Fig. 7.10) is divided into two eras, before the flood of 1927 and afterwards. The flood of 1927 was a great national disaster, but from it developed engineering concepts, economic thought, and a new political policy. In the United States prior to 1927, the control of floods was considered largely a local responsibility; after this great flood it became a Federal responsibility.

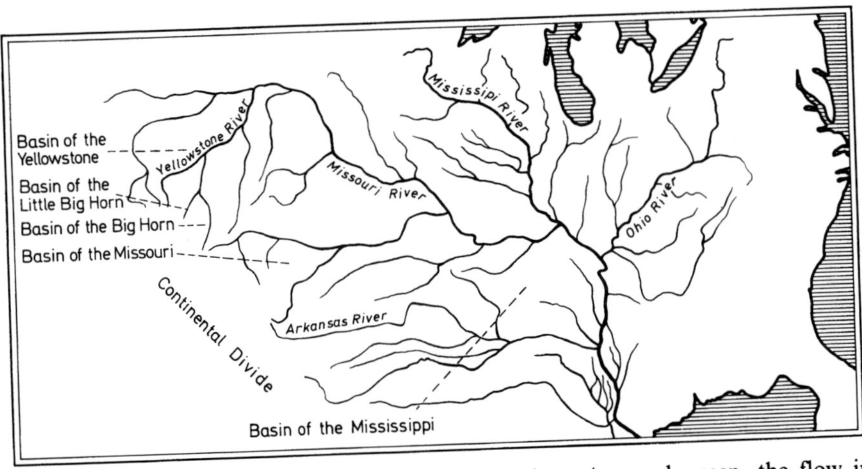

Fig. 7.10. The drainage basin of the Mississippi River. As can be seen, the flow in the Mississippi is the sum of the flows from tributaries having their own drainage basins

The Mississippi flood of 1973 was a significant test of the planning and construction effort that culminated in four major structural elements: levees, floodways, channel improvement, and major tributary storage facilities. This flood set new record flood elevations and tested all elements of the flood control plan—structural features, forecasting, operational actions, and flood fighting capability. The flooded areas covered 5,300,000 hectares in seven States with an estimated damage of 420,000,000 dollars. Without flood control, it is estimated more than 4,000,000 additional hectares would have been flooded, including major industrial centers of the lower valley, losses which would have dwarfed the 1927 disaster.

It is not easy to define the cause of a flood as large as that of 1973 because such a broad area was involved. The Mississippi River drains 40 per cent of the contiguous 48 States, plus parts of the Canadian provinces. It consists of three major tributary areas: the upper Mississippi, the Missouri, and the Ohio, in addition to the lower main channel of the Mississippi from Cairo, Illinois to New Orleans.

As a result of persistent rainfall in the summer and early fall of 1972, flood control reservoirs on the tributary streams began to fill and the watersheds became saturated. The stage readings at Cairo, Illinois, at the head of the valley, indicated that the pre-flood pattern was somewhat ahead of developments preceding the great flood of 1927. On March 11, the Mississippi was at flood stage at St. Louis and flooding continued on the river through April and May. For eight months the Mississippi valley continued to experience heavy rainfall, which continued until April 26, 1973, in the upper half of the Mississippi valley. By this date, the stage at Cairo was 16 m; flood stage is 12.2 m.

Although flood control facilities of the Mississippi River and its tributaries are designed to regulate the worst possible runoff conditions, in a major flood much work is needed to keep them intact and to augment them in critical locations. Levees are subject to river scour and wave wash (see Plate 7.9). The main Mississippi containment system has almost 3,500 km of levees and flood walls; so protection requires a major effort. Thus, in 1973, despite the flood control facilities and flood fight, losses were enormous. North of St. Louis, 39 levees were breached or overtopped, inundating thousands of hectares of farm land and unprotected communities. Three hundred buildings in historic St. Genevieve, Missouri were heavily damaged. St. Louis itself, protected by a flood wall, escaped damage.

Extensive tributary flooding occurred in the States of Arkansas, Tennessee and Mississippi. Four reservoirs in the Yazoo River basin, with a combined capacity of 5 cubic km were filled to overflowing and many months were needed to lower the levels of these reservoirs. More than 1,000,000 hectares of this river basin were flooded and the river remained filled to its banks into 1974.

The 1973 flood was one of the greatest in the history of the river. The maximum flow was about 56,800,000 liters per second, enough to provide the city of New Orleans with its daily water supply in less than 10 seconds!

The Low Countries

In the Netherlands, a great deal of land had been wrested from the sea, beginning before the thirteenth century. Old documents tell a history of land reclamation and flood disaster going back to the early Middle Ages. Chronicles in 1014 reported that waves "rose to heaven". Systematic drainage was not undertaken, however, until the fifteenth century, when windmills were used to pump internal waters over levees into the sea. The reclaimed land, or polder system, now amounts to a large portion of the Netherlands, including Zuid-Holland and Zeeland. (A polder is a basin of land taken from the sea that remains either at sea level or below it and is protected by dikes.) Serious storm floods in Zeeland in 1682 flooded 27,000 hectares and in 1808, hectares 14,000 hectares of land.

Because various land areas are not at the same height, the level of water in every polder is different. In reclaimed lakes in Holland, such as the Beemster, the surface of land may be about 3.5 m below sea level. In the reclaimed part of the former Zuyder Zee the soil is approximately 6 m below sea level. Large tracts of the soil are maintained in a habitable condition by a complicated system of dikes and polders, windmills and pumping stations.

In January 1953, two tragic weeks in the national life of the Netherlands became known as the "battle of the floods". Heavy seas from storms in the North Sea battered the sea defenses of Britain and the Netherlands for days. The raging gales were added to a combination of the Earth and moon, which gave a particularly high tide. On January 31, turbulent seas cascaded over the dike tops and in a matter of hours large parts of the Netherlands were separated from the mainland, communities being isolated so completely that it took several days to reach lost villages and drowned farms and to complete the evacuation of over 100,000 people.

Communications were affected during the early hours; pilot service into the coastal ports was suspended; radio calls for help were received from 35 ships off the coast and ocean cutters, coasters and cargo vessels were washed ashore; fishing fleets were torn adrift from their moorings. In the seaside resorts of Scheveningen and Noordwijk the massive boulevards were turned into rubble and hotels facing the sea were severely damaged by the high winds.

Many dikes, designed to withstand the quiet pressure of wind and water, were breached by the turbulent storms. In Stavenisse, a dike broke and a rolling wall of water 4 m high demolished the prosperous farm houses lying in its shelter, and over 200 people were drowned. In Rotterdam (see front map), the port struggled against the highest tide that had ever been recorded; oil refineries and installations around the waterways were awash. For many days, anxious eyes were turned toward the great Hoge Zeedijk which protects the central part of the country. Fortunately, the dike held (to the relief of about 3,000,000 people who would have been affected) even though one small breach occurred.

On some of the larger islands like Walcheren, the damage was particularly heavy and the death roll long. Members of families disappeared, one by one, as a house gradually crumbled, or a child or parent could hold out no longer against the bitter wind and icy seas. At the village of Tholen, a man saw his wife and 12 children drowned as he was trapped helplessly in an upstairs room. Here, too, a young engaged couple spent 36 hours marooned on a dike. When rescuers arrived the girl had died from exposure and the young man had lost his reason. One man hung on telegraph wires for over 48 hours. Another, caught under the arms by a radio antenna, was still alive 27 hours later when rescued.

An area over 100,000 hectares went under salt water, which is about 6 per cent of the cultivated land of the Netherlands. The population of the flooded area was 660,000 out of the total population of 10,500,000, the number of persons killed was 1,490. Livestock losses were heavy, 60,000 animals drowning. About 5,000 houses were destroyed or badly damaged in Zeeland (see Plate 7.11) and thousands in other places. The total damage was put conservatively at $250,000,000, about double the damage done by floods in England during the same storms.

Following the disaster special studies were made of the emergency response and communications in the stricken area. Generally speaking, citizens behaved well. Boats and helicopters were used for evacuation. As telephone and electricity communications failed over a large part of the flooded area within half an hour, people reverted to informal and age-old methods of communication such as signaling from roofs. Experience during wartime stringencies with flooding in

Plate 7.11. Desolation across flooded polder at St. Philipsland in Zeeland Province in the great January 1953 inundation from the North Sea. (Courtesy of Aero-Photo "Nederland" and Netherlands Consulate General, San Francisco)

1944–1945 led authorities sometimes to hurry away from places where danger was most acute and become isolated from the centers of responsibility.

The use of radio communications was important, being particularly effective when used by fishermen who were familiar with it on their fishing vessels. On the other hand, difficulties arose when radio messages were sent without clear indication of authorization and sometimes with an unwarranted sense of urgency. The greatest problem was the lack of familiarity with modern communication processes and the lack of a flood emergency plan.

A Flash Flood in Black Hills, South Dakota

In terms of lives lost and damage done, the worst natural disaster in the history of South Dakota occurred on June 9 and 10, 1972. Torrential rains of up to 40 cm fell in less than six hours between 18 : 00 h and midnight of the evening of the 9th. Rains over the east slopes of the Black Hills caused *flash floods*

Plate 7.12. The town of Klamath on the Klamath River in northern California was destroyed in a flash flood in 1964. The town has since been relocated at a higher elevation. (Photo by Eureka Newspapers Inc.)

along streams including Rapid Creek, ordinarily a pleasant mountain stream flowing through Rapid City. Total rainfall during this storm nearly equalled the yearly average, a deluge unlikely to occur more than once in several thousand years.

Typically, localized damage in such flash floods is severe (see Plate 7.12). There were 237 deaths and a total of 2,932 persons were injured, with flood damage of over $120,000,000. 750 homes were destroyed and between 1,000 and 2,000 cars were damaged by the flood, over 6,500 families suffering losses. There were at that time no flood protection works of significance in the flooded area, but Rapid City is now planning to spend almost $50,000,000 to acquire private property on the Rapid Creek flood plain and convert it to a park.

A Tropical Storm over Arizona's Desert

Arizona is known as a "fair weather State". It shares the southwest deserts, where the weather usually takes the form of scorching heat, violent thunderstorms, strong winds and cloudbursts of short duration, and it is extraordinary for a Pacific hurricane or tropical storm to visit. Yet this is why on September 5 and 6, 1970, heavy rains caused more loss of life (23) than any other storm of Arizona's recent history. Many dwellings, roads, bridges and other structures were damaged by record flooding.

Tropical storm Norma, centered just west of the southern tip of Baja California, moved northward on September 3, carrying with it copious amounts of moist air. To the north, a cold front was centered off Vancouver Island, while an intense cold front extended from northwestern Montana southward across Idaho, eastern Oregon and northern California. The confluence of streams of moist tropical air as they continued eastward resulted in thunderstorms and orographic rainfall over the mountains associated with the cold front.

New precipitation records for 24 hours were established for many stations, the most spectacular being for a gage in the Sierra Ancha mountains, measuring 28 cm during a 24-h period, double the previous record. The statistics of the local rainfall indicate that this amount of precipitation occurs only once in 500 years. Intensities of rainfall exceeded infiltration rates on the watersheds which had only shallow storage over bedrock and led to high-peak stream flows.

The flooding was reported as spectacular and terrifying, eyewitnesses describing rates at which streams rose as unbelievable. Rivers that drained hundreds of square km rose 2 to 3 m per hour. Uprooted trees, huge boulders, fences, automobiles and small buildings were swept downstream some 50 to 60 km.

A crucial lesson for human safety is that in desert or mountainous regions small remote streams utilized for recreation may become hazardous, and the unwary camper or cabin dweller may not appreciate the vulnerability of his location should a rapid rise in water levels occur. Living sites should be chosen well above the normal level of the stream.

Hurricane Agnes, USA

Termed the greatest natural disaster ever to befall the United States, hurricane Agnes was unique in that she caused disastrous flash floods in a short period over a large area, from Georgia to as far north as New York State.

Agnes, the first Atlantic hurricane of the 1972 season, was not an unusual storm in the beginning. Formed from a depression off the coast of Yucatan on June 15, the storm developed and moved slowly northward, dumping large amounts of rain on western Cuba and spawning tornadoes over Florida. The winds in Agnes never exceeded minimal hurricane intensity, but the area of the storm circulation was exceptionally large, its slow development permitting the large quantity of moisture to be transported from the deep tropics into the northward-moving storm system. Damage by the storm and subsequent floods caused property damage of more than $3,000,000,000 and 118 persons were killed.

Rivers and streams rose to record stages with devastating results. Storm rainfall for the period June 18–25 varied in amounts from 10 cm to as much as 48 cm in some areas. Although the storm lasted some eight days, there were short periods of extremely heavy rainfall. Washington, D.C. was deluged with over 28 cm in less than 18 hours and, as a measure of its broad extent, it was estimated that 93,000 square km received an average of 28 cm of rainfall.

Although there were numerous levee and flood protection projects throughout the flooded area, many were unable to cope with the intensity of this storm. Water in some instances was trapped behind levees and required major effort to pump and drain the water back into the channels weeks after the flood.

Recovery from this disaster took an unprecendented major effort by the individuals of the area, supported by all levels of government. An official in the clean-up commented, "Agnes left a wake of destruction and red tape of such magnitude that few areas of human frustration remainded untouched."

Venice, Italy

Oh Venice! Venice! when thy marble walls
Are level with the water, there shall be
A cry of nations o'er thy sunken halls,
A loud lament along the sweeping sea!
If I, a northern wanderer, weep for thee,
What should thy sons do?—anything but weep:
And yet they only murmur in their sleep,
In contrast with their fathers....
(Lord Byron)

The Adriatic Sea's inundation of Venice is a tragic cultural drama involving loss of irreplaceable art and architecture. On November 4, 1966, while the Arno river was ruining much of Florence (see Section 7.1), the waters of Venice rose to the exceptional height of 1.9 m above the average level, causing not only extensive damage to the older buildings of the city, but severely rupturing the protecting sea walls. The events of that day revealed to a wide audience that the famous city of the Doges was menaced permanently by flood catastrophe.

As the map in Fig. 7.11 shows, the city is built in the center of a lagoon, and in the 16th and 17th centuries, the magistrates of the Republic of Venice had the rivers that emptied into the lagoon diverted to prevent it filling with silt, while entrances to the Adriatic Sea were left open to permit the flushing of the canals by the sea tides.

In recent decades, inundation of the city has become more and more common (see Plate 8.2), for which there appear to be several reasons. First, there is an almost imperceptible rise in the sea level itself, about $1^1/_2$ mm a year. Secondly, the ground has subsided due to the pumping of underground fresh water in industrial areas along the coast, perhaps another 5 to 10 mm change in level a year (see discussion in Chapter 5). Thirdly, protective sea walls have been neglected; and fourthly, the storage area of the lagoon has been reduced by diked fishing grounds and filled land areas (see Fig. 7.11).

High waters submerging first or ground floors in homes and other buildings have become common. Repeated wetting and evaporation from salt water in columns and walls has dramatically reduced the strength of marble and other stonework. In the beautiful Ca'd'Oro, constructed between 1421 and 1440, high tides now cover the ground floor. The entrance ways to many buildings such as the Palazzo Giustiniani on the Grand Canal, dating back to the 15th century, have become seriously damaged or unuseable.

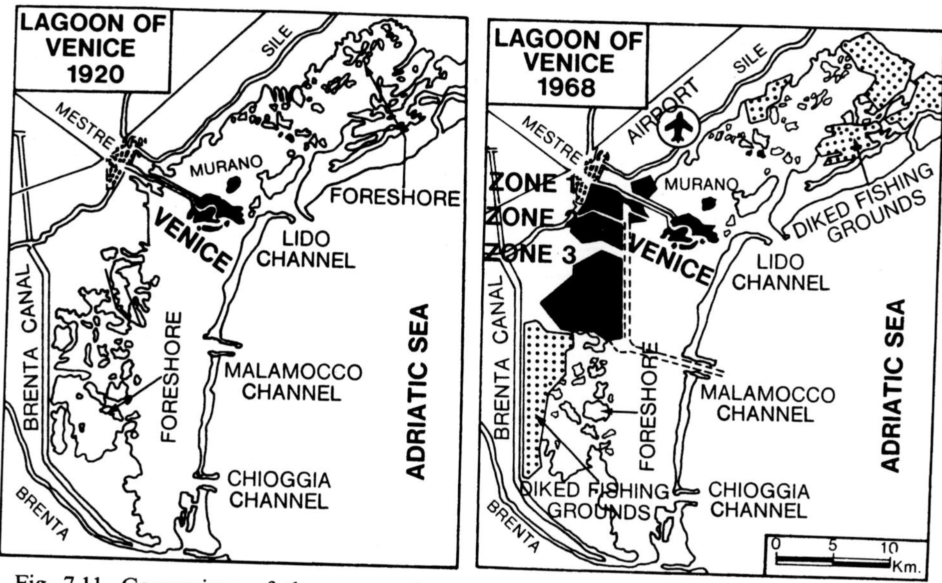

Fig. 7.11. Comparison of the extent of the Lagoon of Venice between 1920 and 1968. Zones 1, 2 and 3 are areas of land fill for housing development. (From "Venice in Peril", International Funds for Monuments)

Population of the city has fallen by over 60,000 in the last decade, and the abandoned buildings accelerate the decay and destruction of the ancient city. Fortunately, in 1966 a committee for the defense of Venice was set up by the Italian government and a considerable sum of money appropriated to combat the pollution and inundation. Many disciplines have been involved in the fight for preservation, including town planning, biology and hygiene, geology, geophysics, zoology, meteorology and hydrology. The struggle to protect this great cultural heritage will, however, be difficult and costly.

A Levee Breach on Andrus-Brannan Island, California

At about 13:00 h in the early morning of June 21, 1972, a levee of the Andrus-Brannan island in the Sacramento River delta in California failed. The delta lies at the extreme eastern side of San Francisco Bay (see Plate 7.6). The breach occurred in the vicinity of a small-craft marina and widened very quickly. Seven hours later, the break had widened to about 90 m.

Many boats at the marina, accompanied by miscellaneous debris, were swept through the breach and stacked up toward the interior of the island. Miraculously, there were no deaths caused by this almost instantaneous flooding.

Andrus-Brannan island is one of several hundred man-made islands within the delta that were created for farming in the late 1800's by dredges that constructed levees to protect the land from winter flooding. Subsidence and oxidation of the peat lands have lowered the islands below sea level, so that now the levees protect against flooding the year around. Although still used for farming, proximity of the islands to the San Francisco Bay area has made them popular for recreation purposes. 1,100 km of waterways now support numerous marinas, and on the islands themselves are mobile homes, trailers and small communities.

Although the flooding was disastrous for anything near the breach, the rapidly rising waters also quickly affected anything at lower elevations in the interior of the island, including homes and trailer parks and it was soon apparent that the town of Isleton would be flooded. Isleton was evacuated and a major flood fight was undertaken in an attempt to protect the town by building a temporary levee. The effort was not successful, however, and the town was inundated. The flooded area was about 5,000 hectares and depths of water ranged from 1 m to nearly 6 m.

This form of flooding differs from that normal in a river overflow. The water remains until the levee can be repaired, pumps installed and the water pumped out, over several months. Much equipment, homes and farm development that was flooded was, therefore, a complete loss. Accounts report damage estimates ranging in excess of $40,000,000. The cost alone of closing the break was close to $1,400,000.

A critical side effect was the deterioration of water quality of normally fresh water released from storage dams upstream of the delta (see Plate 7.6). This fresh water flows through the delta channels and is pumped into conveyance canals. When the Andrus-Brannan levee failed and water poured onto the island, salt water from San Francisco Bay moved in to fill the void. To make the water useable again for irrigation, industry and drinking, tremendous quantities of fresh water had to be released from upstream storage to reestablish a hydraulic barrier. Even though in 10 days, over 3.7×10^8 cubic m of fresh water was released to dilute the salinity, large quantities of salt water remained trapped in the delta and could only be removed by pumping into conveyance canals to be spread on irrigated land. It was estimated that 50,000 tons of salt were spread on agricultural lands in this manner, thus bringing losses not only to those flooded, but to others not directly involved.

This disaster, unlike those imposed by the fury of storms, was wholly man-made. Levees built on peat soils are structurally unstable and subject to failure at any time. This case history illustrates well the risk that humans sometimes take with their lives and property by relying on unsound engineering structures.

Baldwin Hills Dam Failure, California

Although not a widespread form of flood disaster, occasionally flood damage results from the failure of a dam. On Saturday, December 14, 1963, a water-regulating dam between the communities of Beverly Hills and Inglewood, Los Angeles county, California, failed. Although the amount of water in storage was only about 1.06×10^6 cubic m, the damage was severe because there was intensive urban development below the reservoir. Scores of homes were destroyed, and total damage was estimated at $15,000,000.

Water surging from the breached dam swept through the residential subdivision immediately below, washing homes from their foundations. At the base of the hills, the flow fanned outward swiftly, engulfing the heavily populated urban area.

Regular inspections were made of Baldwin Hills dam and the Saturday of the failure was no exception. At 08:00 h the caretaker began his routine observations. At about 11:15 h, while standing over the spillway intake, he heard a light but unusual sound of rushing water, and saw that about four to five times the usual amount of water was flowing in the spillway pipe and the water was muddy. The next two to three hours brought engineers and other officials to the site. They attempted to staunch the flow with sandbags, but the leakage steadily increased. By 13:30 h, a crack across the crest had started to open, and widened from 6 mm to 75 mm in a period of 15 minutes. Leakage through the downstream face of the dam was first observed at about 14:00 h. By 15:15 h, the breach, now clearly visible on the upstream face, was 3 m wide and a heavy discharge was issuing from the widening breach. Suddenly, there was an eruption of muddy water near the bottom of the breach and the upper areas of abutment and embankment collapsed. By 15:38 h, the failure was complete. By 18:50 h, flood water had receded to a point where equipment could be moved in for clean-up.

After excessive leakage from the drainage system was first observed at 11:15 h, a controlled discharge was made from the reservoir, evacuation was ordered, teams of police raced from home to home warning residents, and alerts were broadcast over radio and television. Although all this began at mid-day, five persons lost their lives.

Most of the water supply for Los Angeles is imported from distant points through aqueduct lines. Therefore, a large close-in storage capacity must be maintained to assure sufficient water for both peak demand and emergency purposes. Because of the great area to be served, long lines and strategically located storage, such as at Baldwin Hills dam, are required for local water distribution.

What caused the failure? Opinion is divided, but investigation by the Mayor's board of inquiry indicated that the watertight lining of the reservoir broke along a geological fault line through the reservoir soon after the first filling of the reservoir. Later studies indicated that the breakage was caused by differential compaction of the soil on each side of the fault (see Chapter 5), rather than by fault movement. Throughout the life of the reservoir, water continued to leak into the fault zone, eroding the adjacent material and developing pipes and cavities under the reservoir and dam structures. These developed to such an extent that the floor of the reservoir collapsed along the fault line shortly before the final failure, permitting water under the full reservoir head to enter the permeable zone which had been created through the years. The increase in erosion rate rapidly led to the final collapse.

7.5. References

Alley, R.: Man Against Flood. Peking: New World Press 1956.

American Society of Civil Engineers: Hydrology Handbook (1949).

Burton, J., Kates, R.W., Smead, R.E.: The Human Ecology of Coastal Flood Hazard in Megalopolis. Chicago: University of Chicago, Department of Geography. Research Paper No. 115 (1969).

Freeberne, M.: Natural Calamities in China 1949-61: An Examination of the Reports Originating from the Mainland. Pacific Viewpoint **3**, 33–72 (1962).

Horn, W.L.: Snow Pack and Water Supply. Journal of American Water Works Association **60** Dec. (1968).

Hoyt, G., Langbein, W.B.: Floods. Princeton University Press 1955.

Judge, J.: Florence Rises from the Flood. National Geographic, Washington, July 1967.

Kates, R.W.: Hazard and Choice Perception in Flood Plain Management. Chicago: University of Chicago, Department of Geography. Research Paper No. 78 (1962).

Lambrick, H.T.: The Indus Flood Plain and the 'Indus' Civilisation. Geographical Journal **133**, 483–494 (1967).

Leopold, L.B.: Water. San Francisco: W.H. Freeman 1974.

Leopold, L.B., Maddock, T.: The Flood Control Controversy. New York: Ronald Press Co. 1954.

Linsley, R.K., Franzini, J.B.: Elements of Hydraulic Engineering. New York: McGraw-Hill 1955.

Mensching, H., Gressner, K., Stuckman, G.: Die Hochwasserkatastrophe in Tunesien im Herbst 1969. Beobachtungen über die Auswirkungen in der Natur und Kulturlandschaft (The Flood Catastrophe in Tunisia, Autumn 1969. Observations of the Effects upon the Natural and Cultural Landscape). Geographische Zeitschrift **58**, 81–94 (1970).

Munk, J., Munk, W.: Venice Hologram. Proc. American Phil. Soc. **116**, 415–422 (1972).

Netherlands Booksellers and Publishers Association: The Battle of the Floods. Amsterdam: Netherlands Flood Relief Fund 1953.

Oya, M.: Land Use Control and Settlement Plans in the Flooded Area of the City of Nagoya and its Vicinity, Japan. Geoforum No. 4, 27–36 (1970).

Sansome, G.C.: Venice in Peril. New York: International Fund for Monuments 1970.

State of California: Department of Water Resources Bulletin **69**, California High Water (annual series) (1969).

Todd, O.J.: Taming 'Flood Dragons' along China's Wang Ho. National Geographic **81**, Washington Feb. 1942.

UNESCO: Floods and Their Computations. Proceedings Leningrad Symposium, 2 Volumes, IASH – UNESCO – WMO 1967.

United States, 86th Congress: 1st Session, Committee on Public Works. A Program for Reducing the National Flood Damage Potential. August 1959.

United States Department of Commerce: National Oceanic and Atmospheric Administration. National Disaster Survey Reports (periodic reports).

United States Department of the Army, Corps of Engineers: Water Spectrum (monthly publication).

United States Geological Survey: Floods of (*year*) (annual series).

United States Geological Survey: Notable floods are published as water supply papers—Maps of certain localities showing areas inundated in past floods are published as Hydrologic Investigation Atlas.

White, G.F.: Human Adjustment to Floods. Chicago: University of Chicago, Department of Geography. Research Paper No. 29 (1945).

White, G.F.: Strategies of American Water Management. Ann Arbor: University of Michigan Press 1969.

Young, M.W.: Fighting with Flood: Leadership, Values and Social Control in a Massim Society. Cambridge: Cambridge University Press 1971.

Chapter 8
Hazard Mitigation and Control

8.1. Overall Risk Zoning

Synthesis of Geological Risks

In the last seven chapters a treatment has been given of those geological phenomena which pose the severest hazards to society. Worldwide growth of population, and particularly concentration of man and his works into urban areas, has heightened such threats to levels where large-scale, and often costly, planning to reduce the hazards has become essential in many countries.

Overall assessment of actions needed is complicated in many ways. Indeed, the source of a major geological hazard may be, at the same time, a great asset to the community. A mountain range providing water, irrigation, and recreation may lead to killer floods; rich volcanic soil for agriculture may surround a still lethal volcano; by-products of great active faults or rifts are often minerals, natural resources, beneficial climatic effects and magnificent scenery. Volcanic and geothermal areas may provide geothermal steam for power generation (New Zealand, California, Italy).

In this culmination we focus on geological hazards taken as a whole and consider the relative effects of earthquakes, volcanoes, landslides, flooding, and other geologically related processes. This synthesis requires, first, that basic questions on risk assessment be defined. In the process, as is usually the case in science, the availability (or lack of it) of the fundamental geological and demographic data comes to the fore. Not only must planners rely generally on incomplete and uneven statistics to predict, from past occurrences, future catastrophes, but the available data must be worked into a form that allows some quantitative comparison between various geological hazards. Detailed topographic and geological maps at appropriate scales are usually the starting point for such analysis. Then a detailed study of the separate hazards must be made (see Fig. 8.1).

Many variations on the techniques for studying hazard mitigation can be found. The following account is aimed at establishing (i) the need for interaction between those professions which deal with urban development, such as town and county planners, architects and engineers, insurers, local government and public works officials; and (ii) a direction in which improvement can be made in presenting geologic data for environmental studies so that they are comprehensible, less piecemeal and, at the same time, more open to estimates of uncertainty.

Unfortunately, university degree courses in geology, engineering, geophysics, and so on, have usually not included much critical study of either of these points although something has been attempted in the better geography depart-

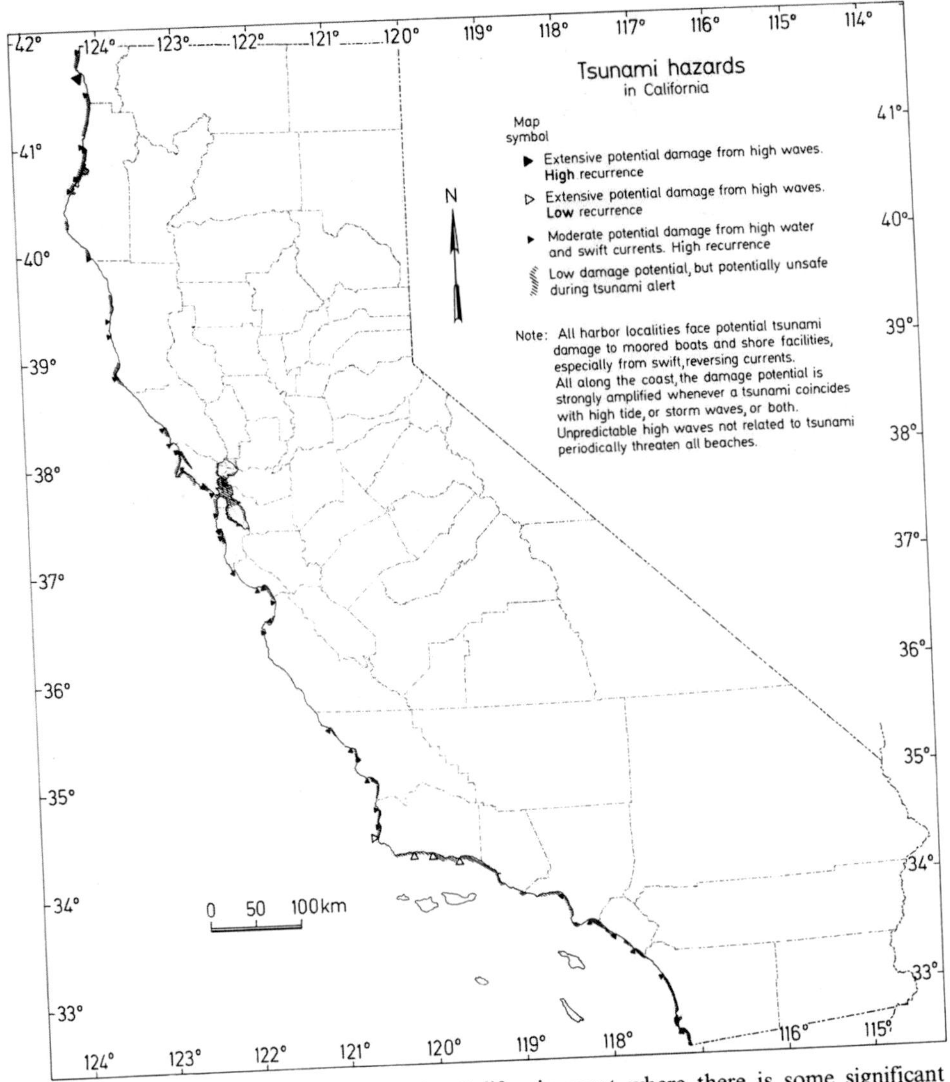

Fig. 8.1. Estimated locations along the California coast where there is some significant tsunami risk (from California Urban Geology Study, 1972)

ments in recent years. Interdisciplinary studies of the environment are becoming more available but they need a solid background in individual disciplines if the work is to avoid the superficial.

In order to link up with a California study to be discussed later, the main geological hazards are considered to be: earthquake shaking, flooding, volcanic eruption, tsunamis, fault displacement, landslides, subsidence, erosion activity and expansive soil (see Fig. 8.4). All have been discussed in previous chapters. Two allied items may be added that are entirely within the control of man—loss of mineral resources and the degradation of ground water resources. Although the latter two do not result in either the loss of life or property, they are hazards to the long-term economic health of a society.

At once a major hurdle arises in most countries. Documentation of life and property losses over a reasonable time-span is poor for many of the eleven problems, the major exceptions being usually floods, earthquakes and tsunamis. Nevertheless, some headway can be made by extrapolation of values from other areas and by comparing relative effects. Another intangible is the material value of a human life; for the sake of longterm planning it may be necessary to assign a rather arbitrary dollar number. In California, recent studies have tested use of life equivalents of $75,000 and $360,000. The first figure was based on the airline industry's limitation of liability for death of passengers on international flights. The international origin of the value is one used at the Warsaw Convention.

The much higher second figure was extrapolated from a sampling of court awards in California for indemnity suits for death and permanent injury, plus hospital costs for associated expectable injuries. Given the often inflated and special nature of such court awards, the latter figure is perhaps an extreme over-estimate in most places. A better procedure altogether is to work in terms of the ratio of benefit to cost. The ratio is dimensionless and would reflect the number of lives that might be saved, say, as a proportion of the number of persons in jeopardy.

In Table 8.1 are listed, as an example, the dollar losses predicted from geological hazards and loss of mineral resources for the next 30 years in California, given a continuation of present practices. The total loss of $55,000,000,000 almost equals the projected loss for urban and forest fires in the same period.

Table 8.1. State of California. Predicted dollar losses[a] from 1970 to the year 2000 (without improvement of existing practices). (From "Urban Geology—Master Plan for California" 1973)

Geologic problem	Dollar losses to year 2000 (in million dollars)
1. Earthquake shaking	21,000
2. Tsunami	40
3. Fault displacement	76
4. Volcanic eruption	49
5. Flooding	6,500
6. Landslides	9,850
7. Subsidence	26
8. Expansive soil	150
9. Erosion activity	565
10. Loss of mineral resources used in urban development	17,000
11. Ground water depletion or degradation	50
TOTAL	55,306

[a] Life-loss factor assessed at $ 360,000 per death.

Use of Probability and Statistics

We must at the outset define what is meant by geological risk. Clearly, a volcanic eruption or flood on an uninhabited island ordinarily poses no risk to man or his works. Risk takes on a cogent meaning only when the geological information is combined with the social and economic circumstances. Two types of risk scale are used.

The first, called *relative risk* (*RR*), merely compares one situation with another and does not take into account probability in an explicit way as the chance or odds of occurrence. The scale is usually an arbitrary numerical or alphabetic one. An example is the earthquake risk map shown in Fig. 1.5 for the United States. Zone 0 has no hazard from earthquakes, while Zone 3 has the highest. In a more restricted case, two or three grades of risk may be established for land use planning and construction along active fault zones. Housing would be banned on the fault strand itself. Another case is the tsunami risk for California as presented in Fig. 8.1.

A more specific scale is that of *probabilistic* risk (*PR*), which is similar to the actuarial concepts embodied in insurance procedures. In this case, the risk of damage or injury from a geological event would be given in terms of the odds or chance of occurrence within a given time interval (design period). This probabilistic risk has been incorporated, for example, in the seismic risk maps for Canada (see Fig. 1.11). Contours give the values of earthquake accelerations which have a probability of one in one hundred of occurring in any year. Another example, for tsunamis, is discussed in Section 3.2.

This kind of statistical risk is not well-developed in the geological context mainly because of lack of actuarial data. It is often held to be too complicated for the general public to grasp but its objectivity would suggest that it deserves more attention. A statistical training is almost a prerequisite for serious students of geological hazards.

Zoning to mitigate risk is also approached in two main ways. One is to treat the risk in a given area as a function of the cumulative severity of damage from earthquakes, floods, and so on, irrespective of the frequency of occurrence of these events. Such a function is a static one in which time is not a factor. Thus, on the seismic risk map of the United States (see Fig. 1.5) the area around Charleston, South Carolina is assigned a risk of 3, equal to that in most of California. This rating is based to a great extent on the single earthquake of August 31, 1886. Yet on present evidence there is a relatively low probability that another severe earthquake will occur within the lifetime of newly-built structures. No clear-cut geological structural evidence has been found even today which marks this as a specially seismically-prone region. For very large-scale problems of this type, Earth satellite photographs may in the future suggest lineations that need ground study.

An alternative rating scheme is to take into account the frequency of occurrence of the geologic hazard. This enables the infrequent but catastrophic events to be given proper weight in comparison with those that are much more frequent but less damaging. For this purpose, recurrence curves which rank the frequency and size of hazardous events per 100 years (say) per unit of area must be constructed.

Sometimes of course the historical information is just not available for anything but the crudest recurrence curves to be produced. Static risk zoning has recently been criticised trenchantly in some quarters in the Soviet Union. It is pointed out that such procedures are so conservative that enormous economic resources are diverted from perhaps more urgent needs of society to satisfy a speculative hazard from the distant past.

8.2. Public Safety

Land Use Planning

At least in many Western countries there is a wide variety of national, local government, and even private organizations having overlapping responsibilities in planning against natural hazards. One of the most effective steps in hazard control is to examine the agency participation and, by means of critical review, generate interagency awareness and cooperation towards the elimination of risk. Few local government officials have appropriate geological or technical training and in most cases the contribution at the local level is chiefly one of balancing financial and aesthetic pressures.

The first questions to be faced in a regional study are (i) which geological events would lead to the greatest losses; (ii) in which areas are the particular events likely to be most severe; and (iii) how the impact of the problems will change as the years go by. Only after these questions have been defined can decisions, for example, on flood control, earthquake zoning and land usage be realistically approached.

State of California Seismic Safety Element

On the provincial government level, legislation requires a general development plan for all cities and counties under the jurisdiction of the State. In California, since 1971, certain geological safety elements have been required, for example, as part of the general plan by the Government Code. The code deals also with other matters such as land use, housing, open space, noise, and geologic hazard mapping. The definition for seismic safety states:

"A seismic safety element [consists] of an identification and appraisal of seismic hazards such as susceptibility to surface ruptures from faulting, to ground shaking, to ground failures, or to effects of seismically induced waves such as tsunamis and seiches."

The effect of the law is to require cities and counties to take seismic hazards into account in their planning. In practice, the effort expended in developing and implementing the geological hazard element depends upon the degree of risk that the community is willing to accept. Often citizens will accept more unregulated risks in the selection of houses than in their hospitals and schools. Although decisions are affected by education and experience, they may change sharply following a calamity and become highly political. It has been said that every damaging earthquake in California has given a boost to legislation on seismic safety; for example, the Field Act followed immediately after the 1933 Long Beach earthquake. Much the same remark is true of other major geological disasters.

8.3. Geological Risk Maps

Structural Maps

The first step in the study of collective geological hazards is the plotting of specific information on maps *at the same scale*. A geological map, for example, presents the areal distribution of rock structure and type. The scale chosen and the emphasis on particular features may be selected to optimize the use of information for a particular need.

In California, a new 1:750,000 scale geological map was produced in 1972 to give an over-view of the geological properties of the State with sufficient detail to be useful for preliminary land-use planning. Published in color, it emphasizes recent volcanic rocks and volcanoes, earthquake faults and the major folds in the layered rocks.

Maps with much more detail than feasible on the usual 1:250,000 to 1:1,000,000 scale maps are needed for specific hazard evaluations. For urban areas, specialized mapping for land-use planning and engineering design must show considerable detail and even include geophysical and borehole studies of local subsurface structure. The required scale may be of the order of 1:20,000. Recent examples are slope maps produced by the U.S. Geological Survey with a scale of 1:24,000. These maps indicate the per cent of slope of hills and mountains by means of a color code so that assessment of hillside erosion and stability conditions can be made. Likewise, U.S.G.S. and Corps of Engineers flood hazard maps at about this scale show the elevations attained by major historical floods and floods of a specific frequency of occurrence.

There are several unsatisfactory features of the usual geological maps published in most countries. First, these maps often emphasize the formations (igneous, basin deposits, etc.) rather than the rock types involved. Alluvium that consists of fine- and coarse-grained material may have depth and horizontal facies changes that lead to major seismic response consequences. Again, it is not sufficient to say that a given formation consists largely of sandstone and shale without mapping bed boundaries. The Geological Survey of New South Wales in Australia has tried to solve the problem by indicating overburden and underlying rock units by appropriate symbols. In this way, the map color defines the underlying rock, while the map symbol tags the type of overburden. In New Zealand, the Soil Bureau of the Department of Scientific and Industrial Research produces maps of soil type that may be read in conjunction with the standard geological maps. In the New England States, USA, one series of maps delineates bedrock and another the superficial glacial deposits.

Another weakness is lack of detail when mapping the *weathered conditions* of the rock types. The depth of weathering may be of considerable importance in estimating the response of the ground to strong earthquake motion. In the same way, locations of unobscured bedrock exposures deserve plotting on the basic geological maps so that when detailed investigations are needed these outcrops can be revisited quickly. Alluvial deposits often require sub-division, appropriate to the scale used (e.g. 1:250,000) showing flood plains, lake deposits, colluvial, residual soils, and so on. In this way, parts of a particular surficial deposit, consisting of fine-grained material with braided stream channels of coarser material, could be identified from the map.

A keen debate continues on whether geologic maps should highlight interpretive features or be restricted to the direct observations of field geologists. Interpretive geological maps contain, by necessity, personal inferences which may have critical consequences for planning and development. An illustration is the construction of maps which show the faults classified as active or inactive.

In a recent California attempt, a three-fold classification distinguishes faults on the basis of recency of movement. The first class consists of those that have moved in historic time (about 200 years); the second class, those that have displaced Quaternary deposits but have no historic record of movement; and the third class of faults (those without recognized Quaternary offsets) are inactive. Immediately we are vexed by the likelihood that faults in the third class may not necessarily be dead or that faults in the early categories may remain quiescent for many thousands of years. How should we classify faults with no visible Quaternary rupture but which have microearthquake foci along them?

Computer Simulation

A recent imaginative development is the use of computers to calculate and draw predictive hazard maps. Once the controlling parameters of the hazard are known these can be combined into a mathematical form and programmed once and for all. The program forms the structural skeleton of the specific hazard (ground shaking, tsunami run-up, flooding, etc.) upon which specific values for particular cases in question must be hung.

The program is of little avail unless it can *reproduce* historical disasters; in practice these are used to test whether the programmed parameters are properly selected and combined. Simply by trial and error, the free coefficients and indices in the program are modified until the historical event is closely reproduced by the program. Only then is it worthwhile to use the calibrated program to predict the hazard map for a future disaster. Such maps can then be used prudently for investigative planning and control.

The simulated intensity of earthquakes provides a straightforward illustration. In Fig. 8.2 the simulated isoseismal pattern for the 1906 earthquake has been calculated and plotted by a computer using seismological factors in ground shaking given in Chapter 1. The mathematical model (program) was built up from largely empirical relations between intensity (Sections 1.2 and 1.4) and earthquake magnitude, fault rupture length, attenuation of shaking with distance, focal depth, duration and local geological and ground conditions.

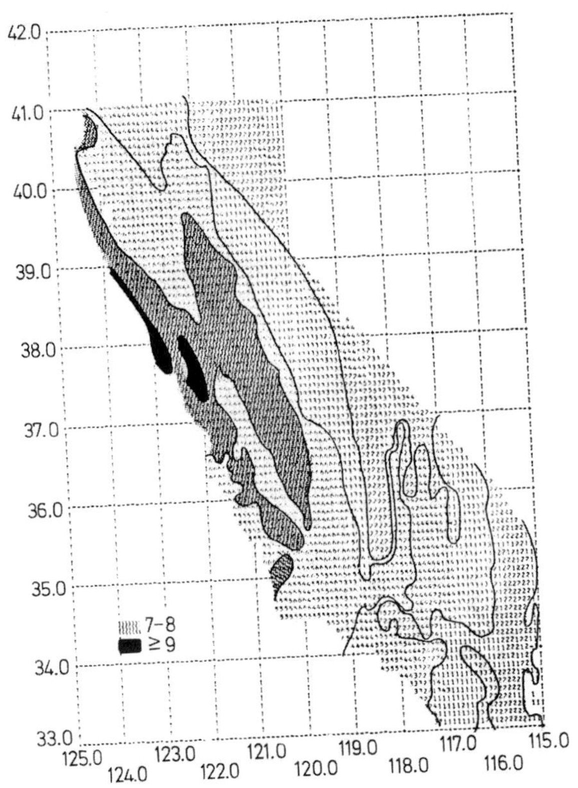

Fig. 8.2. Computer-simulated isoseismal pattern for the 1906 San Francisco earthquake. Numbers representing MM intensities are printed out by computer. Compare the isoseismals with the observed intensity mapped in Figure 1.3. (Courtesy of D.G. Friedman in "Geologic Hazards and Public Problems", 1969)

As Fig. 8.2 shows, the trial and error culminated in rather a close match to the observed intensity for the 1906 earthquake (Fig. 1.3). This calibrated model was then used to predict the largely unknown isoseismal pattern as it might have been in the 1857 Fort Tejon earthquake (Fig. 8.3).

There is still much that is uncertain in such simulation procedures, not the least of which is the use of the subjective intensity scale itself and its narrow applicability to a few types of structures. In addition, some simulation experiments have given reasonably close fits with the intensity of past earthquakes even though intensity is related only to peak acceleration which in turn is strongly related to magnitude. We have given reasons in Section 1.5 why such relations are dubious in general and why such parameters as duration of shaking must be considered. Be that as it may, all indications are that simulated computer models will play a larger role in hazard evaluation as the years go by. Not only is prediction a goal but also the isolation of the contribution of individual geologic factors such as, for example, the role of water saturation in avalanches and seismic ground amplification.

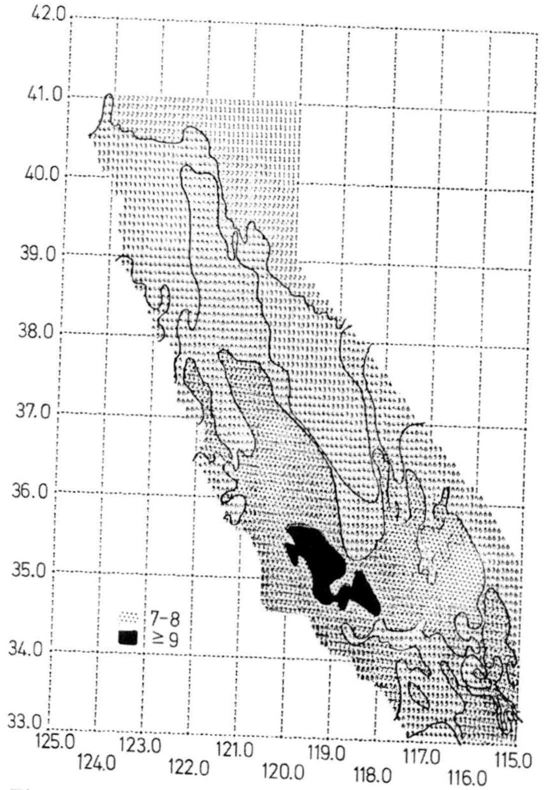

Fig. 8.3. Computer-simulated isoseismal pattern for the 1857 Fort Tejon earthquake in southern California. Numbers representing MM intensities are printed out by computer. The pattern *predicts* in a restricted way the shaking expected in a repetition of this earthquake. (Courtesy of D.G. Friedman in "Geologic Hazards and Public Problems", 1969)

8.4. Population Growth and Priorities

Immediacy Benefit

Geological hazard is a strong function of the population density; St. Pierre Martinique had a population of 30,000 concentrated in a small area when it was destroyed in 1902 (see Plate 8.1). What is more, it is essential in establishing priorities for prediction and control to have also reliable estimates of the population growth rate.

For the estimation of geological hazards the concept of an average urban district is helpful. Such a district can be assigned a fixed area of usable land and an average mixture of dwellings and industrial usage for the province. In demographic studies, the population in an area of study is measured as the number of person-years (*PY*), defined as the average number of residents in the urban district per decade.

Plate 8.1. The ruins of St. Pierre, Martinique, in May 1903. The city was destroyed by the glowing avalanche of May 8, 1902. Mont Pelée, with the dome and great spine at the summit, are visible in the background. (Photo by Alfred Lacroix, courtesy of Masson et Cie, Paris)

In industrial nations the economic growth rate of an urban district will in many cases be reflected mainly by the increase or decline of the population. Because measures taken now to control geological disasters are more valuable over the next decades than *future* control measures, a discount factor which takes into account the benefit of immediate measures also enters consideration. Such weighting factors are somewhat arbitrary but are expressible in terms of a factor called the *immediacy benefit* (*IB*). We might then rate urban districts that are fully developed at present with the immediacy benefit of zero because it is now too late to avoid construction in hazardous spots or to reduce risk by building codes. Actually, the *IB* value is never quite zero because old buildings are constantly being replaced by new ones and there are usually some options which will reduce the threat of loss; thus, a minimum scale value of $IB=1$ is usually assigned. Higher scale values of *IB* are then calculated from the projected growth rate discounted at an appropriate percentage for present worth.

Each urban district can then be considered in terms of a list of geological hazards, from earthquake shaking (*ES*) to ground water degradation (*GWD*). Each geological problem would have to be quantified in terms of the severity for that district. (Of course, the severity may vary within the district and some average would be required.) A relative severity (or risk) scale (running from zero through three, say) is usually adopted, but a probabilistic one could be

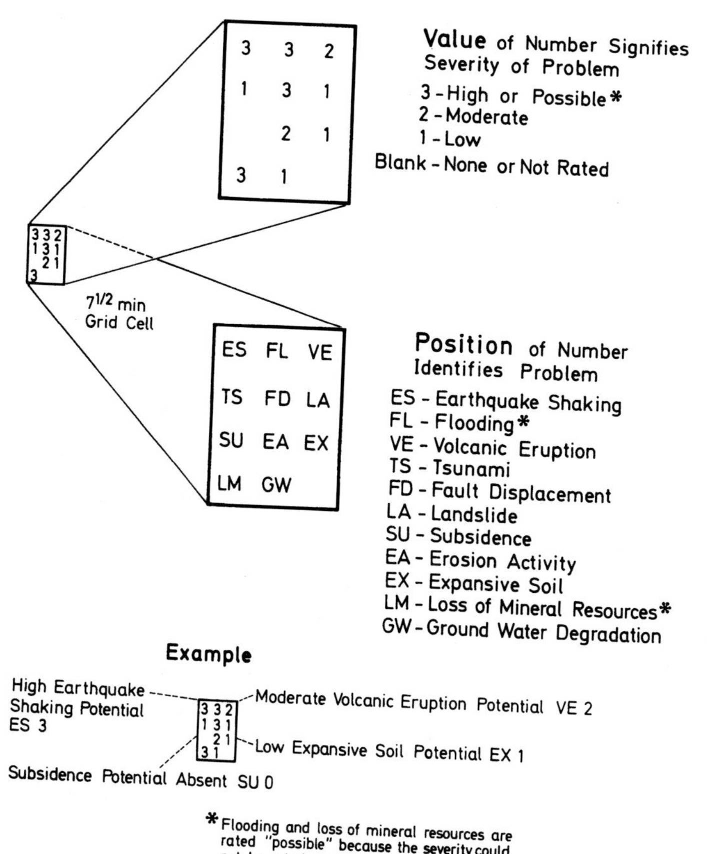

Fig. 8.4. Geologic Problem Code. The position of a number identifies the geologic problem and the value rates the severity of the hazard

used. Fig. 8.4 shows numerical risk estimates coded in a district cell for use in a numerical computer model. At this stage some average probable dollar loss can be assigned if desired (plus, perhaps, also associated socioeconomic loss such as unemployment) for each rating on the severity scale and for each of the eleven geological hazards. These monetary values for a particular district can then be added to provide an additional overall weighting factor called *geological severity* (*GS*).

Hazard Priority

Urban districts may now be compared. The priority to be given to work on mitigation of geological hazards over a whole county or province we may call *hazard priority* (*HP*). In terms of factors introduced already, the hazard priority is obtained by multiplication:

$$HP = GS \times IB \times PY.$$

As can be seen, although a district may have an acute geological hazard problem, urgency in allocating resources may be low if the area is already fully developed or if it has low population density.

8.5. The California Urban Geology Study

The Master Plan

From 1970 to 1973, the California Department of Conservation carried through a project on the needs of California for safe growth relative to geologic hazards and conservation of mineral resources. This Urban Geology Master Plan was conducted by the Division of Mines and Geology and is valuable in its elucidation of the strengths and weaknesses of overall geological hazard studies. The prime goal was to determine the measures necessary to minimize life loss and property damage in urban areas due to dynamic geologic processes.

Urban districts for the whole State were defined and stored in the computer and geological and demographic factors assigned to each district cell. In this pilot study the hazard priority, *HP*, was calculated by computer for all urban districts using the formula given in the last section, each cell marked accordingly and computer maps produced. These computer-produced maps allow verification of the accuracy of the input data and, more importantly, a review of the results of the study in terms of the spatial and geographic relations. The geological severity (*GS*) assessments are helpful in guiding both government and geotechnical agencies to find localities that need their assistance and expertise.

Tests showed, however, that hazard priority (*HP*) values calculated by this method depended most on population buildup (*IB*) rather than the *GS* values. The demographic evidence suggested that in California the urban growth will creep on the fringes of the present urban areas, for the main part, and new urban developments will not be required.

Results

One result of the computer model was to estimate the potential dollar loss in California from the geological hazards considered up to the year 2000. These losses are listed in Table 8.1. The most expensive loss came from local earthquake shaking, with over $21,000,000,000 losses. The figure is very sensitive to casualty estimates and monetary equivalents chosen.

While the actual estimates must be regarded as only rough values, they indicate the scale of problems encountered in a developing a country or province. Further, the study showed that an improvement of current practice to the level now technically feasible might reduce projected overall losses by 70 per cent. The estimated cost of realizing this upgrade might yield a benefit to cost ratio of 6 to 1.

Variations in the Hazard Priority method were not found to lead to much variation in the ranking of either the localities that are either most severely or hardly threatened. It is clear that such an analysis is a great simplification of a most complicated problem. For instance, distant dams prevent local floods and there is much interaction between the population factors (*IF* and *PY*) and the geological hazards (*GS*) of neighboring (and sometimes distant) regions.

8.6. Interdisciplinary Decisions

Opposing Benefits

In the preceding seven chapters measures for control and mitigation of earthquakes, tsunamis, floods, volcanic eruptions and ground surface changes have been suggested. Some of these measures have even been tested in the field. It must be emphasised, however, that control often affects the geological situation itself and the outcome is not clear-cut. Thus, one effect of placing flood-control dams in river systems is often to encourage large developments on down-stream flood plains. In this way the failure of a flood-control dam through some other hazard, such as an earthquake, can have greater catastrophic consequences than if the control had not been established. There are no clear-cut mathematical formulas. What is clear is that an overall approach to geological hazards in

Plate 8.2. The Piazza San Marco at Venice has become a continuation of the Lagoon at periods of high tides and particular weather conditions. Photo taken on November 4, 1966. (Courtesy of Jürgen Schulz)

which the interacting systems are studied together is likely to give a more satisfactory benefit to society than the often piece-meal approach of past years.

A perhaps more intractable problem is balancing the immediacy of benefit to society. A clear example has occurred with land subsidence; sometimes the price of soil settlement, in the long run, is taken to be less important than short-run industrial development.

Two cases in point are severe land settlement in Venice (see Plate 8.2 and Chapter 7) and Shanghai. Settlement in industrial areas in Shanghai was known as early as 1921, yet few measures were taken against it. High tides and storms led to severe flooding of key city areas despite the building of dikes and other protective measures. It was only recently that Shanghai authorities undertook the injection of water into boreholes at the factories to halt further settlement. A beneficial by-product was achieved by having textile factories pump water into the ground in the winter and pump out during the summer. The resultant drop of 10° in available groundwater temperature gives more efficient cooling in the summer.

Inspection Teams

Throughout this book the assumption has been that, for a proper specification of geological hazards, contributions of many disciplines are required. Few individuals are qualified to make a complete assessment of risk. Knowledge must be drawn from geology, geophysics, geography, engineering, land-use planning, statistics, and computer programming.

Geotechnical teams to inspect post-disaster conditions have been invaluable in many instances. In earthquake studies, the practice goes back decades, even centuries (see Section 1.6). In the superb Report of the State Earthquake Commission on the 1906 earthquake, for example, A.C. Lawson and G.K. Gilbert were eminent geologists; H.F. Reid was in applied mechanics. Key advances on the mechanism of the 1964 Alaska earthquake came from many disciplines.

In time, interdisciplinary studies at universities may train specialists in both pre- and post-analysis of natural disasters. For the present, the best check against uninformed opinion outside a field of competence is to rely generally on group decisions.

8.7. Geological Hazard Insurance

Flood Insurance, USA.

Insurance coverage for natural disasters such as floods, earthquakes, volcanoes and avalanches presents special problems for the insurance business. In the first place, there are no means to foretell very closely when, and often where, the disaster will strike, nor how severe it will be (see Fig. 8.5 for the extent of volcanic hazard under certain circumstances). Therefore, useful predictions of damage costs are difficult. Secondly, from an actuarial viewpoint there is not the wide

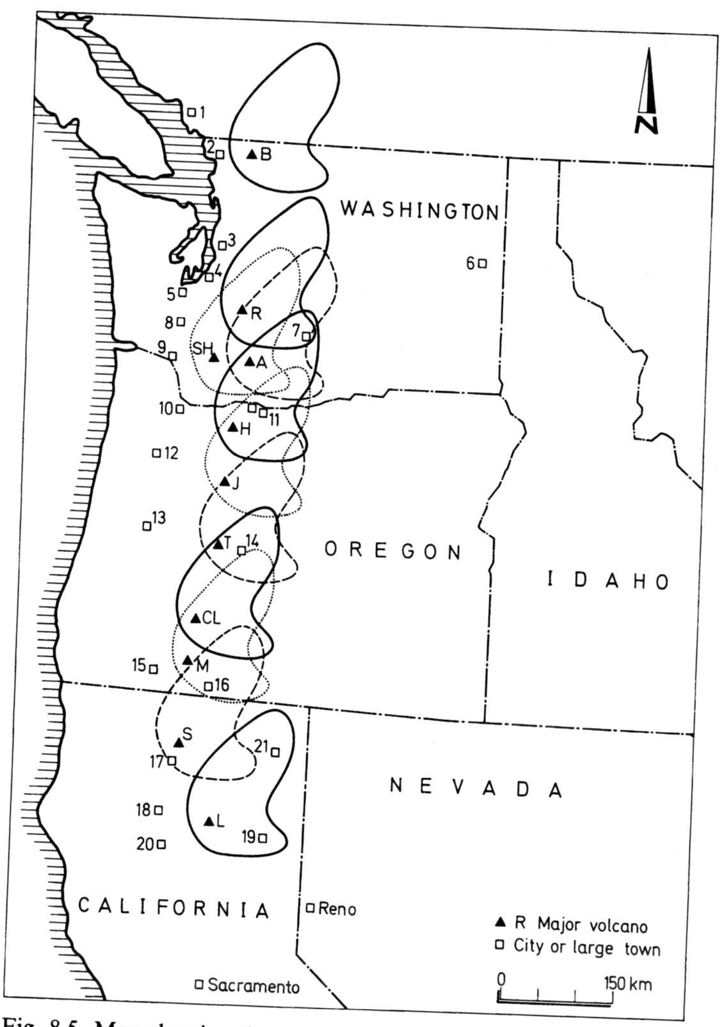

Fig. 8.5. Map showing the area covered by 15 cm or more of ash and pumice during the eruption of Mt. Mazama U.S.A. about 5,000 B.C., superposed on the other major volcanoes of the Cascade Range, indicating the area of devastating ash fall to be expected from each volcano as a result of a similar eruption there under similar wind conditions. (After Crandell and Waldron, 1969; see Chapter 2). The letters and numbers indicate respectively volcanoes and cities, as follows: *A* Mt. Adams, *B* Mt. Baker, *CL* Crater Lake (Mt. Mazama), *H* Mt. Hood, *J* Mt. Jefferson, *L* Lassen Peak, *M* Mt. McLaughlin, *R* Mt. Rainier, *S* Mt. Shasta, *SH* Mt. St. Helens, *T* Three Sisters; *1* Vancouver, *2* Bellingham, *3* Seattle, *4* Tacoma, *5* Olympia, *6* Spokane, *7* Yakima, *8* Centralia, *9* Longview, *10* Portland, *11* Hood River and The Dalles, *12* Salem, *13* Eugene, *14* Bend, *15* Medford, *16* Klamath Falls, *17* Dunsmuir, *18* Redding, *19* Susanville, *20* Red Bluff, *21* Alturas

geographic spread of risk that makes mass market rates possible in other types of property insurance. The coverage is purchased, generally, only by those owners who live in areas where the geological hazard is known and is acute. (In Section 2.4, insurance problems for destruction by lava are discussed.)

Low-cost flood insurance, for example, has been for these reasons generally unavailable to property owners in the United States and most parts of the world although replacement costs, particularly in cities (see Plate 8.2), can be very high. In 1956, the United States Congress passed the Flood Insurance Act, but failed to appropriate the funds to put it into operation. The idea did not die, however, and in 1968 the National Flood Insurance Act, specifically titled XIII, was incorporated into the Housing and Development Act.

The Act provides that a limited amount of flood insurance will be made available if the community involved exhibits a willingness to set up and enforce standards for land-use control in those areas subject to flood and mudslides.

The flood coverage was developed jointly by the insurance industry—represented by the National Flood Insurance Association—and the Federal Insurance Administration. In the twelve months from June 1972 through May 1973, the number of communities in the United States and Puerto Rico qualifying for flood insurance jumped 89.4 per cent, from 1,201 to 2,275; the number of policies in force increased by 158 per cent. The dollar amount of insurance skyrocketed, from $148,000,000,000 to more than $400,000,000,000. The catastrophic floods of mid-1972, including those which devastated Rapid City, South Dakota, early in June and submerged portions of the eastern seaboard (in Hurricane Agnes; see Section 7.4) later in the same month are credited with this sharp increase.

Nevertheless, just as was the case after the Rapid City flood, thousands of persons whose properties were carried away or damaged by floods during the first half of 1973 had failed to take advantage of the protection available through flood insurance. In California, insurance has been available, under the 1968 act, since 1970 for the floods and mudslides that are widespread in the State (Fig. 8.6). The servicing company is Fireman's Fund American. After a community participates in the program, policies may be purchased from any agent. The policy covers losses resulting from partial or complete inundation of normally dry land areas from overflow of inland or tidal waters, rapid accumulation or run-off of surface waters, and mudslides caused by accumulations of water on or under the ground.

Insurance may be purchased in amounts up to $17,500 for a single-family residence and up to $30,000 for all other structures. The policies have a deductible clause applicable separately to structure and contents of $200, or 2 per cent of the amount of loss. The rate per year per $100 of structural coverage is 35 cents for a single-family residence.

Earthquake Coverage, USA

Similar actuarial problems apply to earthquake insurance. Although roughly two-thirds of the earthquake insurance written in the United States is in force in California, the coverage is not widespread in any State. After the San Fernando earthquake, effort to broaden insurance coverage was increased, but little progress was made in State Legislatures at that time.

For rating purposes, California is divided into three zones, according to the likelihood of severe earthquakes:

Zone 1. All the coastal counties and the counties east of the Sierra Nevada,

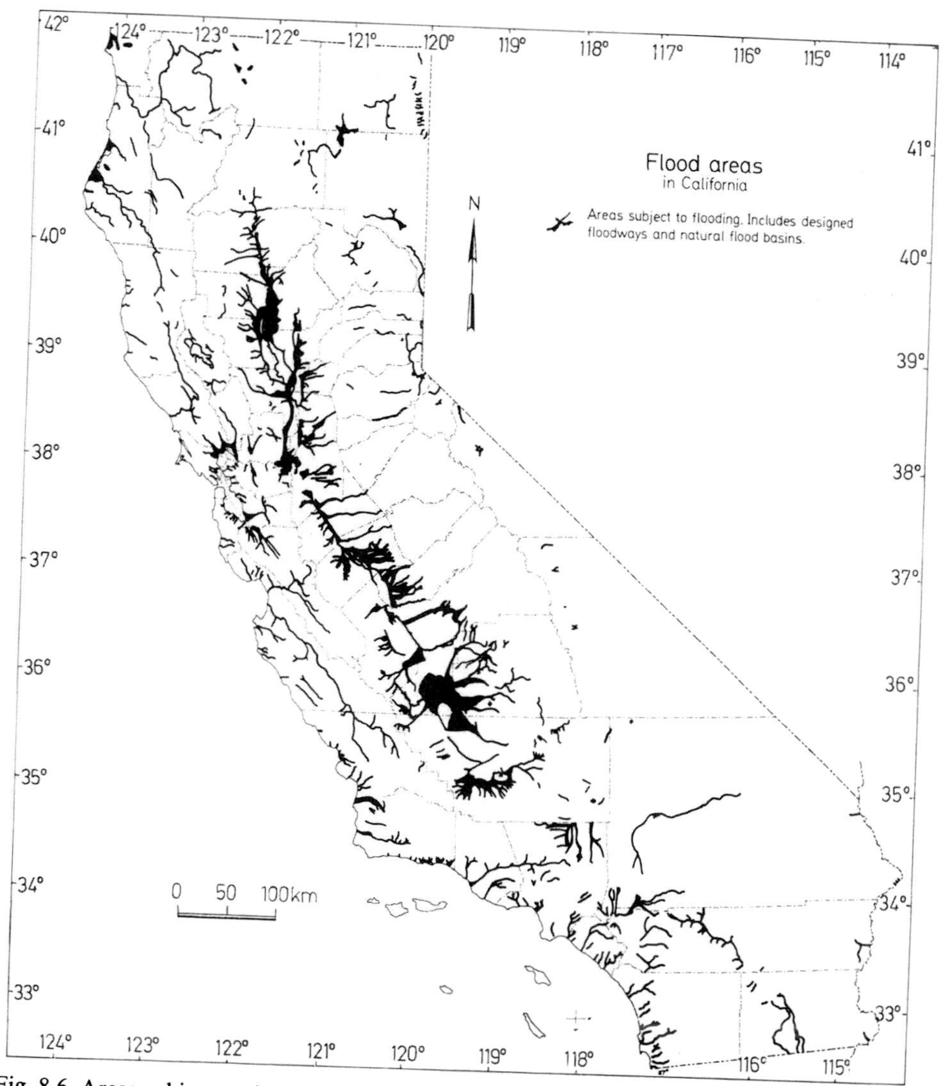

Fig. 8.6. Areas subject to significant flooding in California including constructed floodways and natural flood basins. (From "Urban Geology Master Plan for California", 1973)

Zone 2. Central section of the state from Kings and Tulare counties north to the Oregon border,

Zone 3. Imperial County.

Zone 2 carries the lowest earthquake insurance rate and Zone 3 the highest. For a wood frame dwelling and for the contents, the cost of earthquake insurance generally ranges from 15 cents per $100 in Zone 2 to 30 cents per $100 in Imperial County. The coverage is written usually as an endorsement to a fire or homeowner's policy, usually with 5 per cent deductible applicable to the dwelling and the contents separately. Earthquake coverage for older construction, such as adobe and unreinforced brick, is virtually nonexistent because of the likelihood of total loss.

In order to spread the risk from geological perils, it has been suggested that such disasters be dealt with in one insurance package. Recently in the United States, at least one company has proceeded in this way. In 1973, Wells Fargo Bank offered disaster insurance coverage to owners who have mortgages with the bank for an annual cost of $24 for $25,000 in coverage with $500 deductible. The bank set up a disaster relief insurance trust for interested homeowners. The coverage is provided against direct damage not only by earthquakes, but surface water, mudslides, landslides, tsunamis, tidal waves, wave wash, as well as collapse, breakage or overflow of dams, whether attributable or not to earthquakes.

The maximum limit of liability of the underwriters for a single occurrence is $20,000,000 and arrangements have been made through reinsurers to increase limits to $100,000,000 or more. If the limit of liability in effect at the time of a single occurrence is exhausted there will be a pro rata distribution of the available funds amongst those sustaining loss.

Earthquake Coverage, New Zealand

Elsewhere in the world, availability of insurance coverage against geological hazards is very variable. Of particular interest is an earthquake insurance scheme established in New Zealand following the second World War. The War Damage Act of 1941 required insurance companies to pay to the War Damage Commission a premium levied on the sum insured under fire insurance policies. The fire insuring company was reimbursed by the insured. Private insurance companies maintained records of risk and collected premiums, but the Government Commission settled claims and carried the responsibility.

In 1944 the Earthquake and War Damage Act replaced the earlier act, and a balance of $8,000,000 (New Zealand) in the War Damage Fund provided the basis for indemnity for earthquake shaking and fire damage. In 1948 cover was extended to include extraordinary storm and flood damage, and later still, volcanic eruption damage. Automatic landslide insurance was introduced in 1970 without any extra premium. Landslide damage covered subsidence of any substantial land mass other than by settlement, soil shrinkage, or compaction, and included movement from hills, banks, and slopes.

The private insurance company under the act must collect a premium of 5 cents for each $100 of cover; one-half of the premium may be claimed from a mortgagee. Property owners, however, must bear 1 per cent of each loss with a minimum of $10 and a maximum of $100. The scheme is not optional but is automatically included in all insurance policies—even those on automobiles. Only those unwise enough to be totally uninsured are not covered.

The largest claims for earthquake damage occurred in 1969 after the 1968 Inangahua earthquake (see Section 1.4). During 1972 3,700 earthquake claims were recorded; payments, including an estimate for claims outstanding, totaled about one-quarter of a million dollars. This was the second largest number of earthquake claims recorded in any year. The largest earthquake in the period was a 6.7 magnitude earthquake on January 6, 1973, which affected an area from Hawkes Bay to Wellington, with a total of 2,300 claims.

The earthquake fund, from which war damage, earthquake and earthquake fire damage claims are met, stands in 1973 at $170,000,000. The fund from which storm, flood, and volcanic eruption claims and landslide damage claims are met stands at $3,000,000. It is clear that a great earthquake, near to one of the four most populated areas of New Zealand, would easily exceed the capital accumulated in the fund. Presumably, the government would have to employ funds from other sources to cover the insured properties.

8.8. References

Alfors, J..T., Burnett, J.L., Gay, T.E.: Urban Geology—Master Plan for California. Calif. Division of Mines and Geology, Bulletin **198** (1973).

Anon.: Town Planning and Earthquake Faults. Town and Country Planning Bulletin No. 7, New Zealand Ministry of Works (1965).

Algermissen, S.J., Rinehart, W.A., Dewey, J.: A Study of Earthquake Losses in the San Francisco Bay Area, NOAA Report for Office of Emergency Preparedness (1972).

Anon.: Proceedings. The International Conference on Microzonation for Safer Construction—Research and Application, Seattle, Washington: The National Science Foundation 1972.

Anon.: Earthquake and Insurance. Center for Research on the Prevention of Natural Disasters, DRC-73-02, Calif. Institute of Technology (1973).

Anon.: Report of the Earthquake and War Damage Commission for Year 31 March 1973. New Zealand: Government Printer 1973.

Alquist, A.E.: Meeting the Earthquake Challenge, Sacramento: Report to the Legislature, Joint Committee on Seismic Safety, California Legislature 1974.

Bowden, M.J.: Reconstruction Following Catastrophe: The Laissez-Faire Rebuilding of Downtown San Francisco after the Earthquake and Fire of 1906. Proceedings of the Association of American Geographers **2**, 22–26 (1970).

Cochran, A.: A Selected Annotated Bibliography on Natural Hazards. Natural Hazard Research Working Paper No. 22. Toronto: Department of Geography 1972.

Freeman, J.R.: Earthquake Damage and Earthquake Insurance. New York: McGraw-Hill 1932.

Friedman, D.G.: Insurance and the Natural Hazards. The Austin Bulletin **7**, Part 1 (1972).

Glaken, C.J.: Traces on the Rhodian Shore: Nature and Culture in Western Thought from Ancient Times to the End of the Eighteenth Century. Berkeley: University of California Press 1967.

Graupner, A., Pahl, A.: The Present State of Engineering Geology Mapping in the Federal Republic of Germany. Bull. Int. Assoc. Eng. Geol., No. 7 (1971).

Hewitt, K.: Probabilistic Approaches to Discrete Natural Events: A Review and Discussion. Economic Geography **46**, 332–349 (1970).

Latter, J.H.: Natural Disasters. Advancement of Science **25**, 360–380 (1969).

McHary, I.: Design with Nature. American Museum of Natural History. New York: Natural History Press 1969.

Olson, R.A., Wallace, M.M. (Eds.): Geologic Hazards and Public Problems. U.S. Office of Emergency Preparedness (1969).

O'Riordan, T.: The New Zealand Earthquake and War Damage Commission—A Study of a National Natural Hazard Insurance Scheme. Natural Hazard Research Working Paper No. 20. Toronto: Department of Geography 1971.

Park, C.R.: Affluence in Jeopardy—Minerals and the Political Economy. San Francisco: Freeman, Cooper and Co. 1968.

Rees, J.D.: Paricutin Revisited: A Review of Man's Attempts to Adapt to Ecological Changes Resulting from Volcanic Catastrophe. Geoforum No. 4, 7–25 (1970).

Smithsonian Institute: Natural Disaster Research Centers and Warning Systems: A Preliminary Survey. Cambridge, Mass.: Center for Short-Lived Phenomena 1971.

Steinbrugge, K.V.: Earthquake Hazard in the San Francisco Bay Area: A Continuing Problem in Public Policy. Berkeley: Institute of Government Studies, University of California 1968.

Wallace, R.E.: Goals, Strategy, and Tasks of the Earthquake Hazard Reduction Program. U.S. Geological Survey Circular 701 (1974).

Wright, R., Kamer, S., Culver, C. (Eds.): Building Practices for Disaster Mitigation. National Bureau of Standards, BSS **46** (1973).

Appendices

Appendix A: Notable World Earthquakes. (Source: NOAA USA)

Year	Region	Deaths	Mag.	Comments
December, 856	Greece, Corinth	45,000		
January 9, 1038	China, Shensi	23,000		
1057	China, Chihli	25,000		
1268	Asia Minor, Silicia	60,000		
September 27, 1290	China, Chihli	100,000		
May 20, 1293	Japan, Kamakura	30,000		
January 26, 1531	Portugal, Lisbon	30,000		
January 23, 1556	China, Shensi	830,000		
February 5, 1663	St. Lawrence River			Max. intensity X. Chimneys broken Massachusetts
November 1667	Caucasia, Shemaka	80,000		
January 11, 1693	Italy, Ćatania	60,000		
October 11, 1737	India, Calcutta	300,000		
June 7, 1755	Northern Persia	40,000		
November 1, 1755	Portugal, Lisbon	70,000		Great tsunami
February 4, 1783	Italy, Calabria	50,000		
February 4, 1797	Ecuador, Quito	40,000		
December 16, 1811	U.S.A., New Madrid, Mo.	Several		Intensity XI. Also Jan. 23, Feb. 7, 1812.
December 21, 1812	Off-shore Santa Barbara, Calif.	Several injuries		Max. intensity X. Reported tsunami uncertain
June 16, 1819	India, Cutch	1,543		
September 5, 1822	Asia Minor, Aleppo	22,000		
December 28, 1828	Japan, Echigo	30,000		
January 9, 1857	Calif., Fort Tejon			San Andreas fault rupture. Intensity X–XI
August 13–15, 1868	Peru-Ecuador	25,000		
August 13, 1868	Peru-Bolivia			
August 16, 1868	Ecuador, Colombia, Guayaquil	Ecuador 40,000 Colombia 30,000		
March 26, 1872	Calif., Owens Valley,	About 50		Large-scale faulting
August 31, 1886	South Carolina, Charleston-Summerville	About 60		
October 28, 1891	Japan, Mino-Owari			
June 15, 1896	Japan	22,000		Tsunami

Appendix A (continued)

Year	Region	Deaths	Mag.	Comments
June 12, 1897	India, Assam	1,500	8.7	
September 3 & 10, 1899	Alaska, Yakutat Bay		7.8 & 8.6	
April 18, 1906	Calif., San Francisco	700	$8^1/_4$	
December 28, 1908	Italy, Messina	120,000	$7^1/_2$	
January 13, 1915	Italy, Avezzano	30,000	7	
December 16, 1920	China, Kansu	180,000	$8^1/_2$	
September 1, 1923	Japan, Kwanto	143,000	8.2	Great Tokyo fire
December 26, 1932	China, Kansu	70,000	7.6	
May 31, 1935	India, Quetta	60,000	7.5	
January 24, 1939	Chile, Chillan	30,000	$7^3/_4$	
December 27, 1939	Turkey, Erzincan	23,000	8.0	
June 28, 1948	Japan, Fukui	5,131		
August 5, 1949	Ecuador, Pelileo	6,000		
February 29, 1960	Morocco, Agadir	14,000	5.9	
May 21–30, 1960	Southern Chile	5,700	8.5	
September 1, 1962	Northwest Iran	14,000	7.3	
July 26, 1963	Yugoslavia, Skopje	1,200	6.0	See text
March 28, 1964	Alaska	131	8.6	Prince William Sound, Tsunami
May 31, 1970	Peru	66,000	7.8	$530,000,000 damage. Great rock slide. See text
February 9, 1971	Calif., San Fernando	65	6.5	$550,000,000 damage. See text
December 22, 1972	Nicaragua, Managua	5,000	6.2	See text

Appendix B: Important Earthquakes of the United States, Canada and Mexico.
(Source: NOAA USA)

Year	Date	Place	MM Intensity	Remarks
1638	June 11	Mass., Plymouth	IX	Many stone chimneys down. Chimneys down in shocks in 1658 and probably other years
1663	Feb. 5	Canada, Three Rivers, Lower St. Lawrence River	X	Chimneys broken in Massachusetts Bay area
1732	Sept. 16	Canada, Ontario	IX?	7 killed at Montreal
1755	Nov. 18	Mass., near Cambridge	VIII	Many chimneys down, brick buildings damaged, stone fences generally wrecked. Sand emitted from ground cracks. Felt from Chesapeake Bay to Nova Scotia
1769	July 28	Calif., San Pedro Channel area	X	Major disturbances with many aftershocks
1790	?	Calif., Owens Valley	X	Major shock with appearance of fault scarps
1811 1812 1812	Dec. 16 Jan. 23 Feb. 7	Mo., New Madrid	XI	Three principal earthquakes. Town of New Madrid destroyed, extensive changes in configuration of ground and rivers including the Mississippi River. Chimneys down in Cincinnati and Richmond. Felt in Boston. Several killed. The 3 shocks had Richter magnitudes of about 7.5, 7.3 and 7.8
1812	Dec. 8	Calif., San Juan Capistrano	IX	Church collapsed killing 40
1812	Dec. 21	Calif., near Lompoc	X	Churches and other buildings wrecked in several towns including Santa Barbara
1836	Jan. 10	Calif., San Francisco Bay area	X	Ground breakage along Hayward Fault from Mission San Jose to San Pablo
1838	June	Calif., San Francisco	X	Fault rupture phenomena along San Andreas rift. This earthquake is probably comparable with the earthquake of April 18, 1906
1857	Jan. 9	Calif. Fort Tejon	X–XI	One of the greatest historical Pacific Coast shocks. Originated on San Andreas fault in northwest corner of Los Angeles County. Buildings and large trees thrown down
1865	Oct. 8	Calif., San Francisco Peninsula	IX	Extensive damage in San Francisco, especially on made ground

Appendix B (continued)

Year	Date	Place	MM Intensity	Remarks
1868	Oct. 21	Calif., Hayward	X	Many buildings wrecked and damaged in Hayward and East Bay. Severe damage at San Leandro and San Francisco. 30 killed. Rupture of Hayward fault
1870	Oct. 20	Canada, Montreal to Quebec	IX	Widespread. Minor damage on coast of Maine
1872	Mar. 26	Calif., Owens Valley	X–XI	One of the greatest earthquakes in Pacific Coast area. 7 m fault scarp formed. 27 killed at Lone Pine out of 300 population. Adobe houses wrecked
1886	Aug. 31	S.C., Charleston	X	Greatest historical earthquake in eastern states. 102 buildings destroyed, 90 per cent damaged, nearly all chimneys down. $5,500,000 damage. About 60 killed. Felt at Boston, Chicago and St. Louis. (Current equivalent $25,000,000)
1887	May 3	Mexico, Sonora	XI	Widespread in border states. Chimneys down in several towns, including El Paso and Albuquerque
1895	Oct. 31	Mo., near Charleston	IX	Felt in Canada, Virginia, Louisiana and South Dakota. Acres of ground sank and lake formed. Many chimneys demolished
1899	Sept. 10	Alaska, Yakutat Bay	XI	Great earthquake. Widely felt. Slight damage because area uninhabited. Shoreline rose 15 m
1899	Dec. 25	Calif., San Jacinto	IX	Nearly all brick buildings badly damaged in San Jacinto and Hemet. Chimneys down in Riverside. 6 killed. Another severe shock in 1918
1900	Oct. 9	Alaska, Kenai Peninsula?	VII–VIII	Felt from Yakutat to Kodiak. Severe damage at Kodiak
1906	Apr. 18	Calif., San Francisco	XI	Great earthquake and fire. About 80 per cent of estimated $400,000,000 damage due to fire. 700 killed. Greatest destruction in San Francisco, Santa Rosa. Horizontal slipping along San Andreas fault, 6.5 m. Greatest damage on poorly filled land
1909	May 26	Ill., Aurora	VIII	Many chimneys down. Felt over wide area
1915	June 22	Calif., Imperial Valley	VIII	Nearly $1,000,000 damage. 6 killed. Well-constructed buildings were cracked

Appendix B (continued)

Year	Date	Place	MM Intensity	Remarks
1915	Oct. 2	Nev., Pleasant Valley	X	Widespread. Adobe houses and water tank towers wrecked. Fault break 35 km with 3.5 m vertical throw in one place
1925	Feb. 28	Canada, Murray Bay	VIII	Felt in many eastern and central states. Damage less than $100,000
1925	June 27	Mont., Manhattan	IX	Landslide blocked entrance to railroad tunnel. Some buildings wrecked and many chimneys fell. $300,000 damage
1925	June 29	Calif., Santa Barbara	IX	$6,000,000 damage. 13 killed. 70 buildings condemned
1929	Aug. 12	N.Y., Attica	IX	250 chimneys toppled
1929	Nov. 18	Grand Banks, off Newfoundland	X	Submarine shock broke 12 trans-Atlantic cables, some breaks 240 km apart. Some deaths by tsunami along Burin Peninsula. Some chimneys in Canada toppled
1931	Apr. 20	N.Y., Lake George	VIII	Chimneys fell
1931	Aug. 16	Texas, near Valentine	VIII	All buildings damaged, many chimneys fell
1932	Dec. 20	Nev., Cedar Mountain	X	In sparsely settled region. Widespread
1933	Mar. 10	Calif., Long Beach	IX	$41,000,000 damage, 120 killed. Fire damage insignificant
1934	Mar. 12	Utah, Kosmo	VIII	Marked changes in terrain north of Great Salt Lake. 2 killed
1935	Oct. 18 Oct. 31	Mont., Helena (strong aftershock)	VIII	$3,500,000 damage, 4 killed, less than 50 injured. More than half of buildings damaged from 2.5 to 100 per cent. Second shock strongest of many aftershocks
1935	Nov. 1	Canada, Timiskaming	IX	Widespread, Landslide near origin
1940	May 18	Calif., Imperial Valley	X	$6,000,000 damage, 8 killed, 20 seriously injured. 65 km fault appeared with maximum horizontal displacement of 4.5 m
1941	June 30	Calif., Santa Barbara	VIII	$100,000 damage
1941	Nov. 14	Calif., Torrance, Gardena	VIII	About $1,000,000 damage. 50 buildings severely damaged
1944	Sept. 5	Canada-New York, Cornwall and Massena	IX	On St. Lawrence River. $1,500,000 damage reported. 90 per cent of chimneys in Massena destroyed or damaged

Appendix B (continued)

Year	Date	Place	MM Intensity	Remarks
1946	Apr. 1	Alaska, Aleutian Islands	?	Great earthquake. Tsunami destroyed a light station and caused severe damage in Hawaii. Estimated damage $25,000,000
1948	Oct. 16	Alaska, Nenana	VIII	Rock slides and damage to Alaska railroad
1949	Apr. 13	Puget Sound	X	$25,000,000 damage. 8 killed directly and indirectly. Damage confined mostly to marshy, alluvial, or filled ground. Many chimneys, parapet walls and cornices toppled
1952	July 20	Calif., Kern County	X	$60,000,000 damage, 12 killed, 18 seriously injured. Railroad tunnels collapsed and rails bent in S-shape. Surface faulting with about a half meter of vertical, as well as lateral, displacement
1952	Aug. 22	Calif., Bakersfield	VIII	2 killed, 35 injured. Damage $10,000,000
1954	July 6	Nev., Fallon	IX	Extensive damage to irrigation canals. Several injured
1954	Aug. 23	Nev., Fallon	VIII	Surface ruptures. Damage more than $91,000
1954	Dec. 16	Nev., Dixie Valley	X	Surface ruptures along 88 km linear distance and up to 4.5 m vertical throw in sparsely populated desert
1957	Mar. 22	Calif., San Francisco	VIII	Damage in Westlake and Daly City area
1958	Apr. 7	Central Alaska	VIII	Severe breakage of river and lake ice, pressure ridges and mud flows
1958	July 7	Alaska, Lituya Bay	XI	Major earthquake. Landslide created water wave that denuded mountain side as high as 540 m. Long fault break. Cables severed. 5 killed by drowning
1959	Aug. 17	Mont., Hebgen Lake	X	Huge landslide dammed river and created lake. Fault scarps with 4.5 m throw. Maximum vertical displacement 6.5 m. 28 killed. $11,000,000 damage to roads alone
1964	Mar. 27	Alaska, Prince William Sound	X–XI	Great (Good Friday) earthquake. Damage to public property $235,000,000; real property $77,000,000. In Anchorage, extensive damage to moderately tall structures (45 m or less) and to poorly constructed low buildings. Landslides and slumps caused total damage to many buildings. Docks in several ports destroyed by submarine slides and tsunami. Sea-wave damage on U.S. coast and elsewhere. 131 lives lost. Shorelines rose 10 m in places and settled 2 m elsewhere

Appendix B (continued)

Year	Date	Place	MM Inten-sity	Remarks
1965	Apr. 29	Wash., Puget Sound	VII–VIII	Property loss $12,500,000 mostly in Seattle. Felt over an area of 350,000 square km. 3 persons killed and 3 died apparently of heart attacks
1969	Oct. 1	Calif., Santa Rosa	VII–VIII	Property loss of $6,000,000. Felt over an area of 30,000 square km
1971	Feb. 9	Calif., San Fernando	VIII–XI	$500,000,000 direct physical loss, 65 killed, more than 1,000 persons injured. Felt over an area of 230,000 square km

Appendix C: Number of Active and Other Geologically Recent Volcanoes In Various Regions. (After Macdonald, 1972)

Region	Approximate number of volcanoes		
	Active[a]	Recent[b]	Total
Antarctica and the Scotia Arc	10	6	16
New Zealand	5	1	6
Tonga-Kermadec Archipelago	14	1	15
Samoa	4	16	20
Melanesia (including New Guinea)	30	28	58
Indonesia	75	58	133
Philippines	15	16	31
China Sea	7	?	7+
Ryukyu Islands	6	7	13
Mariana and Izu Islands	20	3	23
Japan	31	14	45
Kuril Islands	33	6	39
Kamchatka	19	9	28
Aleutian Islands and Alaska	39	21	60
British Columbia	0	2	2
Cascade Range, northwestern United States	7	8	15
Mexico and Central America	42	18	60
South America (Andes)	47	13	60
Pacific Ocean	15	?	15+
West Indies	9	8	17
Atlantic Ocean	22	?	22+
Iceland and Jan Mayen	23	?	23+
Mediterranean	13	12	25
Africa	14	29	43
Asia Minor	8	19	27
Indian Ocean	4	?	4+
Total	513	228+	741*

[a] Volcanoes with certain or very probable eruptions within historic times.
[b] Volcanoes with solfataric activity of very well preserved forms, suggesting that they have been active during the last few tens of thousands of years.

Appendix D: Major Flood Disasters of the World 1963–1974. (After W.L. Horn)

Year	Location	Remarks
1963	Caribbean	Hurricane, Oct. 1–9 killed 4,000 in Cuba and Haiti
	Haiti	Flood, Nov. 14–15 deluges and landslides killed 500
	near Belluno, Italy	Oct. 9 landslide into Vaiont Dam created a wave of water that topped the dam and resulted in death for more than 2,000 persons
1964	Montana	Floods, June 8–9 responsible for 36 deaths
	Louisiana	Hurricane, Oct. 3 and associated tornadoes killed 36
	California, Oregon and Washington	Floods late December responsible for 45 deaths
1965	Southwest U.S.	June 18–19 floods killed 27
1966	Rio de Janeiro	Flood, Jan. 11–13 deluge and landslides killed more than 300
	Caribbean and Mexico	Hurricane, Sept. 25–Oct. 1, fatal for at least 200
	Arno, Italy	Flood, Nov. 3–4 overflow of Arno river killed 113 and destroyed priceless art treasures in Florence
1967	Rio de Janeiro, San Paulo, Brazil	Flood, Jan.–Mar. heavy rain floods killed more than 600
	Lisbon	Flood, Nov. 25 fatality toll 457
1968	Gujarat, India	Flood, Aug. 8–14 killed about, 1,000 persons
1969	Southern California	Floods, Jan. 25–29, Feb. 21–25 killed 115 and property damage of $300,000,000
	So. Michigan and No. Ohio	Flash floods, July 4 killed 33
	Mississippi and Louisiana	Hurricane Camille, Aug. 17 killed about 200
	Virginia	Floods, Aug. 23 killed 100
1970	Cuba, Florida and Texas	Hurricane Celia, Aug. 3 killed 31
	East Pakistan delta region	Typhoon, Nov. 12 storm and floods killed 500,000? Worst natural disaster of the century
	Oradea, Rumania	Flood, May 11–23, 200 killed and more than 225 towns destroyed
	Arizona, USA	Tropical storm Norma killed 23
1971	None	
1972	Buffalo Creek, West Virginia	Flood, Feb. 26 coal mine waste water caused a makeshift dam to fail, 118 dead
	Rapid City, North Dakota	Flash flood, June 10 caused over $120,000,000 in damage, 237 deaths
	East coast of United States	Hurricane Agnes, June 18–25, first of season, killed 118, property damage of $3 billion; considered greatest natural disaster to befall United States
1974	Northern Australia and Brisbane	Monsoon flooding and cyclones in Queensland killed at least 15 persons. Australia's greatest natural disaster

Appendix E: Metric-English Conversion Table

Length		
1 millimeter (mm) [0.1 centimeter (cm)]	=	0.0394 inch (in.)
1 cm [10 mm]	=	0.3937 in.
1 meter (m) [100 cm]	=	39.37 in.
	=	3.28 feet (ft)
1 kilometer (km) [1000 m]	=	0.621 mile (mi)
1 nautical mile [1852 m]	=	6,076.1 ft
Area		
1 square centimeter (cm^2)	=	0.155 square inch ($in.^2$)
1 square meter (m^2)	=	10.76 square feet (ft^2)
	=	1.196 square yards (yd^2)
1 hectare (ha)	=	2.4710 acres (a)
1 square kilometer (km^2)	=	0.386 square mile (mi^2)
Volume		
1 cubic centimeter (cm^3)	=	0.0610 cubic inch ($in.^3$)
1 cubic meter (m^3)	=	35.314 cubic feet (ft^3)
	=	1.31 cubic yards (yd^3)
1 cubic kilometer (km^3)	=	0.240 cubic mile (mi^3)
1 liter (l)	=	1.06 quarts (qt)
	=	0.264 gallon (gal)
1 cubic meter	=	8.11 $\times 10^{-4}$ acre feet
Mass		
1 kilogram (kg) [1000 grams (g)]	=	2.20 pounds (lb)
	=	0.0011 ton (tn)
1 metric ton (MT) [1000 kg]	=	1.10 tn
Pressure		
1 kilogram per square centimeter (kg/cm^2)	=	14.21 pounds per square inch ($lb/in.^2$)
	=	2,046 pound per square foot (lb/ft^2)
Velocity		
1 meter per second (m/s)	=	3.281 feet per second (ft/s)
1 kilometer per hour (km/h)	=	0.9113 ft/s
	=	0.621 mile per hour (mi/h)

Appendix F: Geologic Time Scale

Relative Duration of Major Geologic Intervals	Era	Period	Epoch	Approximate Duration in Millions of Years	Millions of Years Ago
Cenozoic	Cenozoic	Quarternary	Holocene	Approx. the last	10,000 years
			Pleistocene	2.5	2.5
		Tertiary	Pliocene	4.5	7
			Miocene	19.0	26
			Oligocene	12.0	38
			Eocene	16.0	54
			Paleocene	11.0	65
Mesozoic	Mesozoic	Cretaceous		71	136
		Jurassic		54	190
		Triassic		35	225
Paleozoic	Paleozoic	Permian		55	280
		Carboniferous: Pennsylvanian		45	325
		Carboniferous: Mississippian		20	345
		Devonian		50	395
		Silurian		35	430
		Ordovician		70	500
		Cambrian		70	570
Precambrian	Precambrian			4,030	

Scale (millions of years ago): 0, 50, 100, 150, 200, 250, 300, 350, 400, 450, 500, 550, 4,600

Subject Index

The Geology of Continental Margins

Editors: C.A. Burk, C.L. Drake
730 figures. XIII, 1009 pages. 1974
ISBN 3-540-06866-X Cloth DM 85,30
ISBN 0-387-06866-X
(North America) Cloth $34.80

Contents: Geological Significance of Continental Margins. – General Bathymetry and Topography. – Transition from Continent to Ocean. – Recent Sedimentation. – Deformation at Continental Margins. – Geology of Selected Modern Margins: Atlantic Region. – Geology of Selected Modern Margins: Pacific Region. – Geology of Selected Modern Margins: Indian Ocean Region. – Geology of Selected Small Ocean Basins. – Ancient Continental Margins. – Igneous Activity and Ancient Margins. – Resources at Continental Margins. – Continental Margins in Perspective.

The continental margins of the world constitute the largest and most impressive physiographic feature of the earth's surface. Because of their fundamental geological significance, continental margins have been the subject of increasing attention in recent years and the resulting body of new data has provided further insights into their character. This interest was further stimulated by the realization that, in addition to abundant living resources, continental margins contain petroleum and mineral resources that are accessible with existing technology. This practical concern, coupled with basic geological questions, has fostered further research into the nature of continental margins throughout the world. A summary of these findings, related to both recent and ancient continental margins, is the subject of this book.

**Springer-Verlag
Berlin
Heidelberg
New York**